Advanced Stimuli-Responsive Polymer Composites

Advanced Stimuli-Responsive Polymer Composites

Guest Editor

Lucas Kreuzer

MDPI

Basel • Beijing • Wuhan • Barcelona • Belgrade • Novi Sad • Cluj • Manchester

Guest Editor
Lucas Kreuzer
German Aerospace Center
(DLR)
DLR-Institute of Material
Physics in Space
Köln
Germany

Editorial Office
MDPI AG
Grosspeteranlage 5
4052 Basel, Switzerland

This is a reprint of the Special Issue, published open access by the journal *Polymers* (ISSN 2073-4360), freely accessible at: https://www.mdpi.com/journal/polymers/special_issues/Stimuli_Responsive_Polymer_Composites.

For citation purposes, cite each article independently as indicated on the article page online and as indicated below:

Lastname, A.A.; Lastname, B.B. Article Title. *Journal Name* **Year**, *Volume Number*, Page Range.

ISBN 978-3-7258-4927-7 (Hbk)
ISBN 978-3-7258-4928-4 (PDF)
https://doi.org/10.3390/books978-3-7258-4928-4

Contents

About the Editor

Lucas Kreuzer

Dr. Lucas Kreuzer is an instrument scientist at the neutron spin-echo instrument RESEDA at the Heinz-Maier-Leibnitz Neutron Source (FRM II) in Munich, Germany. He studied Chemistry in Bayreuth, Germany, and Easton, USA, and went on to earn his Ph.D. in Physics at the Technical University of Munich in 2021. He then worked as a postdoctoral fellow at the German Aerospace Center (DLR) in Cologne and a visiting scientist at the Institut Laue-Langevin (ILL) and Université Grenoble Alpes in Grenoble, France. His research focuses on cellulose-based nanocomposites and their interactions with water. He also hosts the monthly science podcast "Your Friendly Physicist and Other Nerds".

polymers MDPI

Review

Smart Polymer Surfaces with Complex Wrinkled Patterns: Reversible, Non-Planar, Gradient, and Hierarchical Structures

Mauricio A. Sarabia-Vallejos [1], Felipe E. Cerda-Iglesias [2,3], Dan A. Pérez-Monje [2], Nicolas F. Acuña-Ruiz [2], Claudio A. Terraza-Inostroza [4], Juan Rodríguez-Hernández [5] and Carmen M. González-Henríquez [2,6,*]

1 Facultad de Ingeniería, Arquitectura y Diseño, Universidad San Sebastián, Sede Santiago, Santiago 8420524, Chile
2 Departamento de Química, Facultad de Ciencias Naturales, Matemáticas y del Medio Ambiente, Universidad Tecnológica Metropolitana, Santiago 7800003, Chile
3 Programa PhD en Ciencia de Materiales e Ingeniería de Procesos, Universidad Tecnológica Metropolitana, Santiago 8940000, Chile
4 Research Laboratory for Organic Polymer (RLOP), Facultad de Química y Farmacia, Pontificia Universidad Católica de Chile, Santiago 7810000, Chile
5 Polymer Functionalization Group, Departamento de Química Macromolecular Aplicada, Instituto de Ciencia y Tecnología de Polímeros-Consejo Superior de Investigaciones Científicas (ICTP-CSIC), 28006 Madrid, Spain
6 Programa Institucional de Fomento a la Investigación, Desarrollo e Innovación, Universidad Tecnológica Metropolitana, Santiago 8940000, Chile
* Correspondence: carmen.gonzalez@utem.cl

Abstract: This review summarizes the relevant developments in preparing wrinkled structures with variable characteristics. These include the formation of smart interfaces with reversible wrinkle formation, the construction of wrinkles in non-planar supports, or, more interestingly, the development of complex hierarchically structured wrinkled patterns. Smart wrinkled surfaces obtained using light-responsive, pH-responsive, temperature-responsive, and electromagnetic-responsive polymers are thoroughly described. These systems control the formation of wrinkles in particular surface positions and the reversible construction of planar-wrinkled surfaces. This know-how of non-planar substrates has been recently extended to other structures, thus forming wrinkled patterns on solid, hollow spheres, cylinders, and cylindrical tubes. Finally, this bibliographic analysis also presents some illustrative examples of the potential of wrinkle formation to create more complex patterns, including gradient structures and hierarchically multiscale-ordered wrinkles. The orientation and the wrinkle characteristics (amplitude and period) can also be modulated according to the requested application.

Keywords: wrinkles surface; surface microstructuration; smart interfaces; non-planar structuration; hierarchical surface structures

1. Introduction

The design, development, and testing of stimuli-responsive materials are critical for several application areas. Thus, the dynamic and reversible control of surface topography has been extensively studied due to their use, such as smart optics and diffraction grating for tunable light manipulation [1], wearable tactile devices, micropatterns to control friction between solid surfaces separated by a thin fluid film [2,3], refreshable Braille surfaces [4,5], anti-biofouling coating [6], adjustable wettability [7], smart adhesion [8], e-skin and stretchable devices [9,10]. Compared with traditional static surfaces, these microstructured responsive surfaces are highly desired due to their capacity to tune responses, creating more flexible devices suitable for soft electronic and haptic engineering. These applications need the materials to actively change their characteristics, geometry, or function to fulfill each case's requirements. Polymers have been widely used to fabricate these devices because they present several advantages against other materials (ceramic, alloys, or hybrids), like easy and low-cost fabrication; besides, polymers are simple to employ. In general, their properties can be adjusted using a straightforward synthetic methodology. Also, various stimuli-responsive polymers are biocompatible or biodegradable, making them suitable for several biomedical applications [11,12].

1.1. Smart Polymers

Smart materials have intrinsic properties that allow them to experiment with physical or chemical changes in response to external stimuli, including slight variations in the surrounding environment, such as temperature, light, electric/magnetic fields, mechanical stress, pH, humidity, ionic strength and existence of metabolic chemicals [13].

According to Jingcheng et al. [14], in the last two decades (since 2001), there have been significant progress and publications in the area of smart polymers, increasing from 4000 articles published in 2001 to more than 14,000 in the year 2020. The first scientist to coin the name "smart polymers" was Dagani et al. in 1995, who highlighted the similarity of these materials with some biopolymers that present "responsive" natures. Other authors compare smart polymers with machines, which can convert an energy input into mechanical work (shape-shifting). The decisive benefit of smart materials for medical device design is the simplicity of the resulting devices compared to complex mechanical engineering solutions [15].

Smart polymers could change their structure at different length scales ranging from macro- to micrometer sizes. These morphological changes or transitions are usually reversible and return to their initial state when the external stimulus is removed. Recent studies demonstrated that these compounds could change their shape and other characteristics like solubility or modify properties such as wettability or adhesion when placed on surfaces. Similarly, sol-gel transitions or molecular self-assemblies can occur spontaneously on these materials. From the macroscopic level, the adaptive performances of smart materials are achieved by the autonomous behaviors of molecules or atoms at a microscopic or nanoscopic level. When stimuli are introduced, aggregation, rearrangement, and directional movement of the molecules/atoms can occur, thus triggering different reactions in the macroscopical behavior of the material [14].

According to Raquez and Peponi [16], there are two main types of smart materials: self-evolving and self-healing. The first can, for example, reconfigure themselves "on the fly" to react to changing environmental conditions. Some cases of these kinds of materials are shape-memory polymers, adaptive hydrogels, and actuators. In addition, intrinsic self-healing polymers are synthesized on reversible bonding, such as Diels-Alder reactions.

Zhou et al. [17] described a different categorization for self-evolving materials, which can be subdivided into four groups: self-assembly [18], deformation mismatch [19], bistability, and shape-memory effect [20–22]. This last category can be classified into five different families, where shape-memory polymers (SMPs) and shape-memory alloys (SMAs) [22–24] are the two most important.

In summary, the unique features of smart polymers enable their use in different technological areas, for example, for biomedical purposes such as drug delivery systems [25] or for being used in cardiovascular implants [15], or as part of microscale machines [26] such as devices for data storage, smart textiles, or robotic e-skin.

1.2. Instability-Based Surface Patterning: Formation of Wrinkled Polymer Surfaces

The observation of wrinkles on polymer surfaces has been widely studied. Thus, the first reports evidencing the formation of wrinkled coatings using China wood oil treated at high temperatures under atmospheric oxygen. During World War II, airplane wings were fabricated using crosslinked polymer foams and rigid composites, which, under bendings and compressions, could generate these wrinkled patterns on the surface. Despite the initial objection to using these coatings, their use was extended, giving to wrinkled finished coatings [27].

In the last two decades, the interest in this type of surface has increased considerably, being the study of Mahadevan et al. [28] and Genzer et al. [29], the ones who kicked off the renaissance of microstructured wrinkled surfaces. Mahadevan et al. set the physical foundations surrounding the formation of these patterns and explained how the accumulation of tension in the film triggers the formation of ordered wrinkles patterns on the surface of the material. On the other hand, Genzer and their group analyze the processes involved in forming nested wrinkled patterns using elastomeric substrates, extending its dimensions over five orders of magnitude (from a few nanometers to millimeters). Their experimental results were contrasted with both computations and a simple scaling theory, demonstrating that it is possible to predict and model the wrinkling pattern formation. These studies are considered the cornerstone of recent methodologies used to create wrinkled patterns, which implies that none of the studies, applications and technologies reviewed in this article would be possible without these two articles.

Recently, several material scientists have been focused on reproducing some of the properties or behaviors that natural/biological materials present. For instance, biomimetic materials are used repeatedly in tissue engineering (TE) [30] to fabricate pieces or devices that resemble natural tissues or organs. These materials can alter their designs by stimulating specific cellular responses associated with new tissue formation mediated by biomolecular recognition.

The great effort carried out in elaborating these materials is justified by the wide variety of applications that largely depend on the surface properties of a particular material. For instance, both patterning and functionalization of polymer surfaces have been demonstrated to play a crucial role in the preparation of substrates for different applications in the biomedical [30], soft electronic [31], optical [32], or micromechanical [33] areas, among others.

In this context, many of the strategies reported to reproduce the superior properties of surfaces in nature failed or resorted to intricate methodologies that require the use of expensive and sophisticated equipment. In this sense, some advances in the development of polymeric surfaces have been associated with technological improvements.

However, more recently, several alternative approaches have been developed in which the origin of surface patterns is not the result of advanced fabrication techniques but takes advantage of the instabilities produced in these polymeric surfaces [34]. Two alternative routes can be used to create surface instabilities; firstly, they could be fabricated spontaneously in small-scale compact systems, which are particularly unstable. On the other hand, a specific external stimulus can induce a surface fluctuation in metastable films. These stimuli include mechanical force (stretching or compression), electric field, or heating [27]. After removal, relaxation of the surface towards equilibrium results in modification of the film structure to minimize the free energy in the film. It can be seen that different nano and micrometric size morphologies with particular morphologies and distributions are generated on the surface of the material. In particular, the latter allows the formation of intricate patterns, such as wrinkle or fold patterns, which would be complicated, if not impossible, to fabricate by conventional micro-modification methods.

Accordingly, the general wrinkling system based on surface instabilities is illustrated in Figure 1. For the case of the bilayer system consisting of a rigid top layer over an elastic layer, the first step requires the application of a mechanical tensile force (originating from osmotic pressure, stretching, or heating, to name a few). Next, the rigid layer is deposited on the polymer film. Finally, after the removal of the applied stimulus, the initial shape of the elastic layer relaxes, and a wavy structure identified as "wrinkles" is generated on the surface [35].

This behavior can be explained by considering the soft substrate as an elastic material. The external energy input is partially converted when expanding into elastic potential energy within the substrate. After coating it with a thin hard layer over the soft substrate, followed by the relaxation of the initial stimulus, a new equilibrium state must be relocated due to a mismatch of two equilibrium states between the soft elastomeric substrate and the hard elastic layer [27]. The wrinkle wavelength and amplitude result from the minimization of the total elastic energy in the thin layer and the soft substrate. The wrinkling process relaxes the compressive strain, thus reducing the inner film energy. At the same time, it generates the film's bending, increasing the material's elastic strain energy [35].

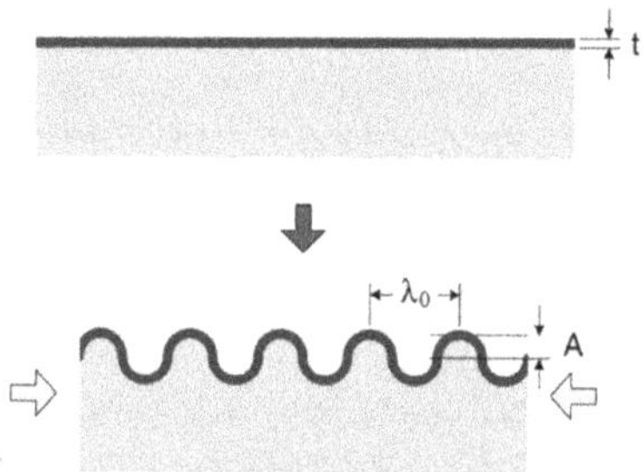

Figure 1. Illustration of the general mechanisms for forming wrinkles in bilayered systems comprising a rigid top layer on an elastic substrate. Reproduced with permission from ref. [35]. Copyright 2010, John Wiley & Sons, Inc.

The wavelength λ_0 depends on the thickness of the thin hard film (t) and the mechanical properties of the thin film and the substrate:

$$\lambda_0 = 2\pi t \left(\frac{\overline{E}_f}{3\overline{E}_s} \right)^{1/3} \quad (1)$$

where $\bar{E}$ is the plane-strain modulus, defined as $E/(1 - \nu^2)$, with E as Young's modulus and ν as the Poisson ratio, the subscripts f and s refer to the top film and the bottom substrates, respectively. It is essential to highlight that the wavelength is independent of the applied strain and stress. On the contrary, the amplitude of the wrinkle pattern depends on the film thickness and the applied strain.

Different micrometric structures could be formed depending on the dimensions of undulated patterns obtained, known as the aspect ratio of the wrinkles (amplitude/wavelength). These structures could be defined as wrinkles (aspect ratio close to 1), ripples (aspect ratio close to 0), or crumples/folds (aspect ratio higher than 1). The mechanisms of how these structures are formed depend mainly on the material's mechanical properties and the rigid layer thickness [36]. Similarly, the type of 2D pattern formed on top of the material depends exclusively on the relaxation/compression directions applied to the systems. Figure 2 shows different 2D patterns (lamellar/stripes, herringbone, labyrinths, circular, radial, checkerboards, etc.) that can be formed just by varying the direction of the applied stress [37]. Mixing these micrometric structures (wrinkled patterns) with the surface responsiveness imparted by the smart polymers could resort in an interesting synergy that should improve the surface performance of the materials.

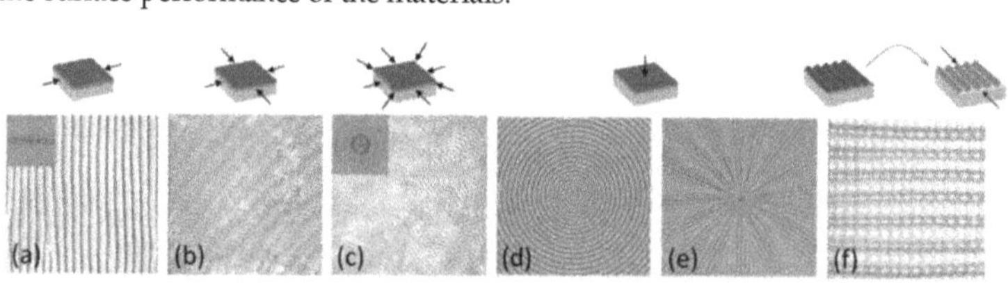

Figure 2. Optical images of various morphological patterns demonstrate that the wrinkling patterns prefer to orient themselves perpendicular to the axis of principal compressive stress: (**a**) stripes (uniaxial mechanical compression), (**b**) herringbones (biaxial mechanical compression), (**c**) labyrinths (isotropic thermal contraction), (**d**) circular, (**e**) radial, and (**f**) checkerboards (uniaxial mechanical compression on a pre-patterned substrate). Reproduced with permission from ref. [37]. Copyright 2011, John Wiley & Sons, Inc.

2. Desing of Adaptive Wrinkled Patterned Surfaces: Reversible Stimuli-Responsive Wrinkle Formation

As mentioned in the previous paragraphs, surface modification has proven relevant for different applications. Within these, homogeneous and reproducible dynamic surface patterns allow the control of encoded surface functions and properties, generating a convenient or stable method to achieve smart surfaces for different types of devices, whether in various scientific areas, such as optical, smart electronics, sensitive microfluidics, switchable wettability or tunable adhesion [38]. However, the fabrication of dynamic surface patterns remains a major challenge, highlighting the various approaches to fabricating surface patterns with different morphologies. The wrinkle phenomenon is an attractive alternative to generate dynamic patterns due to its spontaneous formation, versatility, easy preparation, scaling power, large area production, and sensitivity to different stimuli [39].

The interactions between the material and the stimuli are responsible for the smart behaviors of the material, which commonly occur in the interface between the material and the environment, so increasing this available surface area increases the device's smart capabilities. This last property is why wrinkling morphologies are desirable for fabricating smart devices based on polymeric compounds.

The following section discusses some of the most common materials and methodologies used to fabricate wrinkled patterns with reversible stimuli-responsive behavior. Thus, according to the type of stimulus used, these materials were organized in different topics like light-responsive (UV, visible, and NIR), pH-responsive, temperature-responsive, electro-magnetical responsive, mechanical (stress/stretch) responsive, and chemical reaction/gas/solvent responsive.

2.1. Light-Responsive Wrinkled Patterned Surfaces

The most used methodologies to fabricate spontaneous wrinkled patterns in the materials' surface involve light exposure from UV, visible, or NIR sources to induce surface modification.

Several examples of this strategy have been recently reported. For example, Omenneto et al. [40] presented regenerated silk fibroin to explore wrinkle formation by exploiting silk structure-function relation. More precisely, a bilayer system composed of silk/PDMS (polydimethylsiloxane) was fabricated, depositing the silk fibroin solution onto the PDMS substrates by spin-coating. Afterward, the bilayer samples were heated (and therefore stretched) at 140 °C for 5 min and then cooled to room temperature, thus triggering the formation of wrinkling patterned surfaces. In consequence, the protein conformational changes responsible for these surface changes can be induced by exposure to methanol vapor (MV), water vapor (WV), and UV light irradiation (Figure 3). Upon exposure to MV, the stress release behavior is quite fast, thanks to the rapid penetration of the small water molecules into all amorphous regions of the silk matrix of the composite. On the other hand, the mixture of methanol and water molecules penetrates the non-crystalline areas to induce a local molecular rearrangement without significant conformational transition.

In conclusion, the polymorphic transitions of silk fibroin allow orienting the wrinkled patterns by directly influencing the material's physical properties. The interaction between the silk protein chains and external stimuli controls the protein film's wrinkling. The silk/PDMS bilayer is mechanically firm and allows reversible and repeatable modification in its wrinkled and unwrinkled states in multiple cycles (more than 50 cycles). Figure 3b shows a possible application in which patterns can be revealed under IR imaging, offering the possibility of displaying IR information.

Jiang et al. provide an effective procedure for generating multi-responsive wrinkles based on a supramolecular polymeric network exposed to UV light. The polymeric network consisted of a pyridine-containing copolymer (P4VP-PS-PnBA) and an anthracene-containing carboxylic acid (AN-COOH) dynamically crosslinked by reversible photodimerization of anthracene (AN) and hydrogen bonding between the carboxyl and pyridine groups. The wrinkling pattern was selectively realized and erased by exposure to acid/base gases and light [41]. Cao, Jiang, and Lu et al. [42] have presented a method to control the formation and removal of dynamically wrinkled patterns by visible light exposure, which is generated by an external stimulus on an azo-containing poly(disperse orange 3) (PDO_3) film bonded to a PDMS substrate. Light-induced photoisomerization of the azobenzene units in the azo-polymer films triggers the release of tension, resulting in the elimination of wrinkles. In unexposed regions, wrinkles are also affected and rearranged perpendicular to the exposed area during the ordering.

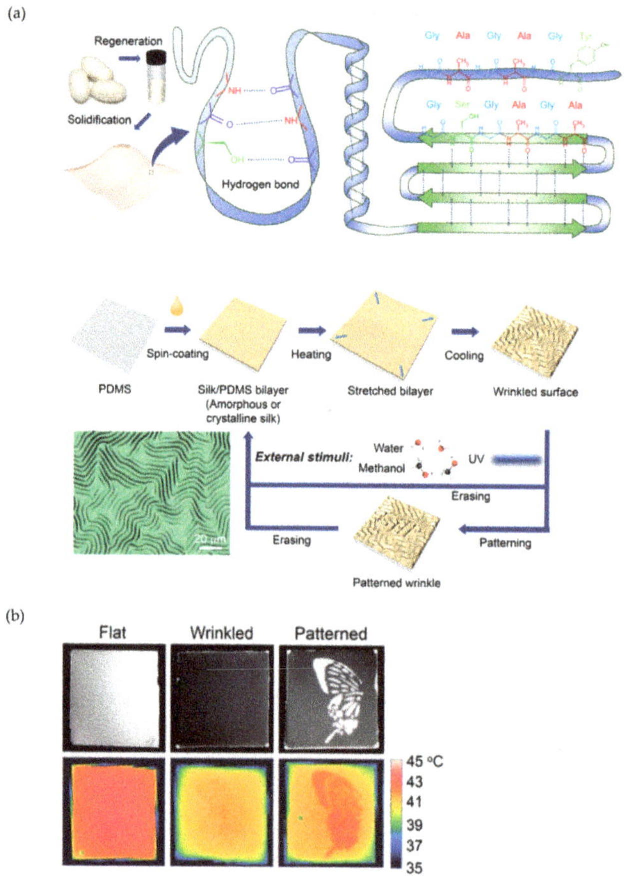

Figure 3. (**a**) Schematic representation of the reconstituted silk-fibroin molecular structure and schematic representation of the process to fabricate the reversible wrinkled patterns. (**b**) Photographs (top) and IR images (bottom) of the flat surface (left), surface with wrinkled structure (middle), and surface with a wrinkled pattern (right). Reproduced with permission from reference [40]. Copyright 2019, Proceedings of the National Academy of Sciences (PNAS).

Cao, Fery, and Lu et al. [43] showed a methodology to trigger various wrinkling patterns on the surface of azo-polymer composite substrate systems capable of switching between flat and wrinkled states via light incidence. The reversible patterning strategy is illustrated in Figure 4a, which is divided into two alternative routes. Route I is based on the exposure of a planar PDO_3/PDMS bilayer system to a high-intensity white light (>0.4 Wcm^{-2}), resulting in randomly distributed wrinkles. Notably, the surface wrinkles are gradually erased using a low light intensity of 15 mW cm^{-2}, allowing the recovery of the initial flat surface. In route II, a selective exposure of high-intensity light is applied by utilizing a mask or a high-intensity 450 nm laser beam with controllable parameters such as speed and movement. Figure 4b shows a schematic illustration of the approaches followed to fabricate dually oriented hierarchical wrinkle patterns. The soft/rigid boundary was generated by the PDMS substrate's selective oxygen plasma treatment (OPT) through copper grids, followed by spin-coating the PDO_3 film. As expected, hierarchically ordered wrinkles are formed after blanket exposure with parallel and perpendicular alignment in the softer D1 and harder D2 boundaries.

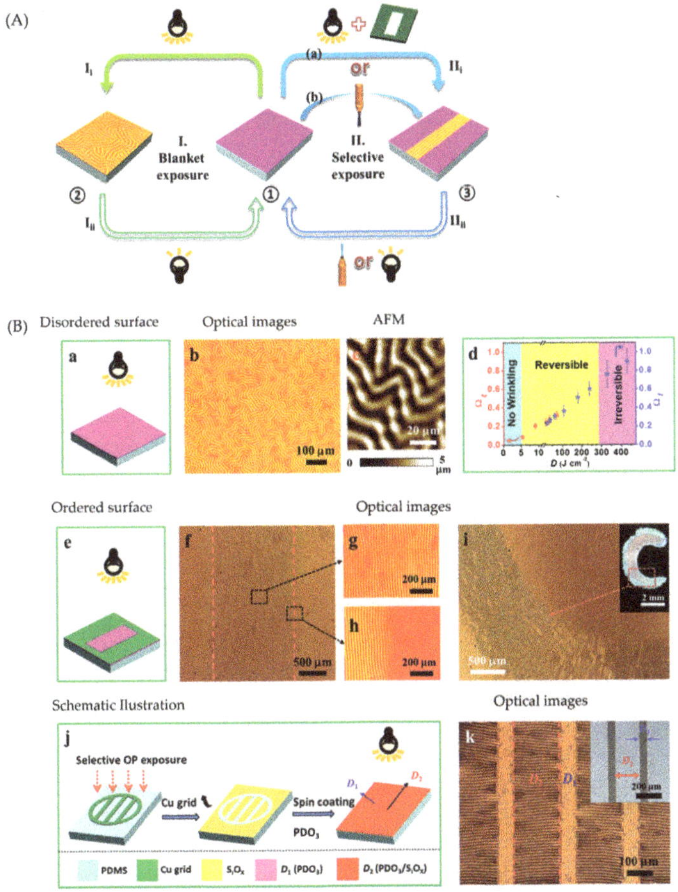

Figure 4. (**A**) Schematic illustration of the all-optical reversible surface patterning strategy and (**B**) Disordered/ordered surface wrinkling induced by the controlled exposure of the PDO3/PDMS bilayer to a high-intensity white light. Schematic illustration (a), optical (b), and atomic force microscopy (AFM) height (c) images of labyrinth wrinkles induced by blanket exposure. (d) The dependence of the aspect ratio (Ω_t and Ω_I) of labyrinth wrinkles on the exposure dose D from different exposure times t and light intensity I employed, respectively. Schematic illustration (e) and optical images (f–i) of oriented wrinkles resulting from selective exposure through an oblong-shaped mask with the width of 1.5 mm and the length of 6 mm (f–h) or "C"-shaped mask with the hole width of 1 mm (i). Schematic illustration (j) and optical image (k) of dually oriented wrinkles induced by blanket exposure, where the substrate was pre-exposed to oxygen plasma through a copper grid (insert of (k)). Reproduced with permission from reference [43]. Copyright 2019, American Chemical Society (ACS).

Another exciting example of reversible stimuli-responsive wrinkled patterns triggered by visible light was reported by Ma and Jiang et al. [44], who reported a reversible dual pattern that responded to light and acid treatment. The material employed is a supramolecular network comprising a copolymer containing pyridine (P4VP-nBA-S) and hydroxyl distyrylpyridine (DSP-OH), which exhibited blue fluorescence on its surface. It is noted that the wrinkle morphology and fluorescence color can be regulated by the isomerization of DSP-OH triggered by visible light or acid treatment (Figure 5a). Experimentally, the polymer blend was spin-coated onto a PDMS substrate, producing a stiff laminate with a thickness of 100 nm at the top. Thermal treatment at 110 °C introduced compressive stress into the system generating the wrinkle surface structure. The resulting hierarchical patterned surface becomes smooth and wrinkled reversible on irradiation with 450 nm visible light or acid treatment (hydrogen chloride gas). Simultaneously, its fluorescence changes dynamically from blue to orange-red due to the chemical structure's strong solid push-pull electronic effect. Figure 5b shows a possible application as an anti-counterfeiting platform related to a simple thermal treatment. After irradiation with 450 nm light through a QR code-shaped photomask, the wrinkled pattern of the QR code with blue fluorescence was carried out. When the sample was exposed to 56.4 ppm HCl vapor for 5 min, the wrinkles flattened, and the QR code fluorescence changed to purple or orange (10 min). After heating to evaporate the HCl, the wrinkles reappeared, and the QR code fluorescence returned to blue.

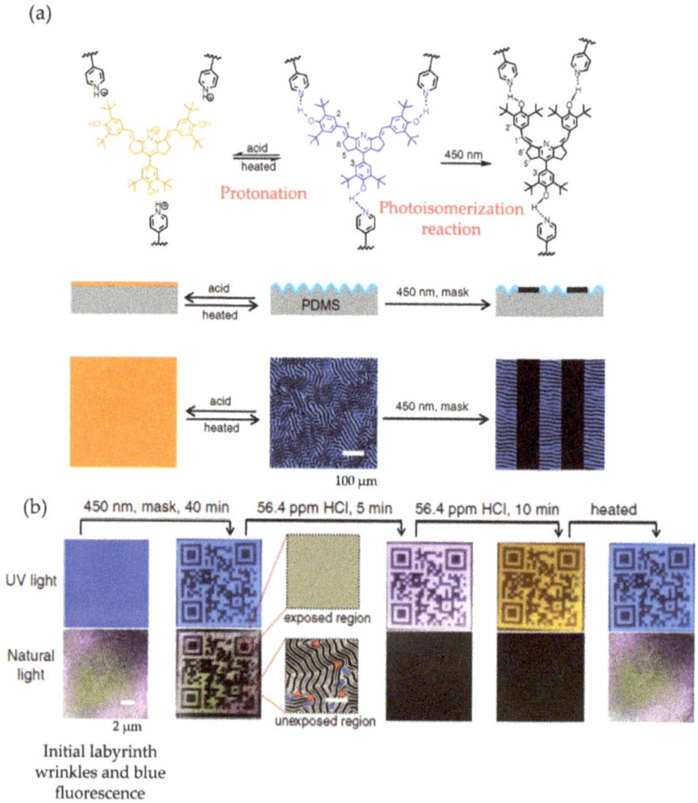

Figure 5. (**a**) Strategy to produce dynamic wrinkled patterns with tunable fluorescence using P4VP-nBA-S/DSP-OH, characterized by laser scanning confocal microscopy (LSCM), and (**b**) photographs of QR code based on wrinkled and fluorescent patterns for anti-counterfeiting applications. Reproduced with permission from reference [44]. Copyright 2020, Springer Nature.

At the same time, Zhang, Jiang, and Lu, et al. [45] have described a simple and effective strategy to control the transformation of the wrinkle structure and the resulting hierarchical micropatterns using an azo-type poly(disperse orange 3) (PDO3) film as a rigid top layer with different rigid interlayers (polystyrene and silica-type SiOx) bonded to a PDMS substrate.

A rigid photoinert interlayer induces a tension-relaxation effect in photomodulated wrinkles. This phenomenon is highly dependent on the material properties of the photoinert interlayer. Consequently, the wrinkles were transformed into highly oriented patterns in unexposed areas after selective exposure. On the other hand, they were either wholly erased or evolved into wrinkles of shorter wavelengths in the exposed part. These studies offer alternative routes for preparing hierarchical surface patterns in multilayer functional devices.

Similarly, Zhang and Zong et al. [46] have published a simple visible light illumination method to fabricate a wrinkle-forming innovative surface with photo-controllable hierarchical surface patterns using an epoxy-based oligomer possessing azobenzene. The epoxy oligomer was synthesized by ring-opening polymerization of bisphenol AF diglycidyl ether (BADFGE) with p-aminoazobenzene (AAB). The elicitation route is shown in Figure 6a. The appearance of surface wrinkling occurred through the circulation of heating/cooling cycles and photo-conformation due to photo softening and the release of stress induced by the photoisomerization of azobenzene. When illuminated with visible light by selective exposure of selected areas, wrinkles could be rapidly photo-eliminated to give rise to hierarchical surface patterns.

Furthermore, the time required to erase the exposed wrinkled regions depended highly on the wrinkles' wavelength and the light's power density (Figure 6b). Finally, as presented in Figure 6c, several patterns could be visualized by employing selective exposure to visible light. Unexposed regions that formed wrinkles showed lower transmittance and intense scattering.

NIR light could also trigger reversible transformations of stimuli-responsive wrinkled patterns. For example, Jiang et al. [47] presented a simple and general strategy for forming NIR-responsive dynamic wrinkle patterns with excellent reversibility and responsive capacity. PDMS elastomers containing a small amount of CNT (CNT-PDMS) served as the substrate for the bilayer system with various functional polymers as rigid top layers. NIR irradiation can again induce thermal expansion of the CNT-PDMS substrate, resulting in a decrease in the compressive stress of the bilayer system and the elimination of wrinkles. After removing the NIR source and cooling it to room temperature, the smooth surface returns to the original wrinkles. This research demonstrates that this strategy can be applied in situ to tune surface properties on demand. The same group [48] reported another method to elaborate a reversible double-dynamic pattern based on a bilayer system that allows simultaneous control of wrinkle topography and fluorescence. A copolymer containing anthracene (AN) and naphthalene diimide (NDI) (PAN-DNI-BA) was used as a rigid top layer. The emerging patterns assume reversible changes between the smooth and wrinkled states upon irradiation with light and subsequent heat treatment.

The fluorescence reversibly changes from red to white and finally to blue-green because of the tunable charge transfer interaction between AN and NDI in the PAN-DNI-BA copolymer. The smart surfaces with dynamic hierarchical wrinkles and fluorescence were achieved by selective irradiation with photomasks. They can be employed for potential applications in smart displays and anti-counterfeiting purposes. Similarly, Jiang, Russell, and Zhang et al. [49] present a promising method for producing long-range ordered patterns by exposing a photomask on a thin polymer film. A thin film of a light-sensitive polymer containing anthracene (LAP), supported on a polyethylene terephthalate (PET) film or glass sheet, was exposed to 365 nm UV light using a photomask for a specific time to initiate photodimerization of the anthracene. Thus, LAP photodimerizes to DLAP, and a gradient in DLAP concentration is generated, where the top layer of the film is highly crosslinked and has a higher elastic modulus than the sub-surface regions. Pattern control was carried out using a photomask, resulting in a compressive force only in areas exposed to UV light. This process gives rise to wrinkling or buckling while the unexposed area remains flat. Similar studies were reported by Jian, Li, and Lu et al. [50], incorporating an amorphous azobenzene-containing PDO_3 film as an intermediate layer to design a photosensitive multilayer-based film/substrate system.

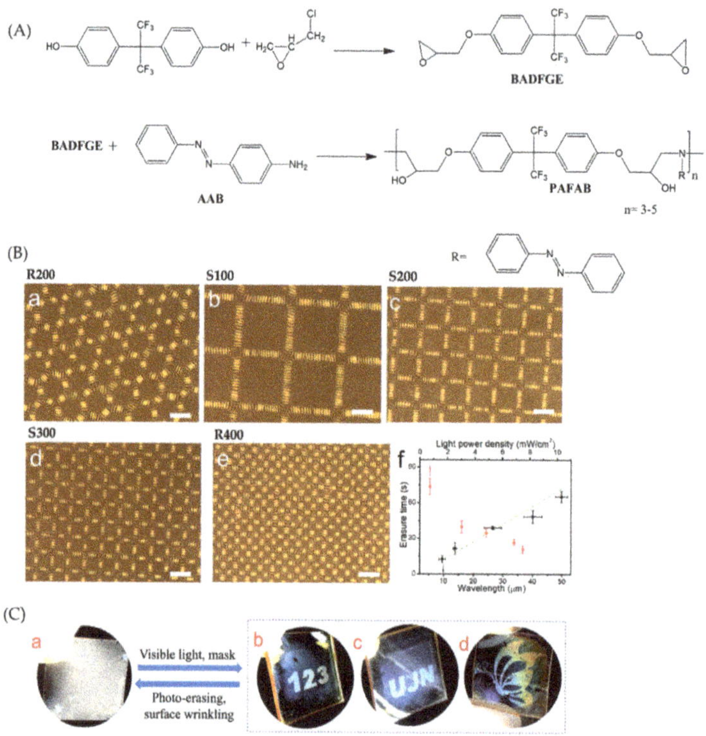

Figure 6. (**A**) Schematic illustration of the process used to synthesize the epoxy-based oligomer (PAFAB), (**B**) optical images of the patterns of wrinkles on the PDMS/PAFAB bilayer system irradiated with visible light through different copper grids (R200 (a), S100 (b), S200 (c), S300 (d), and R400 (e)) and their dependence of the optical erasure time on the light power density and the wrinkle wavelength (f) and (**C**) applications of the PDMS/PAFAB wrinkling system for rewritable information storage as a result of selective exposure to visible light: as-fabricated wrinkle-forming surface (a); various patterns of wrinkles obtained by selective exposure via different photomasks (b–d). Reproduced with permission from reference [46]. Copyright 2020, Royal Society of Chemistry (RSC).

Jiang et al. [51] reported a simple but effective anti-counterfeiting strategy using NIR-driven dynamic wrinkle pattern memory surfaces that can be used as dynamic biomimetic fingerprints. The generated wrinkle pattern works similarly to a fingerprint, which allows a higher level of anti-counterfeiting to be implemented, thanks to the non-deterministic process, unpredictability, and randomness of the wrinkle formation process. These dynamic wrinkles were fabricated using a bilayer system with CNT-PDMS (small amount of carbon nanotubes-CNTs- in polydimethylsiloxane) as an elastic substrate and a photo cross-linkable copolymer containing anthracene (PAN) as a rigid top layer, which was rotationally coated onto the elastomer. This dynamic fingerprint-like wrinkle's key point is that the wrinkle's minutiae remain unchanged during multiple cycles of wrinkle erasure/regeneration. When using heat treatment directly or IR irradiation, the imbalance between the bilayer's moduli and thermal expansion ratios produces compressive stress, resulting in randomly distributed wrinkled patterns, which possess fingerprint-like minutiae: ridge termination and bifurcation [52]. These stress-controlled wrinkles are reversible and sensitive to the bilayer's thermal expansion caused by IR irradiation.

In addition, they developed a fluorescent label that presents dynamic wrinkles for anti-counterfeiting some difficult-to-recognize objects (ancient Chinese Ruyao porcelain from the Song dynasty is tricky because the porcelain surface is usually smooth). The study was carried out by replacing PAN with anthracene (AN) and naphthalene diimide (NDI) motifs with a copolymer (PAN-NDI) as a top layer. The evolution of wrinkles was recorded by surface analysis using fluorescence microscopy. The same group [53] presents a methodology to regulate the wrinkle topographies by precisely controlling the spatial distributions of the modulus and boundary conditions of the top layer in bilayer systems composed of an anthracene-containing copolymer (PAN-BA) on a PDMS elastomeric substrate. The authors state that by this method, they could fabricate and regulate the wrinkles in 2D (Figure 7a).

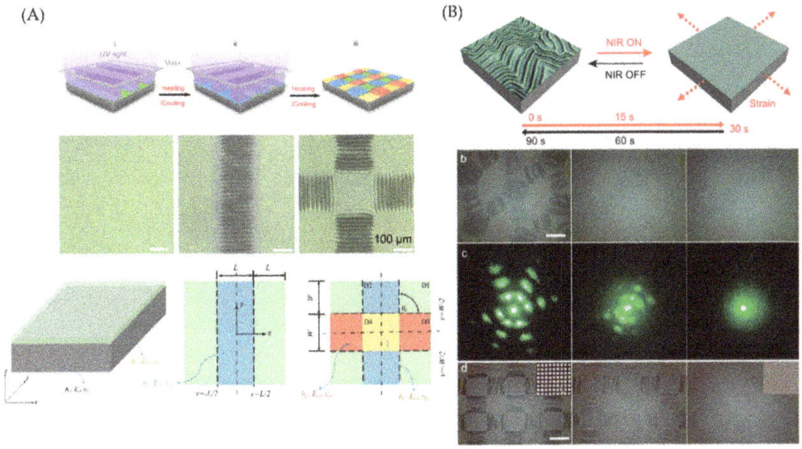

Figure 7. (**A**) Schematic representation of the light-controlled wrinkled pattern fabrication process via a lithographic approach, using 365 nm sequential UV from a 200 μm strip array photomask. (**B**) Illustration of the NIR-driven ordered wrinkles serving as a dynamic light grafting. (b) Optical images of the wrinkle disappearance/regeneration process by the NIR on/off switch. (c) The corresponding light diffraction patterns. (d) Optical images of Au-functionalized wrinkle disappearance/regeneration process by NIR on/off switch. Reproduced with permission from reference [53]. Copyright 2020, Science Publishing & Media Ltd. (Science Press).

By irradiating with 365 nm ultraviolet light for 15 min by a photomask strip, heating it at 70 °C for 30 min, and cooling, 1D wrinkles were spontaneously generated in the exposed domain. Posteriorly, the photomask was rotated horizontally with a certain angle. Thus, the 1D ordered wrinkled surface (one direction on a two-dimensional surface) was again exposed for 15 min, allowing 2D wrinkle formation (multi-direction on a two-dimensional surface) through heating at 70 °C and cooling treatment at 20 °C. This exposure sequence form two strip-like domains (two directions of Poisson's effect in the double exposure domain D4). Parallel, NIR was utilized to tune the resulting wrinkle by controlling the bilayer system's compressive strain; mainly, PDMS-containing carbon nanotube (CNT-PDMS, 0.05 wt%) was used as a substrate for the bilayer system. The formed wrinkled pattern can be erased in situ and in real-time due to the high efficiency of photon-to-thermal energy conversion (Figure 7b). Their thermal expansion can be reversibly controlled by applying 808 nm NIR irradiation to the CNT-PDMS substrate. When the wrinkled surface was irradiated with NIR, the diffraction patterns of 3rd generation disappeared rapidly within 30 s. By removing the NIR radiation, these diffraction patterns were gradually restored to their original state due to the regeneration of the wrinkles.

On the other hand, Bowen Wang et al. [54] reported a simple strategy to create micro/nanopatterns on a surface composed of a film/substrate bilayer based on a non-photosensitive azo-polymer and a soft substrate. In this system, light exposure generated wrinkled patterns on the material's surface. This strategy is illustrated in Figure 8a, which entails two alternative paths. In the first alternative, the planar (POD3-PS)/PDMS bilayer is exposed to a 450 nm laser beam selectively using the copper grating as a photomask, generating labyrinth wrinkles. The second method is based on a dynamic surface exposure that directs the pattern orientation perpendicular to the motion of the 450 nm laser. Therefore, the unexposed surface areas will maintain their basal state. As a result, micropatterns can be formed by writing the surface with the laser, in which the trajectory of this light visualized in Figure 8b is controlled, observing these parallel patterns in the drawn letters.

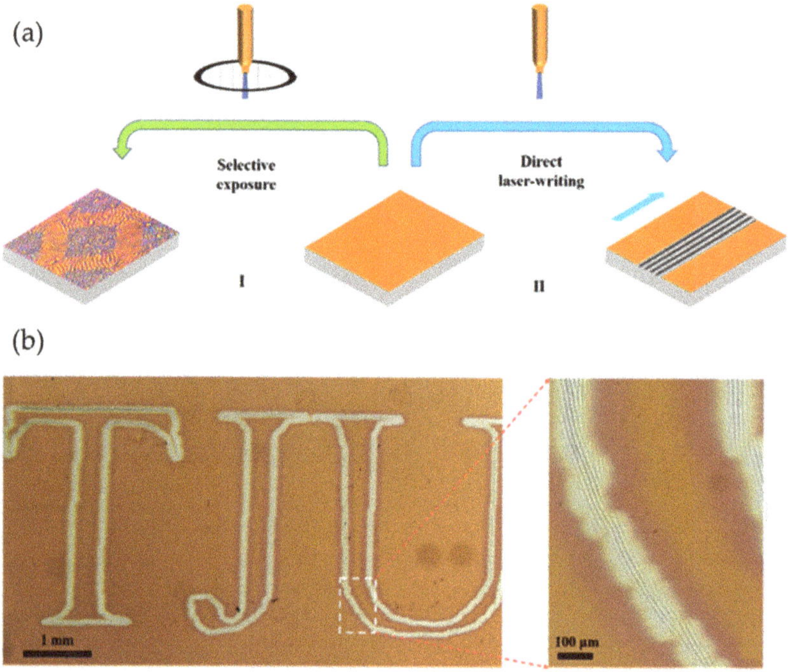

Figure 8. (**a**) Representative schematic of the wrinkle-inducing pathways upon exposure of a (PDO 3-PS)/PDMS bilayer to a 450 nm laser with a photomask (I) or a moving laser (II) and (**b**) Optical image of trajectory-guided oriented wrinkling (TJU) of (2% PDO3 - 2% PS)/PDMS induced by a moving high-intensity laser beam (e.g., 16.20 W-cm^{-2}/1.0 mm-s^{-1}). Reproduced with permission from reference [54]. Copyright 2022, MDPI.

Bai et al. reported the design of a viscoelastic monolayer composed of photosensitive poly(styrene-block-butadiene-block-styreneSBS-An) [55]. The viscoelastic role of the polymer layer is fundamental to understanding the wrinkle growth mechanism. The substrate does not have enough capacity to store energy, which is why the wrinkle formation process, being a kinetic process, is time-dependent. Figure 9a illustrates the wrinkle formation process, in which SBS-An was coated onto a rigid substrate and irradiated with 365 nm UV light. As a result, a crosslinking gradient was formed on the film surface. Figure 9b shows the evolution of the wrinkled morphology when the material is exposed to high temperatures (120 °C). First, a smooth surface is obtained; at 30 s, the first spherical cup patterns are observed, rapidly growing into laberinth-like patterns. These patterns increase amplitude and wavelength with time (from 60 to 600 s). Figure 9c shows the evolution of wrinkles with different temperatures (100, 110, 120, and 130 °C), showing an increase in amplitude with treatment time.

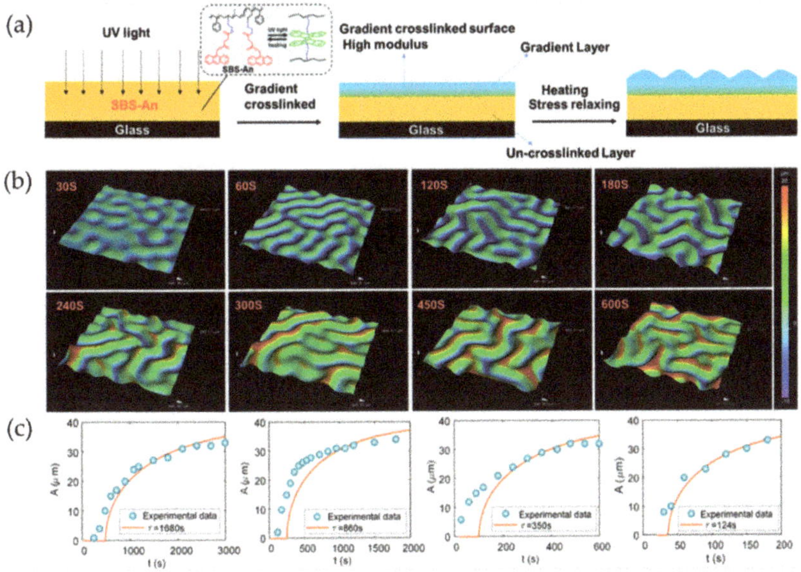

Figure 9. (**a**) Schematic representation of the self-wrinkling process, (**b**) Evolution of wrinkle morphology at 120 °C for different times, and (**c**) Morphology evolution at different temperatures (100, 110, 120, and 130 °C). Reproduced with permission from reference [55]. Copyright 2022, American Chemical Society (ACS).

Um et al. reported a one-step strategy to induce wrinkle formation in hydrogels using oil-water interfaces. The preparation method of wrinkled particles is illustrated in Figure 10a, where an oil phase containing the photoinitiator 2,2-diethoxyacetophenone (DEAP) flowed through microchannels. Subsequently, the oil phase flows to an aqueous phase containing polyethylenglycol diacrylate ($PEGDA_{700}$) polymer, forming droplets that travel to a reservoir to be exposed to UV light, producing crosslinked hydrogels. The morphology of the photoreticulated hydrogel particles will vary depending on the volumetric concentration of $PEGDA_{700}$ in the aqueous solution (Figure 10b). Solution concentrations lower than 20% v/v tend to form thin capsules with wrinkles at the surface, while those containing 30% v/v were similar to coffee beans with folds. Spherical particles without liquid content and prominent surface wrinkles are observed for concentrations higher than 30%.

Moreover, the authors observed that labyrinth wrinkles were produced for concentrations of DEAP at 2% (Figure 10c). After washing and drying, either small folds or ridges were created on the hydrogel surface. For concentrations of 6% to 10%, the labyrinth shape of the patterns was preserved (Figure 10d). However, for concentrations of 20%, no labyrinth patterns are obtained, but rather flower-like patterns are depicted (Figure 10e). The effect of DEAP concentration on the wrinkle wavelength (Figure 10f) and amplitude (Figure 10g) was also determined.

In the following table (Table 1), most of the examples for light-responsive materials previously mentioned are summarized, emphasizing the base material, methodology used to fabricate the wrinkled patterns and the possible application of the device. The examples were ordered according to the type of light used in each case (visible, UV, or NIR light). As is possible to observe, in general, independently of the material used, the process used to fabricate the wrinkles is similar in all the cases.

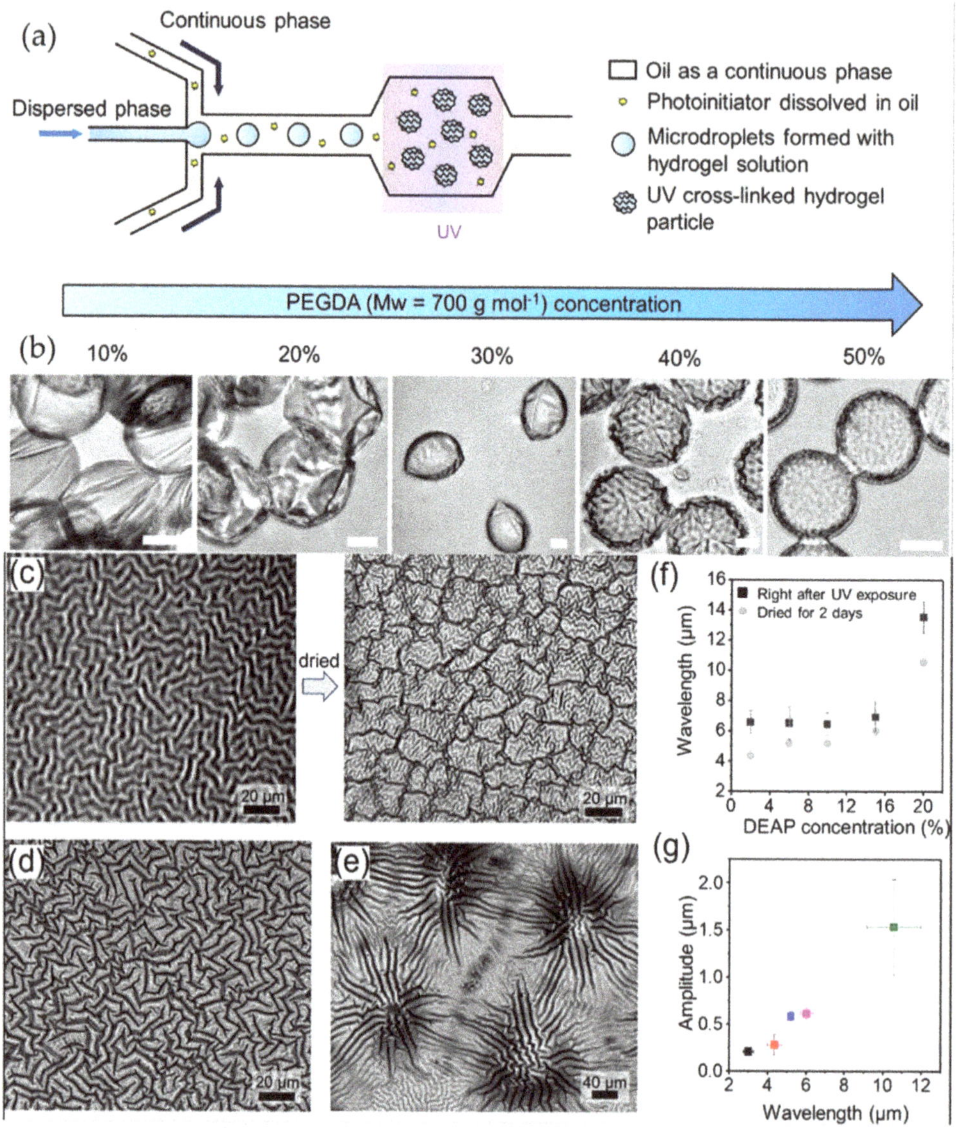

Figure 10. (**a**) Schematic representation of wrinkled hydrogel particles formation, using an oil-water interface, (**b**) Effect of polymer concentration on hydrogel morphology, (**c**–**e**) Labyrinth wrinkle formation and its contraction, and (**f**,**g**) Effect of photoinitiator in wrinkles dimensions. Reproduced with permission from reference [56]. Copyright 2021, American Chemical Society (ACS).

Table 1. Summary of light-responsive materials listed in Section 2.1.

Base Material	Responsive Light Range	Wrinkling Methodology	Posible Application	Ref.
Pyridine-containing azopolymer	Visible light	- Exposure to high-intensity white light - Selective exposure with a mask - Oxygen plasma exposure	Rewritable data storage	[41–43]
Pyridine-containing polymer		- Thermal treatment at 110 °C - Compressive stress - Light exposure	Anti-counterfeiting	[44]
Epoxy-based azobenzene		- Heating to 90 °C for 1 h - Cooling to RT - Light exposure	Rewritable data storage	[46]
(POD-PS)/PDMS		- 450 nm laser exposure - Cooling to RT - Selective exposure with a mask	Rewritable data storage Anti-counterfeiting	[54]
PET/LAP	UV light	- Heating to 90 °C for 1 h - Cooling to RT - Light exposure	Photonic devices Cell culture patterns Flexible sensors	[49,50]
Silk/PDMS		- Heating at 140 °C (5 min) - Cooled to RT - UV Light exposure	Rewritable data storage Anti-counterfeiting	[40]
CNT/PDMS/PAN	NIR light	- Heating to 70 °C - Cooling to RT - NIR exposure	Anti-counterfeiting	[51–53]
CNT/PDMS		- Heating to 80 °C for 3 min - Cooling to RT - UV light exposure for 40 min	Smart displays Dynamic gratings Light control electronic	[47,48]

2.2. Alternatives for the Fabrication of pH-Responsive Wrinkled Surfaces

Several examples have been reported where the interface responds to stimuli like pH or humidity. Sun et al. [57] reported on creating an intelligent film that can reversibly change its wrinkling pattern in response to changes in ambient humidity. The films were formed by depositing hydrophobic SiO_2 nanoparticles (NPs) on wrinkled polyelectrolyte surfaces. Experimentally, the substrate (glass and silicon wafers) was treated and bathed in an aqueous solution of poly(diallyl dimethyl ammonium chloride) (PDDA) to achieve a positively charged surface. Subsequently, the PDDA-modified substrate was immersed in an aqueous solution of poly(acrylic acid) (PAA) and poly(allylamine hydrochloride) (PAH), thus generating a multilayered film. On top of the films (PAA/PAH), a SiO_2 NP layer was formed via immersion. This cycle was repeated to obtain multilayers of (SiO_2/PAH)*m films (where m refers to the number of deposition cycles and a middle number means that SiO_2 NPs are in the last layer). The multilayer was then exposed to thermal crosslinking and subsequent deposition of 1H,1H,2H,2H-perfluoro-octyltriethoxysilane (POTS) on (SiO_2/PAH)*m (PAA/PAH)*30 films by a chemical vapor deposition (CVD) process.

Figure 11a shows wrinkled surface patterns formed based on the ionization ratio of carboxylate groups of the bottom PAA layer of the (PAA/PAH) films at different pHs. The PAA and substrate interaction seems to be reduced when the films are introduced into an acidic aqueous solution (pH < 4.5). After water adsorption, the compressive stress of the film exceeds the confining stress of the substrate, leading to a complete exfoliation of the film from the substrate. Likewise, by modifying the solution to pH 6.5, the interaction between PAA- and PDDA-modified substrates cannot be weakened by immersing the films in a more basic aqueous solution. In this case, the compressive stress generated by the film's swelling is lower than the confining stress of the substrate, which prevents the formation of wrinkles. When the crosslinked (PAA/PAH) films are in a water solution with 5.5 pH, the interaction of the films with the PDDA-modified substrate is moderately weakened. These interaction threshold values allow for partial separation of the films from the underlying substrate upon water adsorption. The separated areas of the film tend to crease, while the areas on the left adhere to the underlying substrates. Thus, wrinkles form in the crosslinked films (PAA/PAH). Figure 11b shows the creation and disappearance of the wrinkled surfaces as the films are subjected to different ambient relative humidities (80% RH and 40% RH).

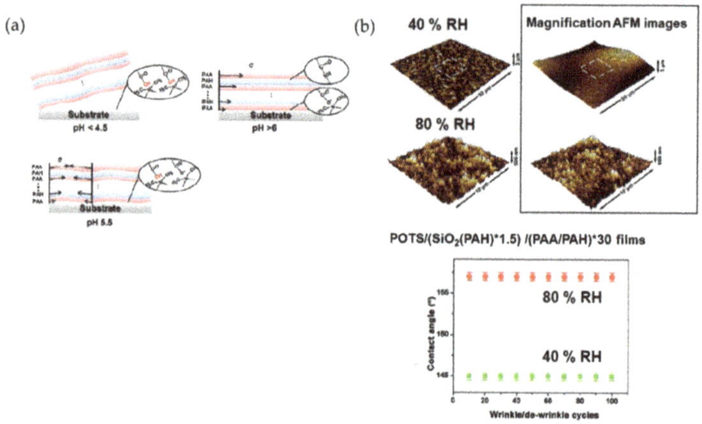

Figure 11. (**a**) Schematic illustration of the interaction of the PAA/PAH films with the PDDA-modified substrate after being immersed in water with different pHs (<4.5; 6> and 5.5.) and (**b**) AFM images and water contact angle changes of the films during repeated de-wrinkle/wrinkle process in 40% RH and 80% RH. Reproduced with permission from reference [57]. Copyright 2016, Springer Nature.

In wrinkled films, structure colors appear due to the interference of reflected light on the wrinkled film surface. The films gradually form wrinkles, and the water contact angle increases with ambient humidity. Nanosized aggregates of SiO_2 NPs are observed on the surface of the non-structured film, and the roughness is monitored by the AFM images of the wrinkles films increase to 256 nm. The researchers indicated that the surface roughness of wrinkled and unwrinkled films significantly influences their wettability and, therefore, can influence their degree of transparency. As a result, the composite can spontaneously change from transparent hydrophobicity to translucent superhydrophobicity through moisture-induced wrinkling/unwrinkling [57].

Similarly, Lu et al. [58] fabricated a film/substrate composed of $(PAA/PEG)_n$/PDMS to investigate the time evolution of swelling/deswelling-induced wrinkle pattern formation. Poly(acrylic acid)(PAA) and poly(ethylene glycol) (PEG) were deposited on PMDS via a layer-by-layer dipping technique, followed by heating-induced covalent crosslinking. Later, two types of systems were fabricated, Pt/$(PAA/PEG)_n$/PDMS and polystyrene (PS)/$(PAA/PEG)_n$/PDMS. In situ optical microscopy imaging monitored the morphological transformation in the covalently bonded (PAA/PEG)n/PDMS system. Thus, when the wrinkled (PAA/PEG)n/PDMS system was immersed in an acidic aqueous solution of pH = 2.5, surface wrinkling patterns occurred, including forming primary wrinkles with a labyrinth-like structure and second-generation wrinkles, which exhibit ordered dimples of hexagonal arrangement and bridging maze patterns. The morphological evolution can be finely tuned considering the thickness of the film $(PAA/PEG)_n$, the additional inert outer layer (Pt and PS), and the swelling directed by the solution pH. The introduction of an extra

Pt layer, resistant to swelling/deswelling, improved the wrinkles' stability, including the heat-induced primary and swelling-induced secondary patterns, leading to increased dimple periodicity. A similar evolution was obtained concerning the PS/(PAA/PEG)n/PDMS system during water swelling. Additional layer deposition was quickly applied to fabricate the desired multifunctional film/substrate systems for tunable stress-relief patterns and external applications.

Another interesting example was reported by Yang et al. [59], who synthesized poly(N-isopropylacrylamide-co-acrylic acid)/copper sulfide (P(NIPAM-co-AA)/CuS) composite microspheres which presented wrinkled zigzag surfaces formed through the in situ biomimetic mineralization reaction between H_2S and Cu^{+2} (Figure 12a). The main objective is to generate wrinkled patterns on a hydrogel film by deposition of a sulfide on the surface, where the amount deposited and the distribution of the sulfide affect the surface morphology obtained by forming these patterns. The effect of acidity and ionic strength of the $Cu(Ac)_2$ solution were analyzed using different solutions of $Cu(Ac)_2$-HAc, $Cu(Ac)_2$-H_3BO_3, $Cu(Ac)_2$-ethylenediamine and buffer solutions of pH = 5.6 and pH = 9.2. In the case of microgels in $Cu(Ac)_2$-HAc solution, small wrinkles are formed due to the degree of acidity of HAc (Figure 12b,g). In contrast, hydrogels in $Cu(Ac)_2$-ethylenediamine solution (Figure 12d,i) and $Cu(Ac)_2$-H_3BO_3 (Figure 12c,h) present a similar structuration between them. When these microgels are under acidic (Figure 12f,k) or basic (Figure 12e,j) buffer solution, they present smaller wrinkles because the buffer influence eliminates the effect of proton release during precipitation. However, the wrinkles in a basic buffer tend to increase their dimension compared to the acidic medium.

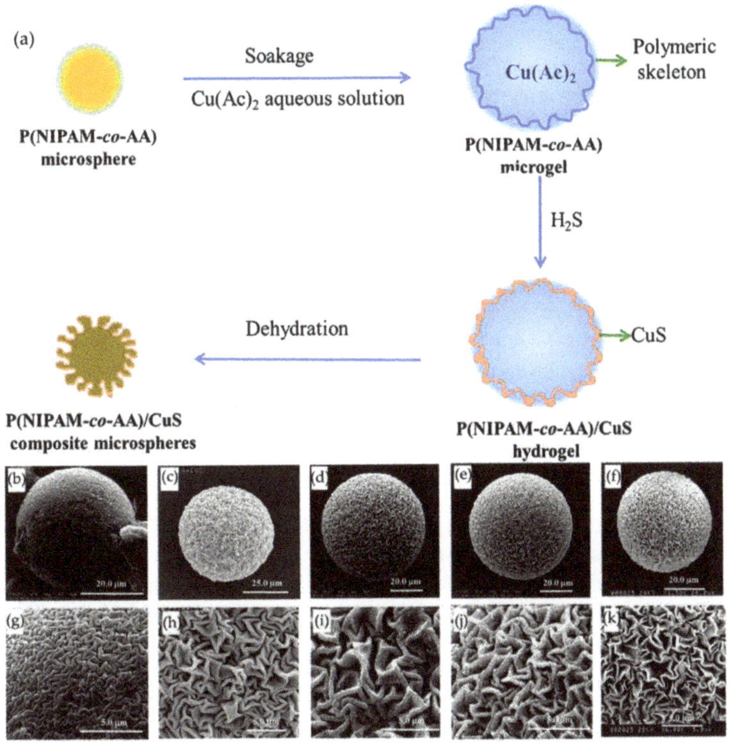

Figure 12. (**a**) Preparation of (P(NIPAM-*co*-AA)/CuS) microspheres, SEM micrographs P(NIPAM-*co*-AA)/CuS by adding HAc (pH = 3) (**b**,**g**), H_3BO_3 (**c**,**h**), ethylenediamine (**d**,**i**), pH = 5.6 (**e**,**j**) and pH = 9.2 buffer solution (**f**,**k**). Reproduced with permission from reference [59]. Copyright 2021, Royal Society of Chemistry (RSC).

Table 2 summarizes the previously mentioned examples for pH-sensitive materials, their methodology for fabricating the wrinkled patterns, and their possible application. The examples were ordered depending on their responsive pH range used in each case. Unlike the examples in the previous section (Table 1), the process of obtaining the wrinkles widely differs in this case.

Table 2. Summary of pH-sensitive materials listed in Section 2.2.

Base Material	Responsive pH Range	Wrinkling Methodology	Possible Application	Ref.
PAA/PEG/PDMS	2.5–7.0	- Heating at 200 °C for 1 h - Immersion in acidic solution pH 2.5	Tunable stress-relief patterns	[58]
PAA/PAH/SiO_2	4.5–6.5	- Immersion in acidic solution pH 2.0 - Immersion in pH 5.5 solution for 5 min - Exposure to 40% RH environment	Morphology-controlled smart superhydrophobic films	[57]
PNIPAM-co-AA/CuS	5.6–9.2	- Deposition of CuS - Freeze-drying for 1 day	Biomimetic mineralization	[59]

2.3. Wrinkle Formation in Systems Incorporating Temperature-Responsive Polymers

Another way to produce stimuli-responsive wrinkled patterns is by using polymers that display temperature-responsive properties. Zhang and Sun et al. [60] studied three types of film-substrate bilayer devices that consist of a stiff hydrophilic polyvinyl alcohol (PVA) film tightly adhered to a hydrophobic soft PDMS substrate (PVA-PDMS film-substrate systems). Some samples exhibit reversible wrinkles forming-disappearing behavior, while others only temporarily show wrinkles formation at the beginning of the moisture exposure, which disappears during the subsequent moisturizing and never reappears in the following drying/remoisturizing. Interestingly, another sample could be prepared that evidenced the formation of permanent wrinkles once exposed to moisture, regardless of the following moisturizing and drying conditions. The design was performed by adjusting precisely the PVA thin film's moisture-dependent responsive properties by controlling the thickness and crosslink gradient and the film-to-substrate thickness ratio. The evolution of wrinkle formation was manifested as the change in the transparency of the bilayer material.

Similarly, Tokudome and Takahashi et al. [61] report the fabrication of adaptive microarchitecture that exhibits a particular peristaltic motion by employing hydrogel actuation on bilayered materials. They designed and fabricated a bilayer structure of poly(N-iso-propylacrylaimide) PNIPAM and hybrid silica. The thickness of the hybrid silica layer was tuned from 680 nm to 1.5 µm using Pluronic F127 as a viscosity improver. 3-(methacryloxy)propyl triethoxysilane was used as a silica source to allow copolymerization with NIPAM and a crosslinker (N,N′-methylenebis (acrylamide), MBA) at the interface of the heterogeneous layers. PNIPAM undergoes shape changes at the lower critical solution temperature (LCST) [62], and therefore, submillimeter wrinkles appear and disappear as a function of the changing environmental temperature. The wrinkles were found to show a peristaltic motion on cooling from 36 °C to room temperature. The periodic length of wrinkles (λ) increases with the crust layer's thickness (t). It also shows that the peristaltic motion was achieved on a patterned hybrid silica layer (triangle-shaped patterning) placed on top of the PNIPAM. Oriented anisotropic wrinkles form on cooling because the patterning can accumulate anisotropic stress. Stress relaxation occurs in the x-direction, and the stress is uniaxially accumulated in the y-direction to form anisotropic parallel wrinkles.

Xiao and Wang [63] studied heat-responsive shape memory polymers (SMP) bonded to PDMS substrates to form bilayer systems able to form surface microstructures. Similar examples were reported by Wang et al. [64], who showed a new strategy for developing moisture and temperature dual-responsible surface wrinkles on PVA/PDMS bilayer films by rational design of the PVA skin layer. The surface wrinkles are produced by uniaxially stretching and subsequently releasing the bilayer films because of the mismatching in elastic moduli of the PVA and PDMS and return to the flattened state when exposed to moisture or high temperature (Figure 13a). For the formation of the initial surface wrinkles, the PVA/PDMS films were maintained under the 70% RH conditions for 10 min before stretching, applying then 15% strain (ε) for 10 s and finally releasing the strain. To decrease the modulus of PVA by moisture or temperature, increase ε_c for the bilayer film, resulting in $\varepsilon < \varepsilon_c$ and the disappearance of surface wrinkles (Figure 13a).

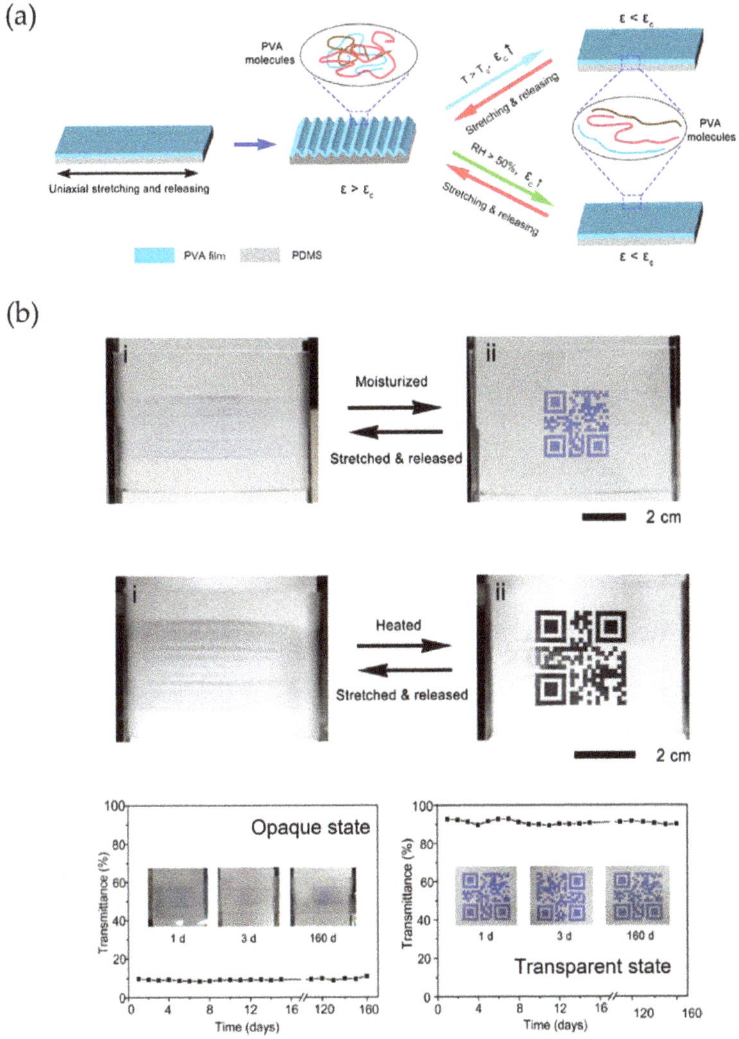

Figure 13. (a) Schematic illustration showing the fabrication process and reversible wrinkling/dewrinkling mechanism of moisture and temperature dual-responsive surface wrinkles on the PVDA/PMDS film. (b) Dual stimuli-sensitive smart windows. Digital photos of humidity-driven smart windows (top, i: stretched and released, ii: moisturized). Digital photos of temperature-driven smart windows (middle i: stretched and released, ii: heated) and transmittance of smart windows as a function of storage times (bottom). Reproduced with permission from reference [64]. Copyright 2019, American Chemical Society (ACS).

Figure 13b shows digital photos of the smart windows actuated by moisture and temperature, respectively. The wrinkling films are highly opaque, which can hide the films' objects; when the bilayer films were exposed to moisture at 80% RH or temperatures over 70 °C, the films rapidly changed to a transparent state to show the object behind clearly. Finally, it shows the wrinkles/dewrinkling film actuated by moisture and temperature stability with high transmittance for over five months at ambient room conditions.

Another example was developed by Xiao et al. [65], who utilized self-assembled surface wrinkling induced by SMPs and localized thermal expansion caused by Joule heating. SMP was synthesized by mixing tert-butyl acrylate (tBA) and crosslinker poly(ethylene glycol) dimethacrylate (PEGDMA) at a ratio of 4:1 and 2,2-dimethoxy-2-phenylacetophenone as a photoinitiator. The solution was injected into a mold with a piece of heating fire (FeCrAl alloy) placed at the center of the mold and cured under a UV lamp. Figure 14a shows a schematic representation of the process followed to obtain temperature-responsive wrinkled patterns; firstly, the SMP substrate was uniaxially pre-stretched with 5% strain (60 °C), using a dynamical mechanical analysis tester. After the sample was quenched to 25 °C, the strains were released, and the temporary and programmed shape was retained. Then, a layer of thin aluminum film of 100 nm was deposited on the programmed SMP substrate's surface using a thermal evaporator. Afterward, the SMP was heated to 60 °C to induce shape recovery in the substrate, leading to wrinkling aluminum film. This behavior can be observed in Figure 14b through optical microscopy images of the surface morphology during the reversible tuning process. According to the profile, the smooth and wrinkled states match well with their correspondents in different cycles, demonstrating excellent repeatability and reversibility.

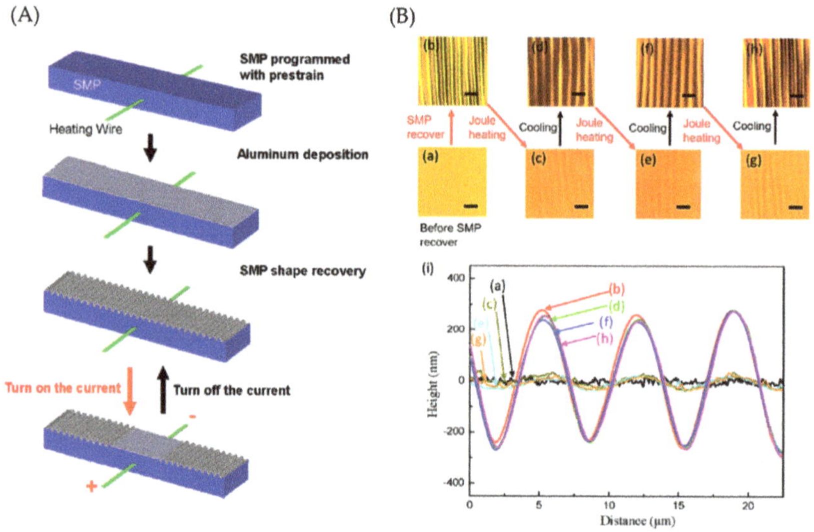

Figure 14. (A) Schematic illustration of the fabrication process to fabricate wrinkled patterns on SMP and locally flatten the surface using joule heating and (B) optical microscopy images of the surface in the reversible and repeatable tuning process (a)–(h) and the corresponding profiles (i) for Joule heating with a current of I = 0.45 A and heating for 3 min and cooling for 5 min. Reproduced with permission from reference [65]. Copyright 2019, AIP Publishing LLC.

Lee et al. [66] presented the fabrication of an interface composed of a polymer electrolyte membrane (PEM) and a hierarchically wrinkled catalyst (CL) layer (PEM/CL) in a PEM fuel cell (PEMFC) by bottom-up methods (Figure 15). Heat-induced repetitive mechanical deformations between a rigid top skin and a soft bottom substrate were used to fabricate these wrinkles. In Figure 15a, three different types of wrinkled PEMs were prepared. Heat-induced mechanical deformations were used to obtain these, with the thermal expansion mismatch between the bilayer causing deformation in the whole area by stress release. In Figure 15b, hierarchical nested wrinkled patterns (1st generation, which they named "G1", similarly 2nd generation "G2" and 3rd generation wrinkle patterns "G3") are fabricated using an argon plasma-treated polystyrene (PS) film that is subsequently heated to a temperature above its glass transition (Tg) and allowed cool at room temperature. Figure 15c shows the process of transferring wrinkled patterns to surfaces. PDMS was used to obtain the inverse morphology of the hierarchical patterns of PS via mask peeling. To perform the Nafion thermal lamination process, the hierarchical PDMS was placed on this layer by applying heat and pressure to the membrane to print the PDMS wrinkles, obtaining a Nafion membrane with a hierarchical nested wrinkled pattern. Finally, this film was coated with a CL layer using spray coating to form the wrinkled PEM/CL interface.

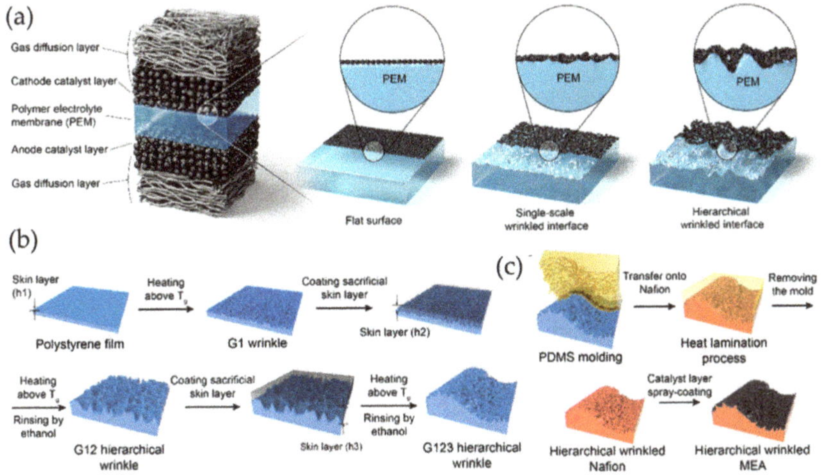

Figure 15. Schematic illustrations of hierarchical wrinkled PEM/CL fabrication. (**a**) The types of hierarchical wrinkled PEM/CL interfaces in PEMFC. (**b**) Fabrication of hierarchical wrinkles through repetitive heat-induced wrinkling processes and (**c**) Transfer of wrinkled hierarchical structures to Nafion membrane by thermal lamination and subsequent CL coating. Reproduced permission from with reference [66]. Copyright 2022, American Chemical Society (ACS).

In Table 3 , it is possible to observe a summary of the examples mentioned for thermo-sensitive materials used to create smart wrinkled patterns. The table also depicts the methodology used to fabricate the wrinkles and its possible application. In this case, and similarly to the one observed in Table 1, independently of the type of material used, the process employed to fabricate thermo-sensitive wrinkled patterns is quite similar.

Table 3. Summary of thermo-sensitive materials listed in Section 2.3.

Base Material	Responsive Temperature Range	Wrinkling Methodology	Possible Application	Ref.
PNIPAM and hybrid silica	RT—36°C	- Sample drying for 15 min - Sample stretching/releasing - Hybrid silica deposition - Heating to 36 °C	Soft micro-actuators	[61]
PEGDMA/tBA	RT—60 °C	- Sample stretching/releasing (5% strain) at 60 °C. - Cooling to RT - Aluminum film deposition	Smart micromirror	[65]
PVA/PDMS	RT—80°C	- Sample stretching/releasing (15% strain) - Exposure to moisture (70% RH) or high temperature (80 °C)	Smart windows	[63, 64]
PS/PEM/CL	RT—100 °C	- PS argon plasma exposure - Heating to 100 °C - Cooling to RT - Deposition of CL layer	Fuel cells	[66]

2.4. Electro-Magnetical Response

Another way to create reversible sensitive wrinkled patterns is by using an electric field stimulus to induce changes in the material surface. The use of electrical sources is attractive due to its ability to be reversible, to have the potential to be miniaturized, to generate an immediate response, to have the scope to integrate functions such as auto-sensing devices, and the possibility of being able to control it in a programmable way. It has been displayed that Van der Waals forces and electrostatic interactions acting through an air gap (<100 nm) between a soft coating and a rigid plate can induce out-of-plane wrinkle instabilities, thus stabilizing stable patterns on top of the material.

Indeed, Van den Ende and Steeneken et al. [67] presented feasibility studies to trigger out-of-plane wrinkle instability by applying an electric field to an elastomer. The acrylate-based elastomeric layer is deposited, and UV cured on an indium tin oxide (ITO) coated glass panel. Thin gold top electrodes are deposited on top of the elastomer layers. When a high voltage is imposed between the upper and lower electrodes, a striking change is observed in the visible appearance of the sample. Thus, the destabilization of the wrinkles generates significant changes in the images reflected or transmitted through the elastomer-coated rigid plates.

Similarly, Kar-Narayan and Smoukov et al. [68] reported the fabrication of controllable regular/irregular wrinkled patterns by a selective electron response system for a responsive polymer coating on underlying counter electrode patterns without requiring mechanical pre-stretch stimuli or the induction of anisotropy in the film. The electrode patterns were created by an aerosol jet printing method, introducing silver nanoparticles to form patterns on the bottom electrode. This research group used a rigid dielectric layer (Novolac bisphenol A epoxide, or SU-8) spin-coated onto the electrode layer (PDMS substrate). As a result, they determined that the polymer acts as a single electrode when there is minimal separation between the electrodes underneath it. The authors established that the regularity and direction of new patterns are influenced by the electrode spacing and defined parameters that regulate the alignment of the wrinkles. Figure 16 shows a schematic representation of fabricating wrinkled patterns with a selective electrical response. These results exhibited that applying this type of customizable wrinkle/electrode patterning with a light reflection-diffusion-grating device works. This technique produces complex responsive surfaces that possess a high electrical capacity.

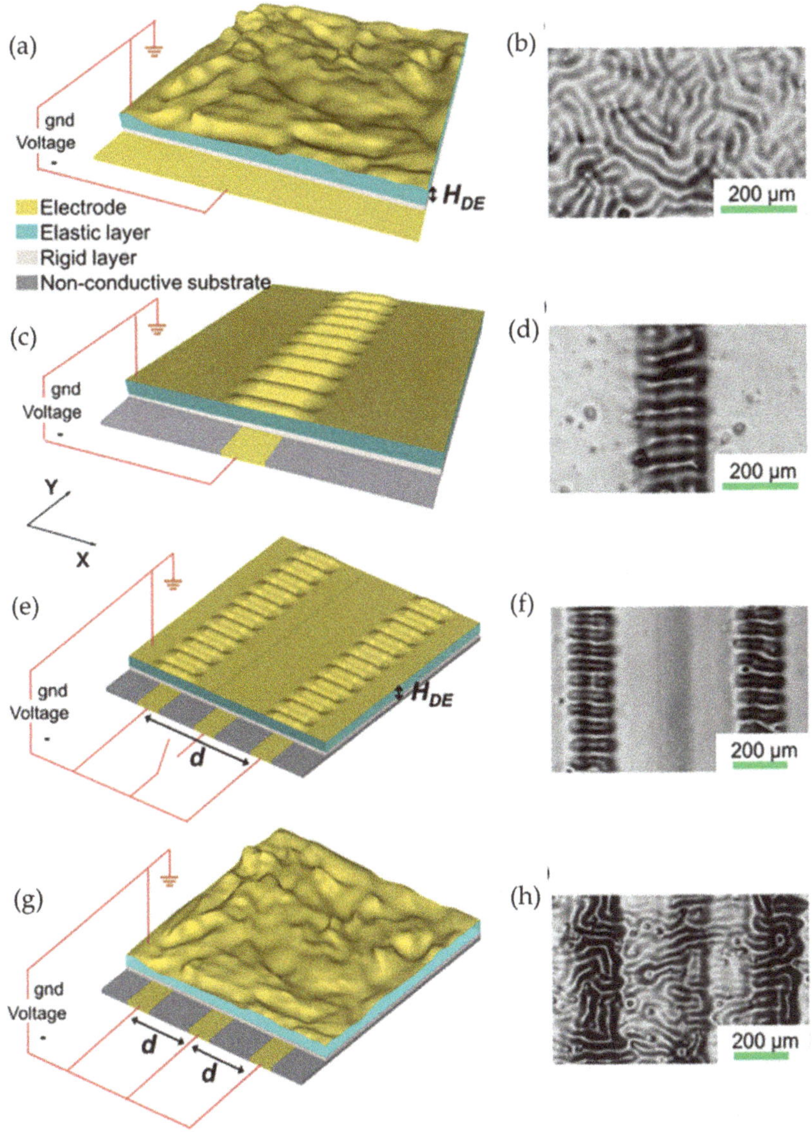

Figure 16. Schematics of the layered structure and operation with different electrode designs and the corresponding optical microscope images. (**a**,**b**) Randomly-oriented wrinkles, (**c**,**d**) When a single narrow electrode and (**e**,**f**) two electrodes spaced by a large distance *d* are used. (**g**,**h**) Randomly-oriented patterns when the spacing between narrow electrodes is smaller than a critical value. Reproduced with permission from reference [68]. Copyright 2020, Elsevier Ltd.

On the other hand, Danas et al. [69] synthesized a polymer with a magnetic response (MRE, magnetorheological elastomer) which is attached to a passive substrate (soft non-magnetic polymer). Additionally, the MRE film includes iron particles isotropically and randomly embedded in the elastomer [70], resulting in no exhibition of magnetic hysteresis, producing that upon removing the magnetic fields, the wrinkled patterns tend to disappear. The analysis demonstrates that due to the robust magnetoelastic interaction, the high pre-compression, and the increasing magnetic field, the wrinkled pattern formed on top of the material tends to evolve from a smooth pattern to a complex one (commonly called crinkling). Thus, by increasing the pre-compression to $\lambda_0 = 0.8$, it is possible to reduce the magnetomechanical wrinkling in the film. Further pre-compression increase generates a decupling between the mechanical and magnetical load effects. Danas et al. conclude that the friction in the film could interact with the magneto-mechanical wrinkling mode produce a decrease in the film thickness.

2.5. Mechanical (Stress/Stretch) Response

Mechanoresponse is an unusual phenomenon that occurs in some materials when a mechanical force is applied to them. This external mechanical stimulus presents several advantages, like manipulating the deformation time, direction, and strain/stress used. Recent studies have corroborated that mechanosensitive surface wrinkles possess dynamically adjustable, repeatable, and practical optical properties, which change their surface topography in response to mechanical stimulus. Indeed, Wang et al. [71] in 2018 reported a facile technique to fabricate PVA/PDMS bilayer films with various materials, including PVA, CS (chitosan), HEC (hydroxyethylcellulose), and the mixture of PVA and CS with mechanoresponsive surface wrinkles. The PVA/PMDS bilayer film surface wrinkled patterns exhibit a high optical response to minor mechanical stress, extensive optical switching cycles, and a wide range of transmittance modulation, attributing excellent mechanoresponsive optical switching properties. They test these systems as smart windows obtaining incredible results.

Some materials, such as mechano-chromic materials under the mechanical stimulus, exhibit a change in their optical properties, such as absorption, emission, and reflection; for example, Yin et al. [72] studied the wrinkling effect of PDMS films as a function of their thickness. In this work, the authors demonstrate that the periodicity and equality of wrinkles formed under mechanical stress are related to the thickness of PDMS-based films. Experimentally, films of different thicknesses were subjected to pre-stretching with a given tension and subsequently coating with a mask, followed by plasma treatment. Afterward, perpendicular wrinkles were obtained on the rigid surface (stretch plasma) after tension-release cycles. (Figure 17a). This new methodology makes it possible to obtain films with bright-colored finishes and predesigned colorimetric responses due to their mechano-chromatic response. Thus, the wrinkle formation mechanism was mapped in two regimes depending on the thickness threshold of around 1 mm. When these films had a thickness of less than 1 mm, wrinkled patterns with defects were formed, and their average periodicity was mitigated when the thickness approached the base value of 1 mm. On the contrary, when the thickness was larger than 1 mm, the wrinkles had a constant periodicity; this value did not vary with increasing thickness. The classical nonlinear finite mechanism can precisely predict it.

Figure 17b illustrates the stress and plasma treatment of PDMS films on the complementary and reversible coloring process. These treatments were performed by applying plasma with and without tension, including the mask. The remaining substrate was oxidized in the last process, obtaining a thin and stiff film. When these films underwent pre-coating, the patterns formed in the first treatment were erased. Meanwhile, uniform and aligned wrinkles were formed in the direction of the stretching during the second plasma treatment.

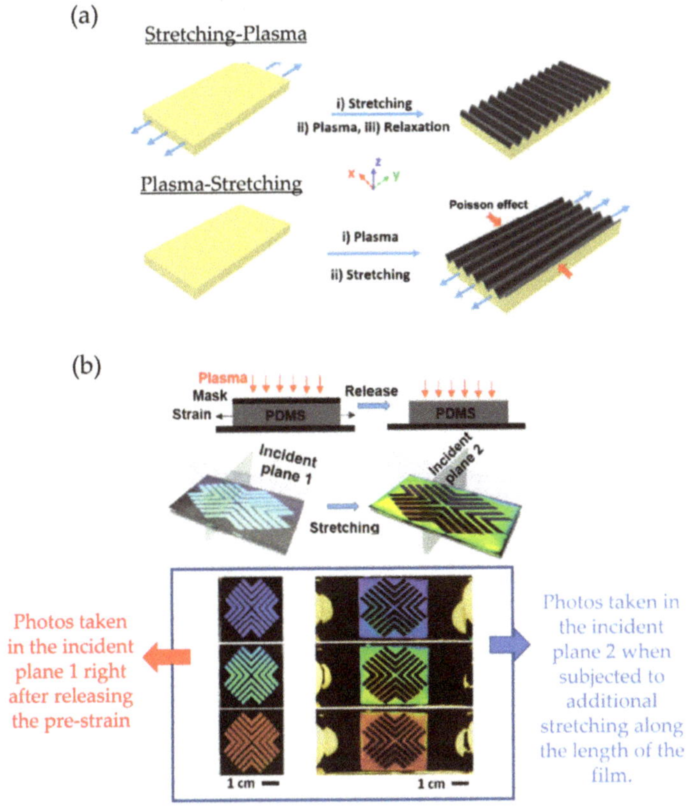

Figure 17. (**a**) Schematic illustration of the fabrication processes and (**b**) mechanochromic response and information encryption. Reproduced with permission from reference [72]. Copyright 2020, Springer Nature.

Similarly, Oyefusi and Chen [73] developed an exciting method to fabricate reversible mechanochromic materials based on optical interference processes. They manufacture stretchable films with excellent low-cost mechanochromic properties through two complementary mechanical strategies based on shear and cracking (kirigami approach). This methodology obtained unique, anisotropic, and load direction-dependent mechano-chromatic properties by including kirigami. This new strategy allows the fabrication of polymeric systems with dynamic reversible mechanical properties with a high degree of sensitivity to small deformations (0–60%) without using external luminous stimuli, making them affordable and scalable. In this case, they select as mechano-chromic material a polyvinylpyrrolidone-iridium-polydimethylsiloxane (PVP-Ir-PDMS) trilayer with the PMDS layer as a stretchable substrate. Figure 18a shows different color patterns formed with PVP-Ir-PDMS trilayer material with and without red dye. With this, stretch-testing photographs are also displayed for these materials, showing interesting reversible mechanochromic properties (Figure 18b).

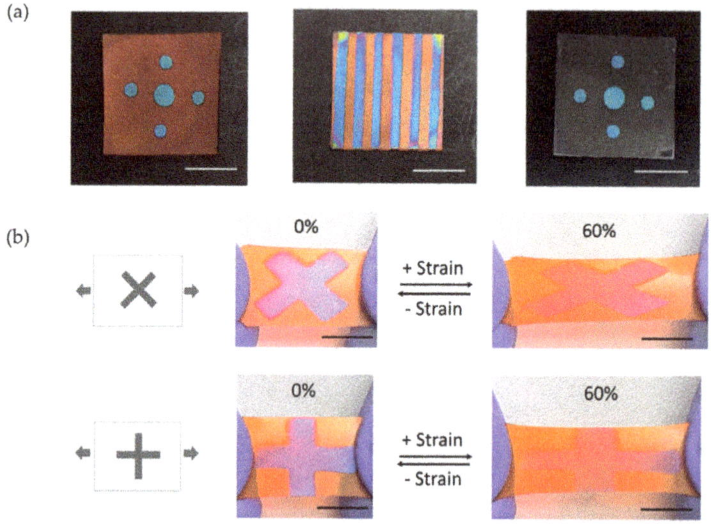

Figure 18. (**a**) Photographs of different color patterns generated in PVP-Ir-Dyed PDMS films with and without red dye and (**b**) stretching tests (from 0% to 60% strain) of the PVP-Ir-Dyed PDMS film, with load direction at 45° or 90° to the blue-colored cross pattern. Reproduced with permission from reference [73]. Copyright 2020, Elsevier Ltd.

However, not just optical changes could be produced by applying external forces. Piezoresponse materials are also commonly utilized for sensing purposes; for example, Jang, Kim, and Lee et al. [74] describe the fabrication of a pressure sensor that posses a large area of microstructured elastomer through a simple and low-cost process, allowing to create a capacitive device with a flexible structure. Experimentally, a soft dielectric Ecoflex film was pre-stretched, followed by UVO exposure. Next, wrinkles on the order of 10 μm were formed on the micro-cut surfaces of the film after stress releases. Finally, electrodes were made from the previously wrinkled film by placing an Au-coated silicon wafer (rigid) and a PDMS (soft) on the top and bottom of the template. Thus, the effect of the films was investigated by comparing the unwrinkled, one-side wrinkled, and two-side wrinkled films on the sensor performance.

When comparing the two-sided wrinkled pressure sensor; with an unwrinkled (smooth surface) one, it improved by 42% and 25% in response and release time. These results minimized the problematic viscoelastic behavior of Ecoflex after the incorporation of wrinkles in the microstructured surface of the sheet, improving its reversible energy release, storage, and sensor performance. A similar example was reported by Xu, Cao, and Lu et al. [75], who fabricated a light-increasing piezoresistive pressure sensor based on a rose petal-templated PDMS. This substrate includes wrinkled nanopatterns on the surface of polypyrrole (PPy)-based films, providing pressure sensitivity, assembling a broad level of pressure detection with low location limit, high sensitivity, fast response/relief, as well as good stability/durability and exceptional pressure detection performance. More importantly, it was possible to achieve a remarkable improvement in sensitivity (70→120 KPa-1, <0.5 KPa) and reduce the detection limit (0.88→0.41 Pa). These new pressure sensors have a potential application in optically controlled fields such as soft robotics, optoelectronics, and artificial intelligence due to their control of light and the incorporation of multiscale hierarchical structures.

Finally, it is worth mentioning that there are some other types of mechanic responses, like the one exposed by Hayward and Ouchi [76], that demonstrated a method in which surface deformations such as wrinkles, creases, and cracks are based on three modes to manufacture stamped surfaces with interchangeable electronic properties. Thus, the regions where the wrinkles are providing stretchability and flexibility. Simultaneously, the patterns were molded into the conductive films' long and straight cracks, presenting large and tunable cracks between one hundred micrometers and centimeters (Figure 19a). Using this method, the authors overcome the limitations of previous switches through mechanically activated electrical switches with adjustable switching voltages from 0.05 to 0.18 and an on/off ratio of high than 10^7. Figure 19b shows the connection of the two patterns in parallel (NAND gate) and series (NOR gate) in the different devices. As a result, they offer an attractive application in the stretch and pressure-sensitive shielding potentials and coatings because the devices respond to in-plane and out-of-plane forces/deformations.

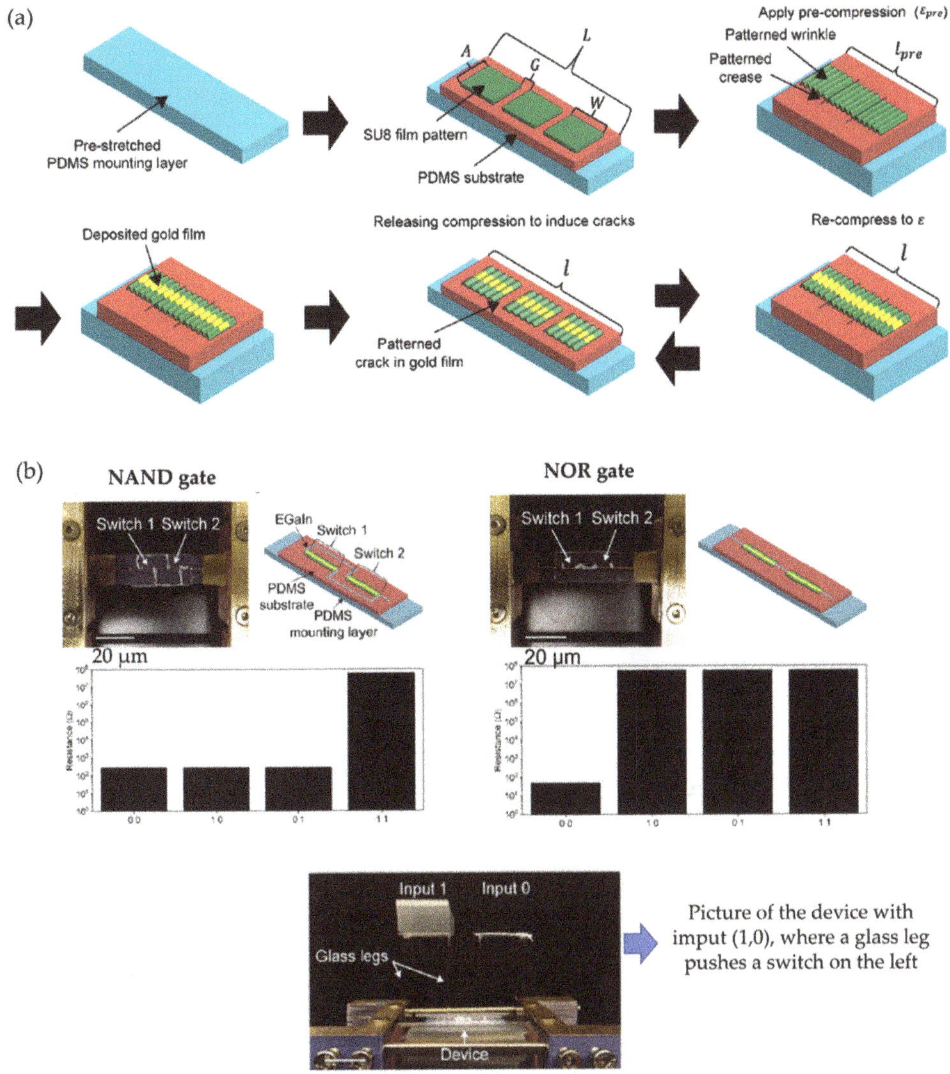

Figure 19. (**a**) Schematic representation of the strategy used to fabricate switchable surfaces. (**b**) Mechanical logic gates (NAND gates, NOR gates). The final image of the device is presented with an input of (1,0), where a glass leg pushes a switch to the left. Reproduced with permission from reference [76]. Copyright 2020, American Chemical Society (ACS).

Similarly, Kashihara et al. [77] fabricated a polyion complex (PIC) in a hydrogel layer through the electrophoresis process, observing the formation of wrinkled patterns using an electric field stimulus that allows wrinkle formation. The representative scheme of Figure 20a explains the wave generation process in the wrinkled areas of the polymer. Through the electrophoresis process, an anionic polymer poly(4-styrenesulfonate) (PSS) was deposited on the surface of a cationic polymer (PDMAAm) and connected between two electrodes. By administrating an electric field, a wrinkled pattern was formed on the material's surface through the cationic gel functioning as an anode and the anionic aqueous solution as a cathode; these patterns are formed due to the voltage mismatch in the PIC layer (Figure 20b). Figure 20c shows the wrinkled patterns obtained after applying the electric field, determining that the wrinkles formed remained static. Still, they returned to their dynamic state when exposed to the electric field. Figure 20d shows the displacement of wrinkles on the material's surface as a function of the speed and time of the electrophoresis process. It was determined that when the first wrinkles have formed, the displacement speed gradually increases, reaching a maximum of ~50 μm/s at 2 s. While higher the electric field, the faster the wrinkles move (increasing speed). Figure 20e shows the trajectories of the wrinkles during the movement provoked in the electrophoresis process.

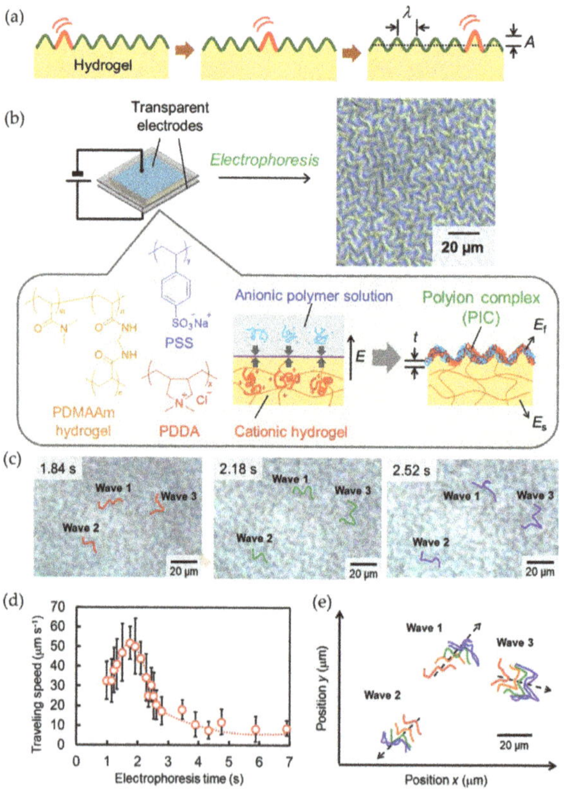

Figure 20. (**a**) Representative diagram of the generation of traveling waves of wrinkles on the hydrogel surface, (**b**) process of wrinkling of the hydrogel surface by electrophoresis, (**c**) images of the wrinkling process on the hydrogel surface during electrophoresis, (**d**) graph of the displacement velocity as a function of electrophoresis time and (**e**) Trajectories of three different waves tracked in the surface. Reproduced with permission from reference [77]. Copyright 2022, John Wiley & Sons, Inc.

Similarly, Wawryk et al. [78] fabricated wrinkled patterns in PDMS using a multi-step procedure. Figure 21a shows a representative scheme of the different manufacturing processes used to create the wrinkled surfaces. The PDMS films were attached to two metal plates, one static and the other in motion, allowing uniaxial movement (stretching up to 16%). Then the films were exposed to air plasma with a power of 30 W for 5 min, followed by a cooling process (1 min) to relax it finally. This series of procedures generate wrinkled patterns on the surface of the material. Atomic force microscopy (AFM) was used to analyze the wrinkled patterns obtained in each one of the different fabrication steps to form the wrinkles (Figure 21b–e). The other surface structures depicted in Figure 21b–e correspond to varying combinations of the processes used to create wrinkled patterns (air plasma (P), strain (S), and relaxation (R)).

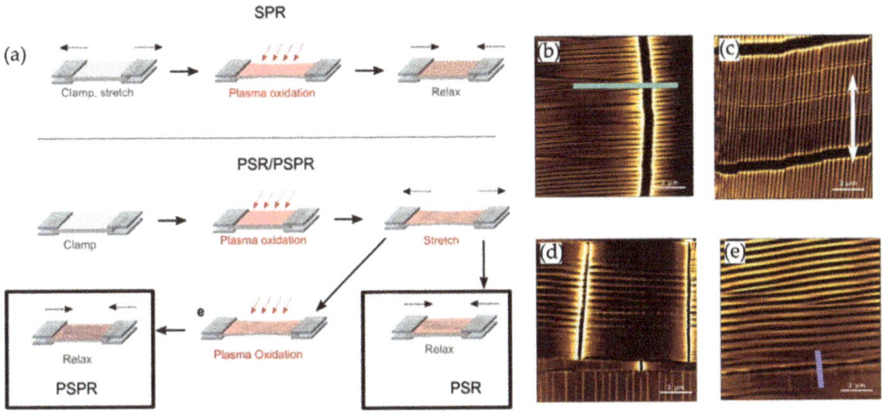

Figure 21. (**a**) PDMS wrinkle manufacturing process, micrographs of wrinkles, (**b**) SPR, (**c**) PS, (**d**) PSR, and (**e**) PSPR. Reproduced with permission from reference [78]. Copyright 2022, Elsevier Ltd.

In Table 4, it is possible to observe a summary of the examples mentioned in Section 2.5 for mechanical-sensitive materials used to create smart wrinkled patterns. The table also depicts the methodology used to fabricate the wrinkles and its possible application.

Table 4. Summary of mechanical (stress/stretch) listed in Section 2.5.

Base Material	Responsive Range	Methodology	Application	Ref.
PVA/PDMS	Stretching	- Oxygen plasma exposure - PVA immersion on PMDS - Drying at RT - Stretching	Dynamic optical switching	[71]
PDMS	Stretching	- PMDS pre-stretch - Mask coating - Vacuum exposure (15 min) - Stress release	Mechanochromic	[72]
PVP/Ir/PDMS	Stretching	- Stretching/relaxation	Structural health monitoring Soft robotics	[73]
PDMS	Stretching	- Stretching (200%) - UVO exposure (30 min) - Stress removal	Portable electronics	[74]

Table 4. *Cont.*

Base Material	Responsive Range	Methodology	Application	Ref.
PPY/PDMS	Pressure	- PPY film deposition on PDMS - Phase oxidation polymerization	Optically switchable microcircuit and photodetection	[75]
SU8/PDMS	Stretching	- -Pre-stretch - Deposit SU8 film in PMDS - System compression - Gold film deposition - Reversible compression	Advanced flexible electronics	[76]
PDMS	Stretching	- Film stretching (16%) - Air plasma exposure (5 min) - Relaxation (30 s)	Microfluidics	[78]

2.6. Chemical Reaction/Gas/Solvent Response

The most simple and common way to fabricate wrinkled patterns is via swelling/deswelling processes; for example, Damman and Vandeparre [79] proposed a mechanism to link molecular diffusion (solvent diffusion) and the stability of a rigid membrane attached to a polymeric layer to generate wrinkled patterns. They produced films with thicknesses between 0.1 and 1 μm made up of a polystyrene (PS) system deposited on a bare silicon (SiOx) substrate. Subsequently, using thermal evaporation, thin layers of titanium (Ti) with a thickness of 10 to 20 nm were deposited on the polymer surface. The multilayer was then immersed in toluene vapors at room temperature to produce wrinkles. These created geometric patterns comprised various fold formats (herringbone, parallel or radial stripes, among others) and dynamically different ones (continuous or discrete) from the diffusion and minimization of wrinkle energy. This work is relevant because it allows them to understand the instability of the different morphological patterns and the formation or elimination of topological imperfections.

Other methodologies, like the one reported by Oscurato et al. [80], involve innovative procedures to spontaneously produce wrinkled surfaces due to naturally occurring chemical transformation processes. The procedure is carried out in smooth thin solid films of 5,6-dihydroxyindole (DHI) over 30 to 60 days in ambient conditions. Due to gradual oxidation in the film, shear stresses were generated in the material inner layers, thus producing a slow swelling/expansion on the material followed by a fast stiffening of the upper layer, exposed to higher oxygen levels. Interestingly, this phenomenon is only observed with layer thicknesses over 250 nm.

However, there are also more intriguing methods to produce and control wrinkles on the surface of the materials; for example, Corté et al. [81] reported a methodology based on the halogenation of rubber to increase the glass transition temperature within a thin layer on its surface that allows creating reversible wrinkled patterns. Thus, the effect of surface bromination of polychloroprene-based rubber films was investigated by examining three different aspects: (1) the spatial distribution of bromine after immersion of a rubber film in a bromine solution, (2) how the glass transition of the material surface is modified by bromine, and (3) how the effect of temperature and mechanical stress are sensitized. Microscopic wrinkling occurs at temperatures below the brominated layer's glass transition, which was visualized in situ in the stretched and unstretched films in the thermomechanical experiments. When the wrinkled layer is submitted to higher temperatures, the wrinkles are eliminated, and it recovers its original magnitudes. This methodology is highly attractive due to its ability to automate surface patterning techniques or design a thermosensitive surface roughness using the gradient of bromine content and the thickness of the modified layer.

Similarly, Zhang et al. [82] report the hysteresis effect produced by cyclic heating and cooling graphene sheets giving rise to reversible wrinkling patterns. The system was constituted by monolayer graphene structure growth on the Pt(111) surface via CVD or surface aggregation. In an ultrahigh vacuum, the substrate was cleaned by Ar+ sputtering, oxidation, and annealing cycles. The surface was observed by low-energy electron microscopy (LEEM). It was found that the oxidative process of graphene and CO was generated underneath the graphene layers. These results suggest that the wrinkles formed may interact as nanometer-sized gas pockets and drive chemical reactions in the graphene structure.

Following the same topic, Jiang et al. [83] reported a simple and effective way to design reversible wrinkle patterns that, through the Diels-Alder (D-A) dynamic chemical reaction, can control wrinkle removal and adjustment in situ. The furan-grated bismaleimide (BMI) and polybutyl acrylate (FBA) top layer system is coated onto the PDMS elastomer. Wrinkle formation was then induced through hardening and crosslinking with a high Young's modulus of the top layer product of the D-A reaction, releasing localized stress to minimize total system energy. Afterward, to produce the decrosslinking of the FBA/BMI top layer, causing wrinkles to disappear and the surface to become smooth, the sample was heated at 120 °C for 20 min to cause a retro-D-A reaction. This mechanism based on D-A reactions provides high versatility in obtaining highly reversible crinkle patterns, providing precise control over characteristics such as wettability, adhesion, and optical properties.

Recently, Qu et al. [84] developed a method for fabricating wrinkled patterns over multilayered materials based on PDMS/Ag nanowire films. Interestingly, wrinkled structures can be formed dynamically in response to volatile organic compounds (such as ethanol, toluene, acetone, formaldehyde, and methanol) due to the multilayer-induced instability caused by their different swelling capabilities. Its multilayered condition could be helpful for responsive and flexible sensor applications as materials present a switchable dual-signal response (transparency and resistance). By controlling PDMS's modulus, composition, and layer thickness, tunable transparency and resistivity could be obtained. As a result, the film with a UVO treatment time of 14 min and a base-to-curing agent ratio of 60:1 can achieve the most sensitive transparency and resistance signal in response to ethanol vapor of different concentrations. The multilayered films also show more selectivity toward ethanol vapor than other volatile organic compounds. Figure 22 shows a schematic description of a wearable device design as a dual-signal sensor that reacts to ethanol vapor.

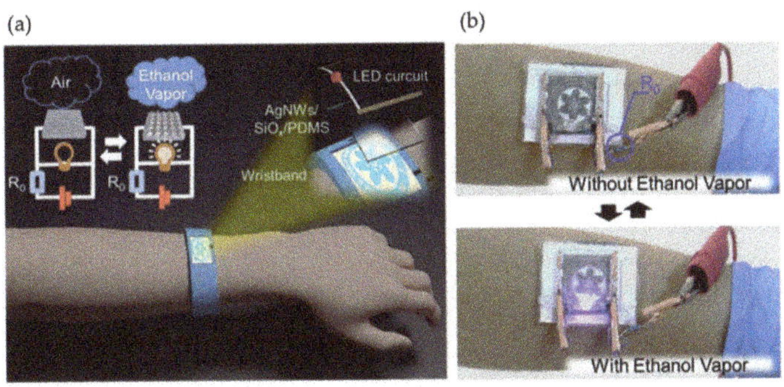

Figure 22. (**a**) Schematic illustration of the smart wearable and flexible electronic device based on the Ag nanowire/PDMS multilayered film and (**b**) a proof-of-concept wearable device that responds to ethanol vapor directly dynamic visual and electrical feedback. Reproduced with permission from reference [84]. Copyright 2019, John Wiley & Sons, Inc.

Similarly, Li and Zeng et al. [85] fabricate functional superhydrophobic materials with stimuli-responsive characteristics. They manufacture these devices by integrating a three-layer structure, a poly(vinylidene fluoride) (PVDF) layer located between two poly(vinyl alcohol) (PVA) layers. The first one tends to absorb solvent vapors to induce swelling, and the second one repulses solvent vapors with finality to maintain their original dimensions (Figure 23a). The samples exhibited superhydrophobicity with a high contact angle of 163° and a low sliding angle of 4.5°. Figure 23b illustrates the potential applications, such as a biomimetic flower, which respond to the absorption/desorption of acetone vapor. Like lilies in nature, the biomimetic glower displayed a dynamic blooming and fading process. Another example was designed as five fingers and assembled into a hand-patterned actuator. Vimineous acetone-absorbing tissue papers were placed under the fingers to activate their bending at ambient temperature.

(a)

(b)

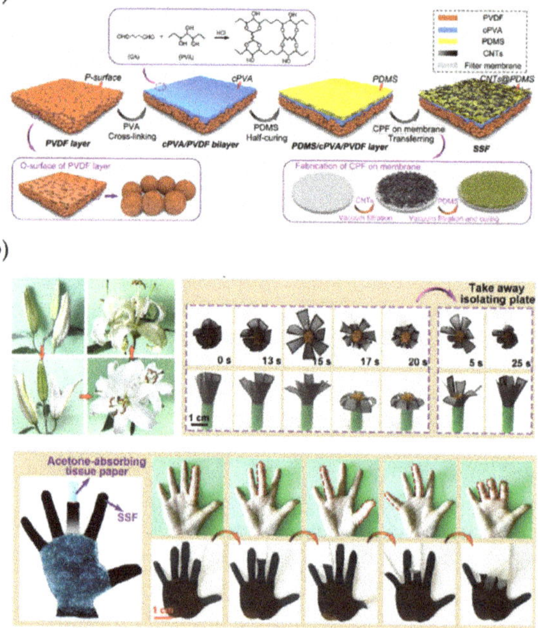

Figure 23. (**a**) Schematic diagram indicating the fabrication of durable stimuli-responsive superhydrophobic films (SSF). (**b**) Illustration of the flower and finger devices. Reproduced with permission from the reference. Reproduced with permission from reference [85]. Copyright 2020, Elsevier Ltd.

Crosby and Kim [86] fabricated stimuli-responsive surfaces based on osmotically driven wrinkling with hierarchical morphology and controlled wavelength. These studies have been based on the PDMS layer's oxidation using UVO supported over the silicon substrate. Subsequently, an ethanol droplet was placed on the oxidized PDMS surface, and wrinkles formed spontaneously, confined within the droplet's trace, and then disappeared upon evaporation of ethanol. Other volatile solvents, such as alcohols, THF, and pyridine, could also achieve this response, affecting the wrinkles' wavelength and amplitude due to their different swelling properties. It can also control the wrinkles patterns' dimension and morphology by changing the oxidation time and the soft layer's thickness. This versatility method allowed the performance of specific functional devices, such as reversible channels, tunable microlens arrays, and facile printing of nanoparticle assemblies [87].

3. Non-Planar Wrinkled Structures

In general, most of the research conducted in recent years by the scientific community has been related to the fabrication and application of wrinkled patterns on flat surfaces. However, numerous applications often tend to be on non-planar surfaces, so the generation and fabrication of these surfaces can sometimes be challenging. Several research groups have studied the experimental and theoretical mechanical framework, models, and constitutive relationships involved in generating non-flat wrinkled patterns to predict the final shape accurately and pattern distribution [88–90]. This section will discuss the most exciting studies on fabricating wrinkled patterns on non-planar surfaces. To organize this section, we will divide it into different topics according to the type of non-planar structure, such as solid core spheres, solid core cylinders, hollow spheres, and cylindrical tubes.

3.1. Solid Core-Shell Spheres

Several of these non-planar wrinkled patterns can modify their adhesion with different surfaces depending on the type of micropattern generated on their surface. Controlled adhesion is essential for developing these materials in many applications, from micro/nanoelectronic systems to bioimplants. Therefore, incorporating geometric features/patterns on the non-planar surface has interested several research groups in recent years. Crosby et al. [91] studied the adhesion of spherical objects with topographical surface patterns created by cover wrinkling/deformation techniques. The surface modifications were fabricated on PDMS lenses with a curvature (R) radius of 7 mm. Surface adhesion was quantified as a function of wrinkle dimensions such as wavelength (20 to 70 μm) and amplitude (0.3 to 5.0 μm). The applied deformation is regulated by varying the vapor pressure and swelling solvents. Ethanol and isopropanol have been used as swelling solvents, and the vapor pressure of these solvents is controlled by mixing glycerol in different proportions.

On the other hand, they describe the effect of adhesion force as a function of wrinkle morphology (wavelength and amplitude) and the dependence of the cumulative effect on the radius of curvature of the surface. Jagota and Yang et al. [92] studied the adhesion of a one-dimensional (aligned) PDMS wrinkled surface using a spherical glass probe. In this study, it was observed that adhesion tends to decrease with increasing wrinkle width. Similarly, Feng et al. [93] reported surface buckling and the morphological transition from a soft-core sphere using a light-tension volumetric growth theory. In addition, they showed that the critical buckling condition and the induced patterns were susceptible to the thickness of the sphere shell and the material's mechanical properties. As mentioned in the introduction section, the increased available surface area due to wrinkling improves several properties related to particles' contact area. For example, the conductive layer with non-planar morphologies becomes more sensitive than planar films. However, ultrasensitive pressure sensors assembled using non-flat conductive wrinkles are rare due to the lack of a wavy arrangement with noticeable surface buckling and adequate adhesion on soft substrates. Wang et al. [94] proposed the formation of wrinkles with electrically conductive behavior by combining polymer swelling and electrodeless deposition. Thus, plasma-treated PDMS films were adsorbed onto a bovine serum albumin layer, followed by swelling in chloroform. A silver film was placed on the swollen PDMS after electrodeless deposition in a mixture of Tollen's reagents. The pressure sensor was fabricated by superimposing two face-to-face wrinkled surfaces, forming a piezoresistive sensor. In this study, the resistance change attributed to the conformal contact change between two conductive layers is more noticeable for wrinkles with longer wavelength and amplitude, which is described as better pressure sensing performance. Thus, the sensor fabricated with silver wrinkles on the patterned pillar was extended to sound-sensing applications.

Similarly, Zhang et al. [95] synthesized core-shelled composite microspheres with a wrinkled micro-structured carbon-titanium dioxide (C/TiO_2) created by taking advantage of microwave absorption properties. Their group established a three-step method to manufacture these wrinkled spheres, based on the polymerization of polyglycidyl methacrylate/divinylbenzene polymer spheres (PGMA/PDVB), a TiO_2 coating of the spheres mediated by hydrolysis of tetrabutyl titanate (TBT) and vacuum carbonization of $PGMA/PDVB/TiO_2$, forming C/TiO_2 composite microspheres. It is worth mentioning that this manufacturing configuration allows controlled regulation of the polymerized wrinkled spheres, their electromagnetic parameters, and their superficial wave reflections.

Other novel applications using wrinkled non-planar (spherical) surfaces are, for example, the one reported by Cao et al. [96], who developed a novel functional interlayer by wrapping dispersed zinc sulfide (ZnS) nanospheres and an ultrathin wrinkled carbon film (ZnS/WCF). Specifically, the wrinkled structure of the porous WCF provides sufficient intermediate spaces to hold the electrolyte and ensure the distribution of lithium ions. In addition, ZnS nanospheres with WCF treatment act as anchor sites for LiPSs (soluble lithium polysulfides) entrapment and promote the breakdown of solid Li_2S with a relatively low surface migration barrier. This composite structure could pave the way for a broad application of Li-S batteries.

On the other hand, Im et al. [97] fabricate a labyrinthine pattern onto uniform-sized microparticles based on a biodegradable polymeric blend using an emulsion-solvent evaporation method. By changing the particle size, the wrinkled patterns switch from a simple labyrinth morphology to a bi-phase pattern where different wrinkle morphologies coexist. The microparticles were fabricated in a tube-type microfluidic device using various amphiphilic block copolymers based on polycaprolactone (PCL), poly(ethylene glycol) (PEG), and poly(D, L-lactide) (PDLA). These copolymers were dissolved in dichloromethane to obtain an oily solution in which polymer microparticles were formed. By regulating the composition and the sizes of the particles, different non-planar wrinkled structures could be obtained. Figure 24 shows some bi-phase wrinkled patterns formed over the spherical microparticles.

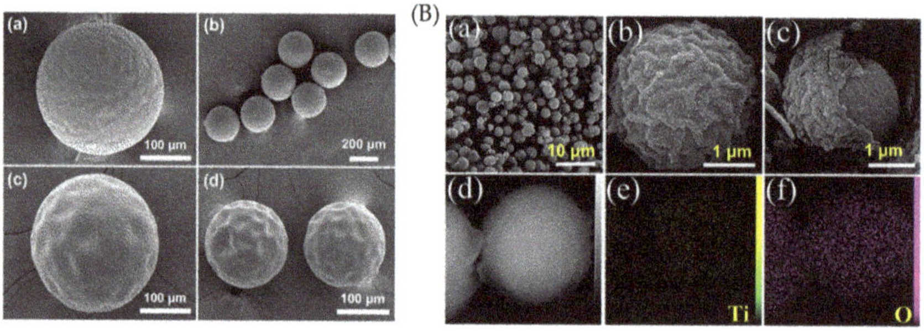

Figure 24. (**A**) SEM images of the dual-wrinkled microparticles from (a,b) PLGA/PDLA-b-mPEG blend and (c,d) PLGA/PCL-b-PEG-b-PCL blend. Reproduced with permission from reference [97]. Copyright 2018, Royal Society of Chemistry (RSC), and (**B**) FE-SEM (a–c), BSE-SEM (d), EDS mapping (e,f) images of the Y@HWS-TiO_2. Reproduced with permission from reference [98]. Copyright 2021, Elsevier Ltd.

Recently, Peymanfar et al. [99] synthesized a wrinkled Ni nanosheet/grapy carbon microsphere (CMS) nanocomposite. The wrinkled Ni nanostructure was prepared using the co-precipitation and hydrothermal method. Similarly, Li et al. also fabricate CMS structures but with some modifications [100]. PS or PVDF was first dissolved in DMF to fabricate the nanocomposite. Next, CMS, Ni nanosheets, and Ni/CMS nanocomposites with 50 wt% (filler/filler + polymer matrix) were separately suspended in the solution using an ultrasonic bath, then heated and molded into rectangular shapes to investigate the microwave frequency absorption properties of the nanocomposites.

In particular, the results manifest that the architecture's microwave absorption and shielding structures illustrated considerable reflection loss, efficient bandwidth, and shielding properties with thin thickness due to the wrinkled morphology of Ni and applied media. An additional novel wrinkle manufacturing method was explored by Jeong et al. [56], using oil-water interphase to fabricate the pattern. The appearance of wrinkles is achieved by the dispersion of the photoinitiator suspended in the oil phase on the aqueous phase containing a prepolymer solution. The explained methodology still requires UV exposure to establish a polymerization. In addition, this oil-water generates tension that promotes the formation of wrinkles on curved structures such as solid spheres or cylinders. Figure 25 describes a schematic representation of this one-step wrinkle-forming procedure using curved surfaces.

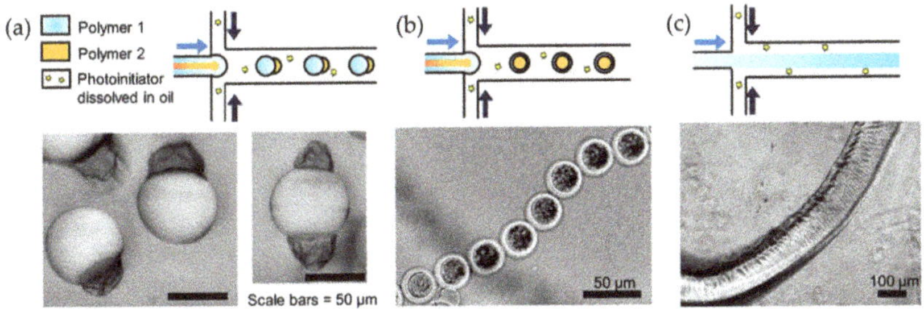

Figure 25. Wrinkled surface patterning formation on various microstructures created using a microfluidic system. (**a**) Janus particles with wrinkled sides, (**b**) Wrinkled core-shell with inner core underneath another hydrogel layer, and (**c**) Wrinkled surfaced fibers. Reproduced with permission from reference [56]. Copyright 2021, American Chemical Society (ACS).

3.2. Solid Core-Shell Cylinders

Another type of structure commonly used to fabricate non-planar wrinkled surfaces is cylinders; for example, Cao and Feng et al. [101] investigated the phenomenon of wrinkle formation on the surface of an elastic substrate subjected to in-plane compression. Thus, two theoretical and finite element analyses of the variations of the substrate modulus along the depth direction (exponential functional and power function) were performed. Feng et al. [102] performed theoretical analysis and numerical simulation to study the surface wrinkling and patterns evolution of cylindrical elastomers with core-shell structure induced by material swelling/shrinkage. Their studies showed that a slight compression on the shell (the shell is supposed to be stiffer than the core) could generate the first buckling characterized by sinusoidal patterns.

Further pressure on its surface can lead to a transformation from wrinkles to folds, resulting in the formation of hairpin patterns that double the first buckling period. The surface modifications that occur at the first and second bifurcation can be adjusted by modifying the mechanical and geometrical properties of the structure. In 2019, Wang et al. [103] investigated a novel method that created concentrical wrinkles on PDMS. The application of UV/O was carried out on pre-stretched PDMS cylinders in constant rotation, which resulted in the formation of annular wrinkles upon release (Figure 26a). Through simulation and experiments, pre-stretching (mechanical strength and stretchability) and exposure time were related to wavelength and wrinkle amplitude. Pearce et al. reported on the surface wrinkling of a thin elastic film attached to a sizeable adjustable cylinder from the perspective of large strain elasticity, concluding that curvature delays the instability of wrinkles [104]. Their findings confirmed the simulation results of Patricio et al. [105], which state that the orientation of the wrinkles can be longitudinal (along the fiber axis), polar (along the fiber perimeter), or a combination of both, depending on the mechanical and geometrical parameters of the fiber. In 2018, Yang et al. developed a highly flexible and stretchable optical strain sensor [106]. In this research line, doped and undoped PDMS was fabricated. Dye molecules were added to the PDMS precursor before curing to dop the PDMS. Thus, the undoped sample showed high transparency, while the PDMS doped with rhodamine B dyes showed higher attenuation, which is attributed to absorption and scattering due to the spatial inhomogeneity of the doped dyes.

On the other hand, PDMS fibers showed excellent mechanical flexibility so that they could be easily bundled and twisted (Figure 26b). The mechanical test responses (dynamic and static) showed stable, reliable, and repeatable performance. The sensor demonstrated that it could detect deformations over a wide dynamic range of 100% with an accuracy of

$\pm$ 0.91%. An example was used to detect dynamic body movements, such as breathing and joint movement, all performed in real-time.

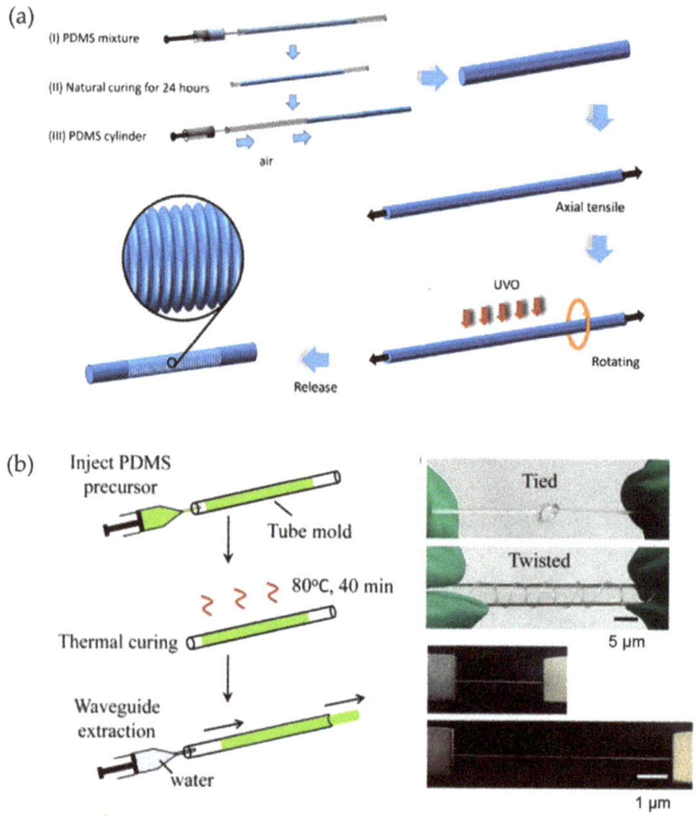

Figure 26. (**a**) Wrinkled PDMS cylinder preparation process and (**b**) fabrication and testing of the mechanical flexibility of the PDMS fibers. Reproduced with permission from references [103,106]. Copyright 2018, American Physical Society. Copyright 2019, Elsevier Ltd.

3.3. Hollow Spheres

However, not just solid structures were developed to form wrinkled patterns over non-planar surfaces, Zhu and Xing et al. [107] synthesized a novel chitosan material (ECS/Ca/CTA) to adsorb SDBS (sodium dodecylbenzene sulfonate) with high efficiency by using hollowed-out spheres with wrinkled patterns on their surface. Glutaraldehyde (GA) was used to crosslink CS (chitosan) into a wrinkled hollow sphere, which increases the specific surface area, chemical stability, and capability of capturing SDBS. The adsorption procedure fits well with the pseudo-second-order kinetic mechanism, and the adsorption equilibrium experiment fits the Langmuir model. A similar example was the pleated hollow-sphere material (CS/Ca/CTA) mixing CS and $CaCl_2$ in acetic acid solution followed by cross-linking with GA and quaternized using (3-chloro-2 hydroxypropyl) trimethylammonium chloride (CTA), which efficiently adsorbed SDBS [108].

Following the same trend, Badiei and Luque et al. [98] reported the fabrication of advanced hydrogenated yolk/wrinkle structures on TiO_2 shell (Y/HWS-TiO_2); these materials demonstrated excellent photocatalytic activity under visible light irradiation for the selective oxidation of alcohols at room temperature. The synthesis includes sequential processing steps, such as solvothermal reactions using polyethylene glycol as soft templates, partial etching, and hydrogen treatment. This research can prove the elaboration of a randomly assembled wrinkled surface on the microsphere's surface (solid core). The photocatalytic oxidation activity of the materials was tested in the selective oxidation reaction of benzyl alcohol to benzaldehyde, obtaining 90% conversion and 97% selectivity. The results showed that these particles possess high porosity and hollow space in the yolk/shell structure, which is beneficial for molecule displacement and attachment, along with a thin, wrinkled shell, which can promote molecule dispersion and create active sites exposed to reagents. In a similar example provided by the same studies, Zeng et al. [109] synthesized hierarchical $CoFe_2O_4$ double-shelled hollow spheres (DHSs) with a wrinkled porous surface. This structure exhibited high sensitivity to ammonia (NH_3) with good stability and fast response-recovery capability at the optimal operating temperature (240 °C), an effect mainly related to two characteristics: its sizeable specific area and, secondly, its penetrable hierarchical shells.

On the other hand, Zhao et al. [110] prepared a hollow tremella-like graphene sphere/ tin dioxide (HTGS/SnO_2) composite via emulsification and impregnation. This structure (HTGS/SnO_2) presents excellent electrochemical stability. The graphene sphere's hollow structure allows the simultaneous insertion of lithium ions from the inner and outer surfaces. Meanwhile, the tin dioxide particles are uniformly dispersed by the graphene's surface's wrinkles, thereby enlarging the space for the volume expansion of tin dioxide to avoid contact with the electrolyte. As a result, the initial discharge capacity reached ~1760 mAh g^{-1}, and the reversible capacity was ~1170 mAh g^{-1}. After 50 cycles, the reversible capacity was maintained at ~940 mAh g^{-1}, and the Coulomb efficiency was consistently above 96%.

Similarly, Davarani et al. [111] have constructed a supercapacitor (GW-MSNGOHS// GW-YS-NFOHS) consisting of an electrode with a high porous nanoarchitecture and superior energy density. It consists of a multilayered $NiGa_2O_4$ hollow sphere wrapped in graphene (GW-MSNGOHS) with an exceptional surface area of ~115 m^2 g^{-1} as a new positive electrode material and a $NiFe_2O_4$ hollow sphere with graphene shell (GW-YS-NFOHS) as a negative electrode material. In 2020, Davarini and Zardkhoshoui [112] encapsulated multi-shelled zinc-cobalt selenide hollow nanospheres as positive electrodes and yolk-double shell cobalt-iron selenide hollow nanospheres as negative electrodes into graphene networks a self-templating technique. The hollow nanospheres, called G/MSZCS-HS and G/YDSCFS-HS, present high electrical conductivity and unique structures, showing exceptional positive and negative capacities (~377 mAh g^{-1} and ~293 mAh g^{-1}), superior rate performances, and robustness after 12,000 cycles.

Likewise, Qi et al. [113] proposed a supercapacitor (electrochemical capacitor) composed of Mn_3O_4/MnS heterostructure building multi-shelled hollow microspheres and active carbon as the cathode and anode, respectively. Figure 27a shows the synthetic route; first, Mn-based coordination polymer spheres (Mn-CPSs) can be synthesized by the solvothermal reaction of Mn^{2+} ions with organic compounds. The multi-shelled Mn_2O_3 spheres were then prepared through a pyrolysis process of the Mn-CPSs (step 1). Posteriorly, it is reacted with sulfur power, and heterogenous Mn_3O_4/MnS multi-shelled hollow spheres are obtained (step 2). The morphology was explored through SEM, TEM, and HRTEM, showing a higher wrinkled degree than the multi-shelled Mn_2O_3 hollow (Figure 27b). The devices exhibited a high capacitance of ~185 F g^{-1} at 1 A g^{-1}, with substantial energy density (~66 W h kg^{-1}) and power density (16,000 W kg^{-1}). The superior electrochemical performances could be attributed to the heterostructures and the stable multi-shelled structure.

A recent implementation for hollow wrinkled spheres was proposed by Wang et al. using SnO_2 hollow microspheres with Au-Ag bimetallic nanoparticles for triethylamine gas detection. This configuration showed high responses, fast response/recovery speed, and long-term stability. These wrinkled spheres were prepared through combinations of Sn_4^+ and resorcinol-formaldehyde (RF) resin polymer spheres, followed by a calcination process [114].

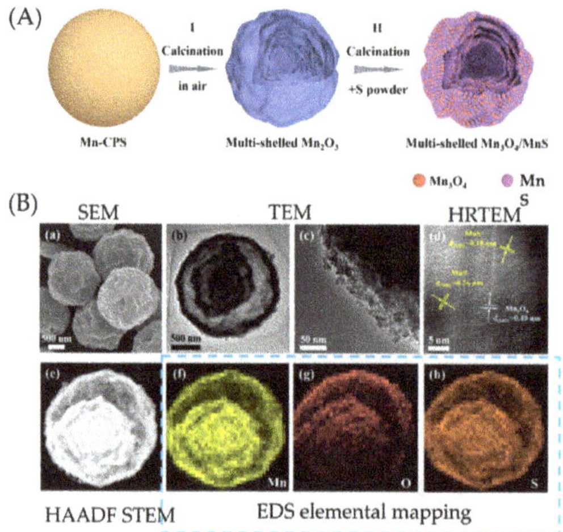

Figure 27. (**A**) Schematic illustration for the fabrication of Mn_3O_4/MnS and (**B**) (a) SEM image, (b,c) TEM image and (d) HRTEM image of Mn3O4/MnS heterostructures building multi-shelled hollow microspheres; (e–h) HAADF STEM and EDS elemental mappings of different elements of Mn, O and S recorded from a single heterogeneous Mn3O4/MnS multi-shelled hollow sphere. Reproduced with permission from reference [113]. Copyright 2022, Elsevier Ltd.

With a similar objective, Li et al. [115] implemented a spray pyrolysis methodology to manufacture wrinkled microspheres for high-performance ethanol sensors. This group studied wrinkled In_2O_3 microspheres, finding a short recovery time, high sensitivity, and up to 100 ppm ethanol detection with a high dependence on atmospheric humidity. Figure 28 presents a schematic description of the process followed in the pyrolysis of wrinkled In_2O_3 microspheres.

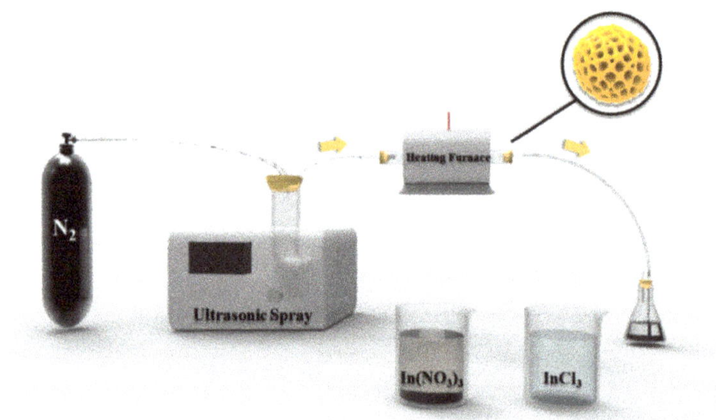

Figure 28. Schematic representation of the experimental procedure followed to fabricate ultrasonic pyrolysis spray-based wrinkled microspheres. Reproduced with permission from reference [115]. Copyright 2021, Elsevier Ltd.

Another exciting application of wrinkled hollow spheres is the use of transition metal oxides as energy storage for future innovations in electronic devices. One of the research related to this feature is that of Naffakh-Moosavy et al., in which the morphological and electrochemical properties of V-doped Co_3O_4 hollow spheres (MS-V-CO) produced by solvothermal/restoration were studied. The fabricated material resulted in low internal resistance, fast kinetics, remarkable reversibility, acceptable capacitance, and extremely high available surface area, which is a viable candidate for implementing a modern capacitor to be used in the future [116].

Finally, Moreno and Carp et al. [117] synthesized an open hollow structure of Co_3O_4 by a three-step procedure, a green alternative to the Pechini method that employs a carbohydrate as a polymerization agent. First comes the initial formation of Co^{+2} (citrus soluble coordination compounds), and the second part comes the hydrothermal processing at (180 °C) in fructose, followed by thermal air processing of the obtained precursors. Notably, the obtained Co_3O_4 consists of crystallites with sizes ranging from ~26 nm in concave hollow shapes with multilayered spheres, which have filigree and wrinkled walls. The lack of citric acid allowed for hollow double-shell morphologies with denser walls. On the other hand, the presence of citric acid imparts high specific capacity, rate performance, and cycling stability after 250 cycles to the structures, compared to the same hollow multilayered spheres synthesized in one ambient in the absence of citric acid. From a broader point of view, the research is an excellent example of how simple synthetic procedures using neither raw materials nor sophisticated equipment can offer feasible access to elegant, elaborate, and high-performance anode nanostructures for lithium-ion batteries.

3.4. Cylindrical Tubes

Cylindrical tubes are the last type of non-planar wrinkled structure in this review. For example, Geng et al. [118] have designed a novel microscale fabrication technique based on structural self-assembly processes to create periodically controlled ring-shaped disturbances (wrinkled patterns). The results show that the wrinkles are formed by stretching a soft polymer fiber made of PDMS at tensions ranging from 10% to 200%, followed by UV/ozone exposure to generate a hard SiOx film on the soft fiber to release the stress of the fiber after its formation. Geng et al. could identify the main parameters that control wrinkle dimensions and distribution over the polymeric fiber.

Similarly, Yang et al. [119] developed an innovative, simple, low-cost method to fabricate periodic wrinkled patterns over elastomeric substrates. PDMS films were prepared over polyethylene terephthalate (PET) substrates using the spin-coating process. Then, the PDMS-PET system was bent until it reached a specific deformation and was exposed to an oxygen plasma treatment during different periods. The plasma exposition generates SiO_x molecules on top of the material, thus producing an oxidized thin layer with increasing stiffness, triggering the formation of a wrinkled pattern after bending release. These materials show good visual performance in transmittance diffraction experiments, enabling the fabrication of highly ordered grating devices, which can further develop portable and cost-effective applications to provide a valuable method for making other optical micro-detection devices.

On the other hand, Sampathkumar et al. [120] fabricate sculpturing wrinkled structures into soft polymers (SU8 photoresist and PDMS) by thermal treatment of graphene flakes. The wrinkles were formed via biaxial compressive stress to the graphene flakes controlling graphene/polymer composites' heating. The wrinkled pattern dimensions are directly related to the graphene's thickness and the young modulus of the polymer substrate. On the other hand, the orientation of the pattern to the shape of the flake. This fabrication process also serves as a simple route to pattern the surface of soft polymers with wrinkling wavelengths that can be as low as ~50 nm.

Notably, Sun et al. [121] studied the spontaneous formation of wrinkled patterns on tubular structures by using poly(vinyl alcohol) (PVA) and Ecoflex®. This bilayered system was exposed to room temperature water mist to plasticize the PVA, followed by immediate stretching to predetermined strain. Then, the bilayered system is dried and released to form the tubular structure. It was possible to predict the elastomeric system's deformation and the wrinkled pattern morphology and distribution formed on top of the material through finite element analysis. The material presents a strong anisotropic constitutive response based on incompressible hyper-elastic neo-Hookean material. Figure 29a shows a schematic description of the process followed to form the self-wrapping tubular structure. Figure 29b,c offer theoretical and experimental results of the system's overall tube shape under different strains.

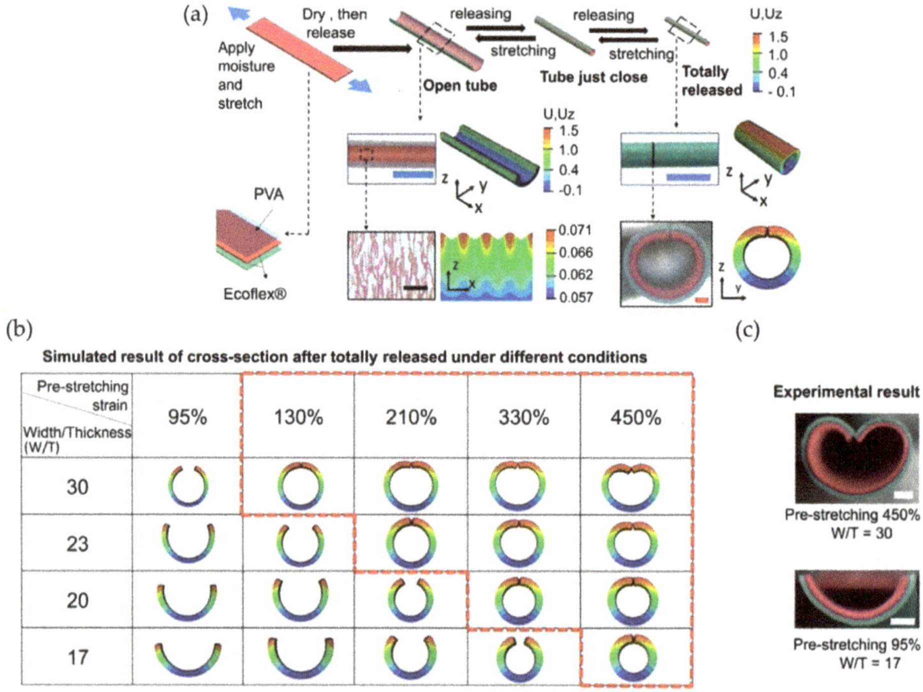

Figure 29. (**a**) Schematic representation of the PVA-Ecoflex® tubular geometry fabrication, (**b**) simulated result, and (**c**) experimental result of the tube cross-section after released under different width/thickness ratios and pre-stretching strains. Reproduced with permission from reference [121]. Copyright 2020, Royal Society of Chemistry (RSC).

Likewise, Sun and Wu et al. [122], following the hollow structure of multilayered blood vessels (lamellar and muscular layers), developed a hydrogel-elastomer hybrid microtube with the quality being elastic and soft. To achieve this, they used a thermoplastic elastomer (TPE) based on a styrene-isoprene-styrene block copolymer as the outer layer and a poly(N-isopropyl acrylamide) (PNIPAM)/clay/Mxene (PCMH) hydrogel which is sensitive to heat and infrared light as the inner layer, as shown in Figure 30a. Due to the mechanical instability presented, the TPE layer has a self-wrinkled inner surface with uniaxial ridges, which is remarkably similar to natural blood vessels. The photothermal effect caused by the NIR is used as a non-contact stimulus model to change the body temperature. It should be noted that the water temperature causes reversible changes in channel size and the appearance/disappearance of microtube wrinkles. Figure 30b a microvalve system connected to a syringe to control the fluid flow, using rhodamine (hot water) and methylene blue (cold water). The channel was opened via NIR irradiation (30 s) due to the rapid contraction of PCMH and then closed for 15 min after the removal of the NIR exposure. In this way, temperature and NIR-induced flow transport and blockage are performed at a distance, such as vasodilatation of the muscle layer and constriction of blood vessels to regulate blood flow. This smart microtubing opens the door to understanding the effect of tissue surface topography on biological functions.

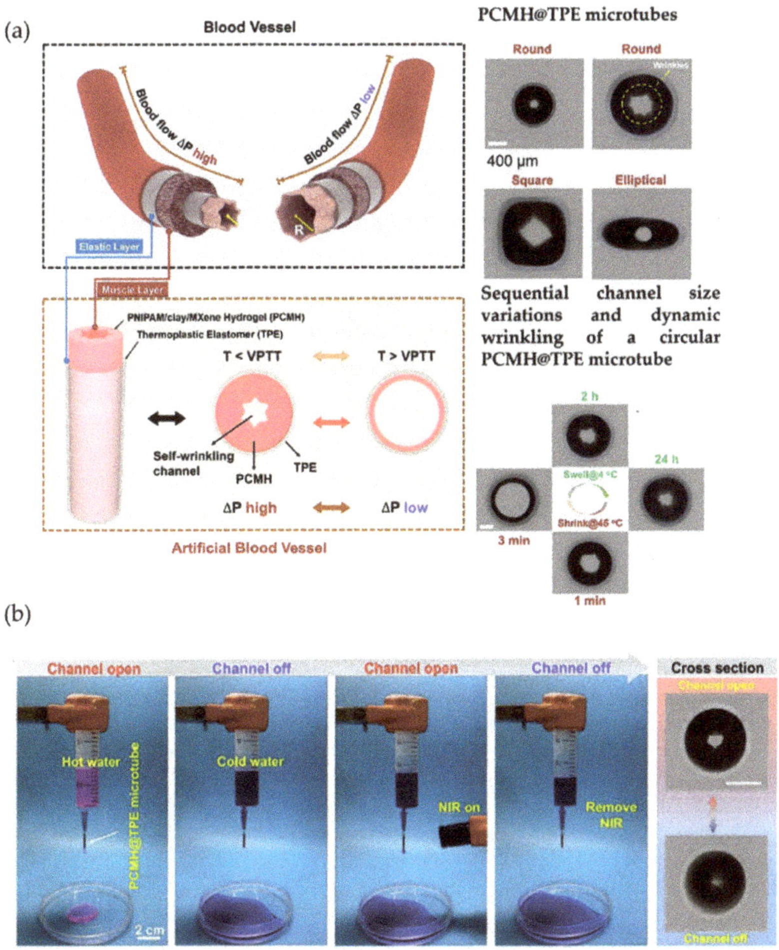

Figure 30. (**a**) Schematic illustration of the layered structure, hydraulic pressure sensing, and actuating functions of artery walls as well as an artificial blood vessel, PCMH@TPE microtube, and (**b**) photographs of PCMH@TPE microtube as a smart microvalve for controlling water flow with the channel open/closed by hot/cold water and re-opened/re-closed by irradiation or remotion of NIR. Reproduced with permission from reference [122]. Copyright 2022, Elsevier Ltd.

Along with the procedures already presented, Shi et al. report a hollow chitosan hydrogel tube that was electrodeposited with controllable wavelength and wrinkle amplitude by immersing the film in acidic sodium dodecyl sulfate (SDS). A more in-depth analysis of this procedure establishes two main steps. Firstly, electrodeposition of chitosan-dissolved solution with a titanium plate via a cathodic neutralization mechanism. Secondly, an SDS soaking process generates the tubular configuration and the controllable surface wrinkles. In Figure 31, the group schematizes the methodology reported to facilitate an understanding of the process. Furthermore, this group reports that the resulting wrinkled hydrogel tube exhibits excellent anti-fatigue capacity, mechanical strength, flexibility, and resilience [123].

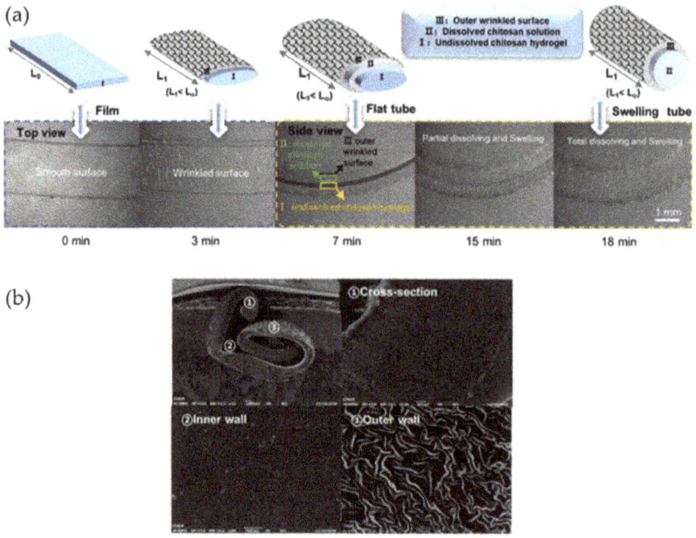

Figure 31. Wrinkled pattern tube formation process. (**a**) Microscopic and scheme of wrinkled tube device and (**b**) Cross-section morphology of the inner and outer wall of the wrinkled tube. Reproduced with permission from reference [123]. Copyright 2019, American Chemical Society (ACS).

4. Complex Wrinkles Patterns

As we already mentioned in the previous sections, the formation of wrinkled microstructured patterns in the material's surface has many advantages for some applications, mainly due to the increase of available contact area between the material and the environment, which could generate several improvements in the development of some particular and exciting devices like chemical/pressure [124,125] and piezoresistive sensors [126], among others. This section will be focused on reviewing the formation of complex wrinkled patterns like gradient and hierarchical patterns and some of their most relevant applications.

4.1. Strategies for Fabricated Gradient Wrinkle Surfaces

As demonstrated in the previous sections, the size and structuration of the wrinkled patterns could be controlled via the alteration of some particular variables like type of compression/relaxation (uniaxial, biaxial isotropic, biaxial anisotropic, etc.), stiffness or thickness of the rigid upper layer, among others. Uniaxial compression in the film results in striped wrinkles, but isotropic biaxial compression generates randomly oriented labyrinth-like wrinkle patterns or ordered herringbone patterns. However, the transition from disordered to ordered wrinkle patterns with particular orientation has been reported to occur from symmetry breaking caused by the anisotropy of compaction in their gradients. Accordingly, and with the finality of studying these symmetry-breaking processes, the fabrication of gradient wrinkled structures could be helpful. Indeed, Yu et al. [127] investigated and reported the wrinkling of patterns in tantalum (Ta) film sputters deposited on the modulus-gradient PDMS substrate. Due to the intrinsic and thermal stresses, wrinkling patterns spontaneously form on the sample surface. The wrinkles' size strongly depends on the substrate stiffness and film thickness (decrease steadily increasing the substrate modulus but increase linearly with the thickness). The gradient sample's hydrophobicity increases with the wrinkling pattern aspect ratio.

Similarly, Yu and Ni et al. [128] report a methodology to control the evolution of folds into wrinkles and create hierarchical wrinkles on the surface of a given soft material by regulating substrate stiffness and sputtering flux, producing a transition from lattice folds to isolated folds and wrinkles, which finally lead to labyrinth-like wrinkles. Furthermore, two distinct wrinkling patterns (G1 and G2) can coexist on the sample surface for a more considerable sputtering flux. Also, Yu and Ni et al. propose an unconventional methodology to form periodic thickness metal films on PDMS foils by masking copper meshes using room temperature DC magnetron sputtering. The thickness of the metal layer is directly related to the wrinkle structure generated; thus, the film surfaces at the mesh centers transition from labyrinth-like wrinkles to herringbone wrinkles and then to stripe-like wrinkles until a wrinkle-free state is reached when the mesh size is reduced or the film thickness is increased.

Likewise, Kim et al. [129] presented a simple process of fabricating a wrinkle pattern with a gradual gradient with a spatially controlled discrete wavelength in distinct zones with clear boundaries on a PDMS substrate. They performed a sequential oxygen plasma treatment for PDMS surfaces, employing a glass plate to prevent this perturbation partially. First, the PDMS substrate was fixed and stretched using a stretching device with 10% pre-strain in a uniaxial orientation. It generated a stable wrinkled groove structure without unexpected bifurcation of the wrinkled groove structure. Subsequently, the PDMS pre-stretched with the glass plate was exposed to oxygen plasma for 12 min. After the first exposure, the glass plate was gently removed to perform additional plasma treatment sequentially for 6 and 3 min, generating distinct regions with three discrete wavelengths. This research further studied sequential and continuous plasma treatment, finding discrepancies in values when the total time was 12 min.

On the other hand, Kato et al. [130] have developed a novel procedure to create wrinkled patterns on the surfaces of hydrogels under aqueous conditions by electrophoresis. A cationic hydrogel was prepared from N,N-dimethyl acrylamide with N,N-methylene bis(acrylamide) in the presence of poly(diallyl dimethylammonium chloride) (PDDA) and subjected to an anionic polymer solution based on sodium poly(styrene sulfonate) (PSS). By applying an electric current through the hydrogel system, it was possible to form a thin layer of polyion complex on top of the hydrogels by free-radical polymerization due to the electrostatic interaction between the cationic PDDA and the anionic PSS, thus generating wrinkled patterns on top of the material (Figure 32a). It was possible to prepare patterned structures (Figure 32c) and gradients (Figure 32d) by modulating the electrode position. The material also shows thermoresponsive characteristics; on the surface of the hydrogel, demonstrating a surface change from (wrinkled to flat) when varying the temperature (above or below the LCST, Figure 32b).

Similarly, Liguori et al. [131] Differentiating stromal cells from adipose tissue was induced using wrinkled gradient models with nano- and microarchitectures on PDMS substrates. Cells were stimulated with TGF-β1 to elicit myogenic differentiation over seven days. Cell alignment occurred at topographies over 1.75 μm in width and 0.63 μm in breadth. Similarly, adhesion alignment increased on surfaces over 0.73 μm and 0.15 μm in width and breadth, respectively. These results make it possible to conclude that adhesion, alignment, and wrinkle differentiation are directly related to signals at the topographical level, suggesting that nanotopography may be an innovative tool to inhibit fibrosis. Similar results were reported by van Rijn et al. [132] but using mesenchymal stem cells to differentiate them into neural cells. Their results indicate that hierarchical structures significantly enhance neural differentiation, providing a new perspective in biomaterial development for tissue engineering applications. These gradient structures are advantageous in some areas, like flexible electronics and soft robotics. Indeed, Gao et al. [126] report the fabrication of an adjustable and highly-sensitive piezoresistive sensor based on 2D MXene material, which possesses wrinkled structures on its surface. The MXene was synthesized via wet chemical techniques; first, LiF was added in an HCl solution, then Ti_3AlC_2 powder was gradually added to the solution, and it was etched at 35° for 24 h, obtaining a dark green supernatant of $Ti_3C_2T_x$ (MXene) with a pH over 6. The sensing device was fabricated by spraying MXene on a pre-stretched polyacrylate tape (50%, 100%, and 150% strain). After releasing the pre-stretched polyacrylate tape coated with MXene, the composite sensing element with a wrinkled structure was fabricated. Figure 33 shows a schematic description of the process used for MXene device fabrication. The MXene composite-based sensor demonstrates high sensitivity, wide pressure range, short response time, and excellent durability; it has been possible to sense physiological signals, monitor robot postures, and spatially map pressure distributions.

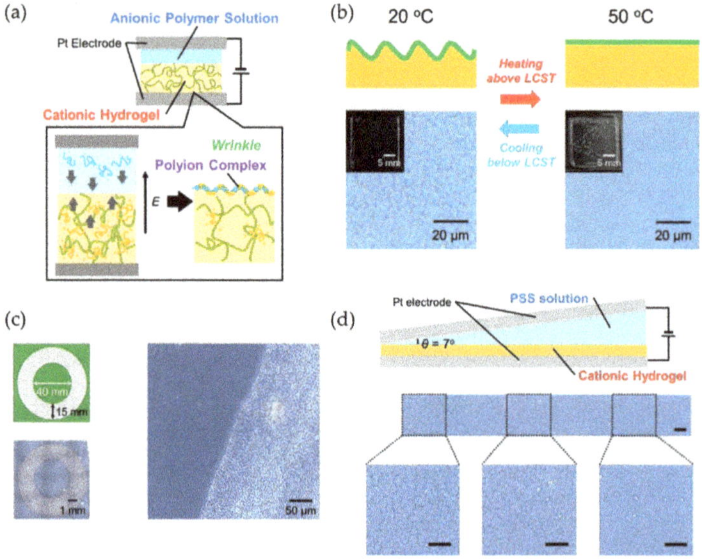

Figure 32. (**a**) Schematic description of the experimental procedure followed to fabricate wrinkled surfaces, (**b**) stimuli-responsive wrinkle structures at the surface of PNIPAAm hydrogel, (**c**) illustration of the Pt electrode's insulting mask patterns, and (**d**) schematic representation of the experimental procedure of gradient wrinkle structure formation using sloped electrodes. Reproduced with permission from reference [130]. Copyright 2019, Royal Society of Chemistry (RSC).

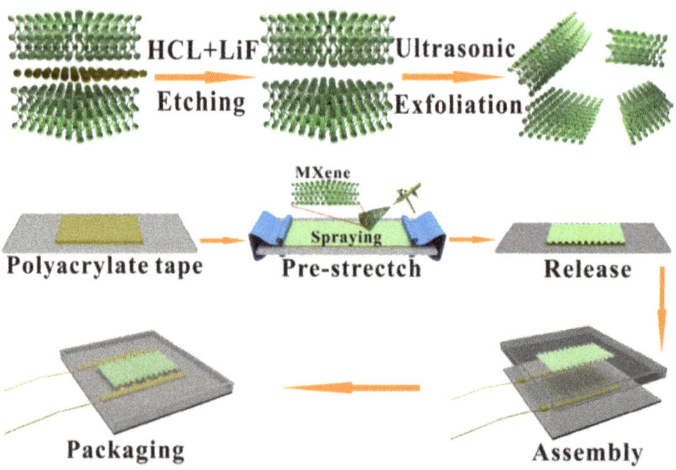

Figure 33. Schematic diagram of the preparation procedure used to fabricate the piezoresistive sensor based on MXene composite with wrinkled structure. Reproduced with permission from reference [126]. Copyright 2018, Elsevier Ltd.

Similarly, the group of Wang et al. [133] fabricates flexible and wearable piezoelectric pressure sensors based on single-walled carbon nanotubes (SWCNT), graphite flakes (GF), and thermoplastic polyurethane (TPU). The wrinkled structures were formed via pre-stretching of the TPU, which was previously exposed to carbon deposition. The material's wavy structure provided high conductivity, lower detection limit, high sensitivity, good flexibility, and short-time response. The pressure sensor was successfully adapted to monitor human activities, like walking, jumping, and pulse measuring. Furthermore, a sensor array is assembled to map an object's weight and shape, indicating its various potential applications, including human-machine interactions, health monitoring, and other wearable electronics.

Chai et al. [134] also developed a methodology to fabricate surfaces with controlled wettability through a series of surface material modifications (chemical gradient, pilar array arrangement, and wrinkled surface) (Figure 34). The chemical gradient is induced via oxygen plasma non-planar exposure over the PDMS substrate. In general, it was possible to demonstrate that the pilar density increases the wettability of the sample. Chai et al. proposed that the spreading length of droplets is regulated in situ by applying external strain.

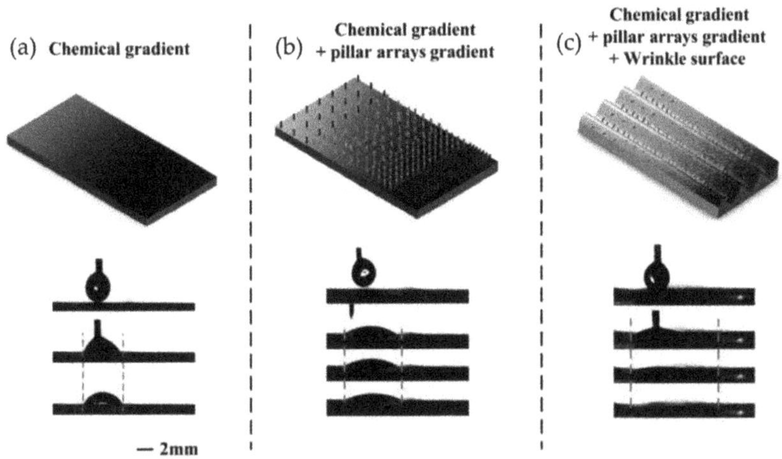

Figure 34. Droplet movement behavior on the surfaces with (**a**) chemical gradient, (**b**) chemical gradient and pillar arrays gradient, and (**c**) double-gradient wrinkled structure. Reproduced with permission from reference [134]. Copyright 2020, John Wiley & Sons, Inc.

Recently, Park et al. presented a strategy with a high level of security against imitation, using wrinkled patterns with various complexities through the development of non-clonable codes. This group fabricated wrinkled gradient patterns utilizing grayscale lithography with a custom mask with different gray levels, generating a controlled photopolymerization of a microparticle/silica-coated substrate. Even though, in most cases, the material implemented in the wrinkle formation has the major part in the matter, these authors propose a gray mask design depending on the factor. This custom gradient gray mask was designed to allow 15 gray levels within the same microparticle mask. It should be noted that the synthesized particles used different wavelengths for each gray level, establishing a total of 21 possible gray levels, each with eight gray level masking units, allowing a controlled gradient wrinkled pattern to be generated [135].

4.2. Hierarchically Ordered Wrinkled Surfaces

In the same way as wrinkled gradient topographies, hierarchical wrinkled patterns could also be generated under particular circumstances. The group of Genzer et al. [29]

reports a multi-step procedure that was employed to generate hierarchical crumpled patterns. With this new method, they could form self-similar strands extending over five orders of magnitude on the length scale, ranging from nanometers to millimeters. Recently, Odom et al. [136] transformed flat polystyrene (PS) plates into three-dimensional hierarchical structures by using sequential cycles of mechanical stretching and trifluoromethane (CHF_3) plasma exposure, followed by a stress relaxation process. Several studies were also carried out to identify the main aspects that dominate hierarchical wrinkled patterns' formation and their topology distribution. For example, Begging and Xu et al. [137] prepared wrinkled surfaces using lithography processes over soft PDMS substrates. SU-8 pillars grown on silicon wafers were used to create holes in the substrate via the spin coating method. The PDMS substrate was then transferred to a soft pre-stretched layer and released to form parallel wrinkled patterns over the surface. The evolution of parameters was studied, like distribution, size and separation of holes, and strain percentage (6%, 15%, 30%, and 55%).

On the other hand, Hu et al. [138] demonstrate that controllable hierarchical surface patterns can be fabricated by combining nano embossing techniques and surface instability of supramolecular hydrogel. Their article used polyethylene glycol (PEG) based hydrogels to fabricate nanostripe arrays upon water exposure. Due to the lateral expansion of the stripes, folded patterns were formed on the material's top. This way, hierarchical structures could be easily assembled, one due to the nano embossing process to create the nanostripes and the other by the surface buckling. The dimensions of the patterns could be controlled via the nanostripe pattern dimensions and by the swelling time.

Likewise, the González Henríquez et al. [139] have proposed a multi-step strategy to form flat, first and second-generation structures using vacuum, argon plasma, and UV exposure steps. To achieve these wrinkle dimensions, different concentrations of monomers such as 2-(dimethylamino)ethyl methacrylate (DMAEMA) and poly(ethylene glycol)diacrylate ($PEGDA_{575}$) were mixed. Besides, Irgacure 2959 was used as a photoinitiator, ammonium peroxodisulfate (APS) as a thermo-initiator, and sodium metabisulfite (SPS) as a catalyst. The APS-SPS was employed as redox pair with the finality to initiate the gelation reaction. To create G1 and G2 patterns, three essential steps must be performed. First, water removal via vacuum generates a gradient in the material, inducing G1 patterns. Second, frontal vitrification is caused by argon plasma exposure, triggering the formation of G2 wrinkles at the top of the G1 pattern. Thirdly, exposition to UV to completely polymerize the surface and fix the pattern to the substrate (Figure 35a). Figure 35b shows three AFM images corresponding to a G1 pattern, an incomplete G2 pattern (lower UV exposition time), and a G2 fully formed.

Recently, Choi et al. [140] also developed a simple method to fabricate multiscale wrinkled patterns with two different structure generations (G1–G2) using fluorocarbon thin films over the PDMS substrate. The elastic properties of the substrate greatly influenced the wrinkled structures when subjected to the fluorocarbon polymer sputtering process; in fact, it is quite possible that some of the wrinkled structures were formed in this time span.

Several of these studies were also performed using metal films to elucidate the evolution of the hierarchical wrinkled morphologies related to several external stimuli. For example, Zhang et al. [141] reported creating hierarchical wrinkles and oscillating cracks in metallic films deposited on scratched silicone oil (or PDMS). Thus, the results demonstrated that wrinkled patterns strongly depend on the stripe width and film thickness. Generally, straight wrinkles were formed near the stripe edge, while labyrinth-like wrinkles appear at the center.

Additionally, as the distance from the edge increases, the wrinkle wavelength increases quickly, and the growth speed slows down gradually. Three distinct wrinkles were observed in the cracked films: hierarchical, wavy (or herringbone), and straight structure. The formation of complex wrinkles in cracked films is attributed to the constrained (or free) edge effect and stress anisotropy. Thus, the wavelength of both hierarchical and wavy wrinkles increases with film thickness.

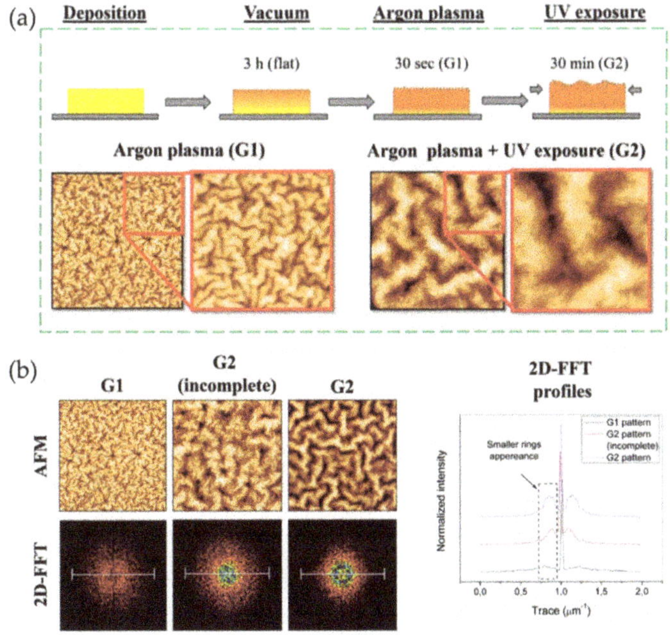

Figure 35. (**a**) Schematic illustration of the methodology used to fabricate hierarchically structured wrinkled hydrogel films, together with AFM micrographs for each sample 15:1 net-poly(DMAEMA-*co*-PEGDA$_{575}$) and (**b**) AFM micrographs of G1, incomplete G2 and fully formed G2 wrinkled patterns, 2D-FFT profile are depicted right. Reproduced with permission from reference [139]. Copyright 2019, Elsevier Ltd.

Another study that elucidates thin metal films' behavior against compressive/expansion stimulus was carried out by Jiao and Yu et al. [142]. It reported spontaneous hierarchical wrinkling in metallic films deposited by sputtering on a liquid gel substrate. Molybdenum (Mo) metal films (from 1.3 nm to 600 nm) were deposited on the non-crosslinked gel substrate (~18 μm thickness PDMS) by magnetron sputtering technique. In this publication, metal atoms and clusters can penetrate the gel substrate in the initial phase of sputtering, and complex surface structures are formed during deposition. The compressive pressure generated by the volume expansion forces the gel surface to generate highly deformed ridge structures, referred to as wrinkles. Thus, the wrinkles have a disordered appearance caused by the isotropic stress field. Also, the wavelength of the wrinkles decreases steadily with increasing separation but is insufficiently sensitive to film thickness relative to the prediction of the elastic theory. The differences in size and length between the spontaneously forming hierarchical structures should benefit various applications, be they photoelectronic devices, surface engineering, and biology-related fields.

There are several applications in which these patterns could be used; for example, structures with hierarchical roughness on both the nanoscale and microscale could help to improve superhydrophobicity. [143]. Indeed, Carter et al. fabricate nanoscale patterns on poly(hydroxyethyl methacrylate) (PHEMA) film by nanoimprint lithography via a reactive methylchlorosilane infusion reaction. The morphology and size of the surface structure are easily controlled by adjusting the initial thickness of the PHEMA films.

Unlike randomly distributed wrinkles, hierarchical wrinkles exhibit interesting orientations, mainly due to pre-imprinted patterns and confinement effects. In general, hierarchically wrinkled surfaces have significantly higher water contact angles than random wrinkled surfaces, showing in some cases superhydrophobicity behavior (water contact angles higher than 160° and sliding angles lower than 5°).

Wu and Chai et al. [144] report a theoretical framework that describes how prestrain and dynamic strain could reversibly tune the wettability of hierarchically wrinkled surfaces. Commercial silicon templates were used to fabricate hierarchical wrinkled surfaces. Then, a PDMS solution was deposited on the material via spin-coating and thermally cured. Finally, a VHB-prestrained film was adhered to the PDMS and heated together to create a strong bond between the films. The system was slowly cooled and released to form wrinkled structures, creating a hierarchical pattern of nano and micro wrinkles on the material. The wetting properties of the material varied with the dimensions and ordering of the wrinkles. It was possible to demonstrate that the influence of prestretch and applied stretching affects wettability. By analyzing the experimental and theoretical results, a relationship between the contact angle and structural size using a thermodynamic equation was established. Furthermore, the surface wettability can be adjusted from superhydrophobic to hydrophobic, even after multiple stretching-release cycles.

Similarly, Yin et al. [145] fabricate smart windows with a variable hydrophobic performance using dynamically tunable hierarchical wrinkles. These devices should present water droplet transport control using self-similar hierarchical wrinkles over PDMS substrate through sequential strain release cycles. They demonstrate that the re-stretching/releasing of the elastomeric material leads to a reversible and repeatable switching between opaqueness and transparency. Yin et al. also show the capacity of the material to modify its water droplet transport; initially, when a water droplet is deposited on the surface, the droplet stays pinned due to the hierarchical pattern presence, then the droplet starts to slide when the surface is stretched and becomes pinned again upon strain release. The sliding process occurs because first-generation wrinkles loosen due to material stretching, leading to a simple wrinkled surface pattern. When the substrate is released, the second-generation wrinkled hierarchical pattern emerges again, thus pinning the drop.

These hierarchical wrinkled patterns could also serve as a base for fabricating pressure sensors; for example, in 2018, Lu et al. [146] manufactured highly-sensitive pressure sensors based on polypyrrole (PPy) films, which are composed of three-scale nested surface wrinkling microstructures. Double-scale nested wrinkles are formed via spontaneous wrinkling after oxidative polymerization growth. The polymerization time is fundamental for this process, allowing the pattern's dimensions to be modulated easily. Heating and cooling cycles induce the third generation of surface wrinkling. The dimensions of the third-generation wrinkles are directly related to the heating temperature and heating cycle duration. These multiscale hierarchical structures, combined with the materials' stimulus-responsive characteristics, allow the fabrication of pressure sensors with high sensitivity, low detection limit, ultrafast response, excellent durability, and time stability, being suitable as an integrative part of wearable electronics.

Accordingly, hierarchical wrinkled surfaces could also be used for different applications in the wearable electronics field; for example, Zhang and Wang et al. [147] proposed a method to fabricate wearable triboelectric nanogenerator (TENGs) based on hydrogel wrinkled electrospun nanofibers. These hierarchical structures form a membrane for effective energy harvesting from human movement. The PDMS substrate was packaged with PAAm-LiCl hydrogel (polyacrylamide hydrogel containing lithium chloride) to fabricate the TENG. A copper belt is embedded inside the hydrogel as an electronic connection. An air layer is added to the middle of the devices as a contact separator; air can enter and leave freely through holes in the walls on both sides of the devices under regular operation. Posteriorly, a polycaprolactam (PA6) solution was deposited via electrospinning onto the PAAm hydrogel's surface. Finally, the films are released by cooling the flask to obtain a hierarchically wrinkled PA6 membrane. This device presents high stretchability and enlarged triboelectric area, improving its performance due to the formation of rough surfaces. These outstanding features may optimize TENGs devices for fitness braces.

Following the topic of flexible electronics, Cui and Liu et al. [148] fabricated a hierarchical $Cu(OH)_2/MnO_2$ core-shell nanorod array on copper foam (CF), utilizing $Cu(OH)_2$ nanorods in situ generated on the Cu foam. Thus, the $Cu(OH)_2/MnO_2$/CF nanorods array display an excellent capacitance of ~709 mF cm^{-2} at the current density of 2 mA cm^{-2} (~283 F g^{-1} at 0.8 A g^{-1}), favorable rate capability (74.65%, from 1 to 20 mA cm^{-2}) and great cycling stability (85.17% of capacity retention after a test for 5000 cycles). Additionally, the assembled $Cu(OH)_2/MnO_2$/CF/activated carbon (AC) asymmetric supercapacitor shows an outstanding energy density of ~18 Wh kg^{-1} at a power density of 750 W kg^{-1}, and excellent cycling stability with 90.03% capacitance retention after 6000 cycles.

On the other hand, wrinkled hierarchical structures could also respond to an external thermal stimulus; for example, Asoh et al. [149] studied aligned wrinkles fabricated via electrophoretic polymerization of a polyion complex (PIC) as a hard-skin layer. The cationic PDMAAm hydrogels with semi-interpenetrated network structures were prepared by free-radical polymerization of N,N-Dimethylacrylamide (DMAAm) with methylenebisacrylamide (MBAAm). The wrinkles on the hydrogel surface were prepared under stretching; aqueous sodium poly(4-styrene sulfonate) (PSS) solution was poured on the top of a stretched cationic hydrogel and then sandwiched between two pieces of Pt electrodes. The stretching ratio of hydrogels during the electrophoresis was changed to control the wrinkle formation. An aligned wrinkled structure was fabricated on the surface of oriented hydrogel using poly(vinyl alcohol) (PVA) and PPA dissolved in water. Then, 25% glutaraldehyde (GA) solution as the cross-linker and hydrochloric acid as the catalyst were added to the solution. Wrinkle formation was performed on the surface of oriented hydrogel, and the applied electric fixed was fixed to 10 V mm^{-1}. The wrinkle geometry was controlled by modifying the electrophoretic conditions and Young's modulus of the hydrogels. Figure 36a indicates the topological modulation of hydrogel wrinkling as a function of stimuli based on the fabrication of aligned and hierarchical wrinkle structures in the hydrogel. The stretching ratio, Young's modulus of hydrogels, and hierarchical wrinkle structures were fabricated after releasing the stretched hydrogels (Figure 36b).

Recently Jiang et al. [150] proposed a simple but robust method to fabricate 3D patterned surfaces with dynamic topographies by taking advantage of the reversible DA reaction. In this article, Jiang et al. synthesize a substrate composed of a copolymer containing furan (PSFB) and bismaleimide (BMI), which can selectively polymerize through photodimerization of BMI by a mask when exposed to UV light (Figure 37a). The spread of maleimide from the unexposed region to the exposed region results in a diffuse relief pattern. By controlling the reversible D-A reaction at different temperatures, orthogonal wrinkles can be sequentially and reversibly generated or erased in the exposed and unexposed regions of the surface, thus fabricating dynamic wrinkle patterns that actively respond to ambient temperature (Figure 37b).

Some applications also involve the use of light-reversible hierarchical wrinkled patterns, like the case of Jiang et al. [151]. A straightforward and efficient strategy has been created to fabricate light-manageable hierarchical patterns with highly reversible morphology and fast turnaround time through dynamic reversible wrinkle-induced dynamic photodimerization.

Jiang et al. reported obtaining wrinkles that deliver a multi-response and surface geometry that can be formed as well as dynamically removed in situ by multiple physical or chemical stimuli, such as UV and an aqueous solution of DL-1,4-dithiothreitol (DTT, reductant) and heat (Figure 38a). The wrinkles were formed by the dynamic cross-linking of the interpenetrating polymeric network (IPN) containing covalent bonds sensitive to redox-type reactions and visible light, designed as the top layer deposited on PDMS elastomers. Given the dynamic reversible covalent -C-C- bond in the anthracene-containing polymer (PAN) and -S-S- bond in the disulfide-containing diacrylate monomer (DSDA), the surface morphology can not only be reversibly and precisely adapted to all kinds of complicated hierarchical patterns permanently, but it can also be controlled by NIR radiation. The authors fabricated a QR code with wrinkle patterns, which is visible to the naked eye due to light scattering from the wrinkles (Figure 38b).

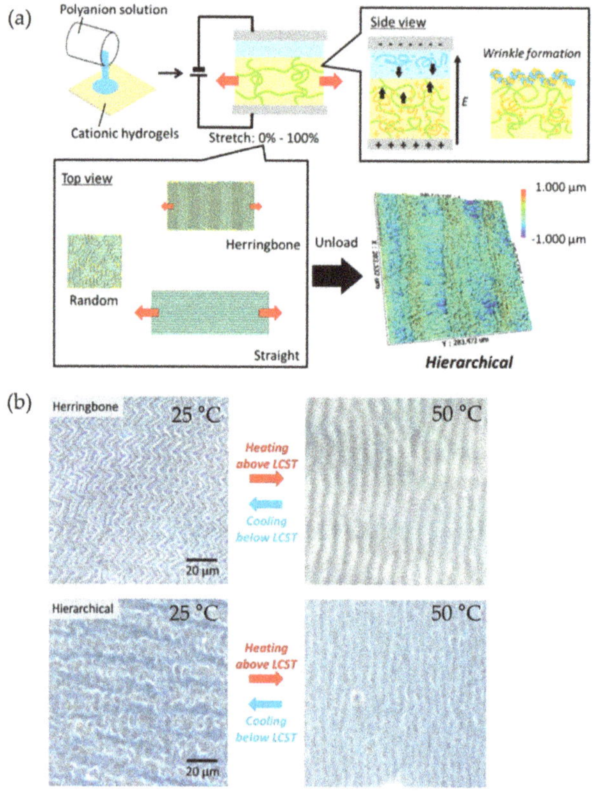

Figure 36. (**a**) Experimental procedure for wrinkle formation on the hydrogel surface by electrophoretic fabrication under stretching and (**b**) thermoresponsive transformation of herringbone and hierarchical wrinkles structures. Reproduced with permission from reference [149]. Copyright 2020, American Chemical Society (ACS).

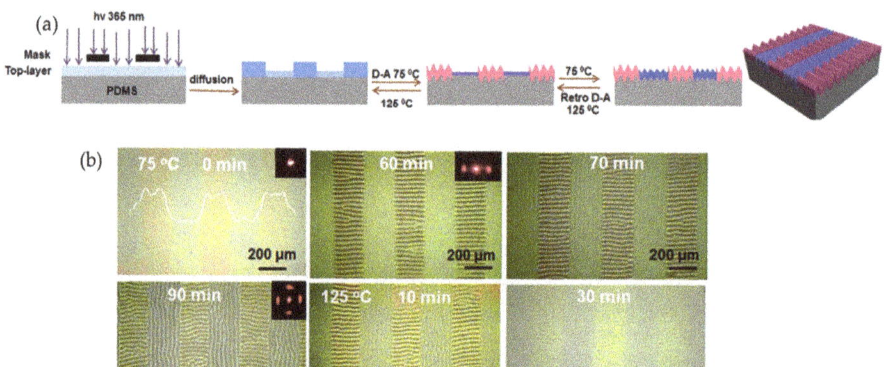

Figure 37. (**a**) Schematic description of the process used for fabricating hierarchical patterns with dynamic wrinkles through a photo-controlled Diels-Alder reaction and (**b**) optical images of the formation/extinction process of horizontally or vertically striped wrinkles on reversible Diels-Alder reaction. Reproduced with permission from [150]. Copyright 2020, John Wiley & Sons, Inc.

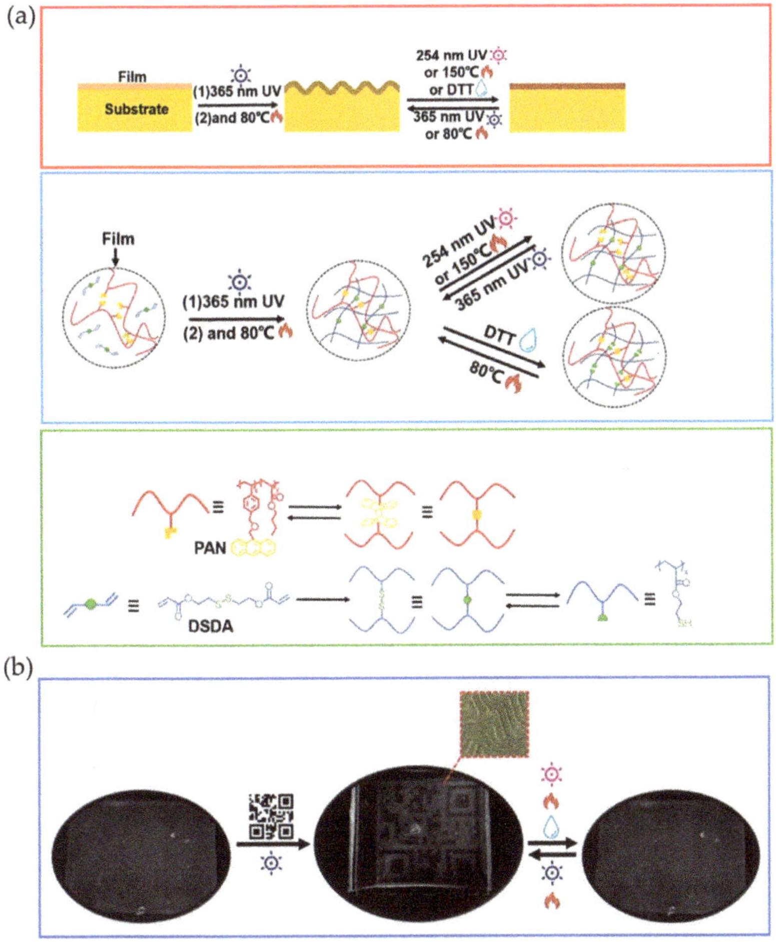

Figure 38. (**a**) Schematic diagram of the IPN fabrication and transformation process and the corresponding wrinkle generation/quenching process. Bilayer wrinkle generation/quenching process (red). IPN formation/quenching in the film by multi-stimuli (blue). The reversible chemical reaction of PAN and DSDA in the skin layer (green). (**b**) Photography of QR code with the reversible wrinkled pattern. This pattern was fabricated using UV irradiation of 365 nm and erased using 254 nm in the UV range. Reproduced with permission from reference [152]. Copyright 2019, American Chemical Society (ACS).

Finally, in 2022, Yang et al. implemented a swelling base wrinkled structure manufacturing process. This group generated stretchable sensors from silver nanowires (AgNW) that exhibit a crumpled hierarchical structure. These structures were fabricated through a two-step process: (1) water-induced swelling and (2) AgNW soaking in a dopamine aqueous solution. It is worth mentioning that the swelling stage endows the sample surface with ultra-hydrophilicity. Subsequently, the dopamine-modified swollen samples fixate the nanowires when immersed in the AgNW/ethanol solution due to the dopamine's highly adhesive layer properties. In Figure 39, it is possible to observe a schematic description of the fabrication of these personalized stretchable elastomer sensors using 3D printing [153].

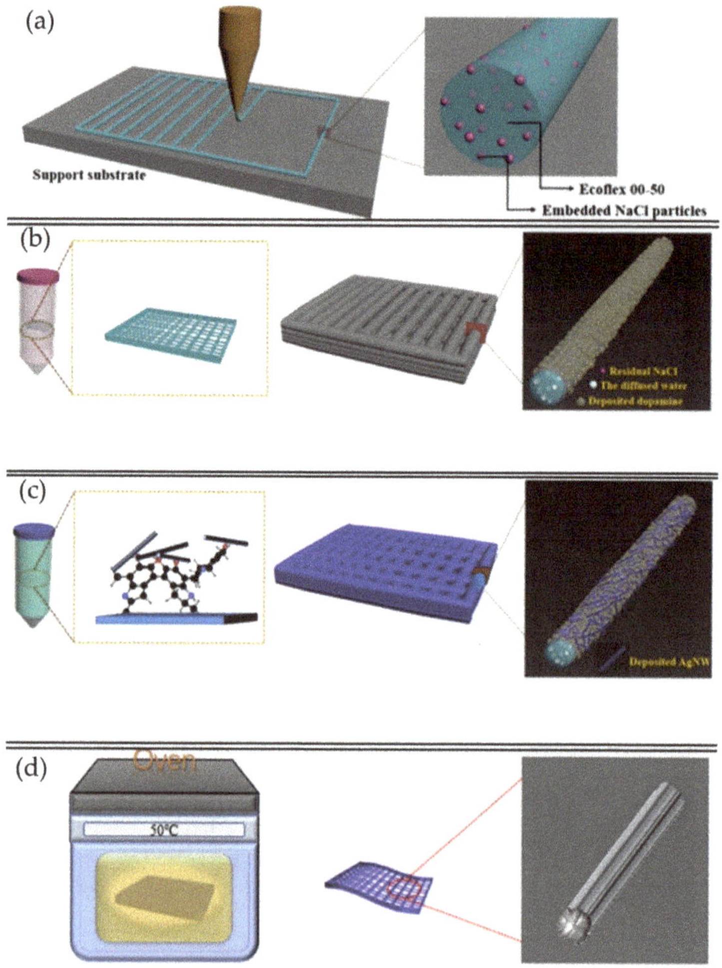

Figure 39. Schematic description of the fabrication process of stretchable elastomer sensors. (**a**) Formation of structures by 3D printing. (**b**) Dopamine treatment via immersion in an aqueous solution. (**c**) Deposition of AgNW on the dopamine-coated sample and (**d**) formation of rough structures after remotion of residual water. Reproduced with permission from reference [153]. Copyright 2022, Elsevier Ltd.

5. Conclusions and Future Perspectives

The microstructure and chemical composition of polymer surfaces are critical aspects that directly or indirectly control the material characteristics and how it behaves with its surroundings; therefore, it influences the final application of the material.

This review mainly focused on wrinkle formation on polymer surfaces using smart materials that can react to slight environmental changes. This type of structuration is obtained by taking advantage of the surface instabilities generated, which result in an out-of-plane deformation. This process can be controlled; thus, the final wrinkle characteristics, including period and amplitude, can be finely modulated.

More recently, a great effort has been focused on elaborating more sophisticated surface structures, including (1) reversible stimuli-responsive wrinkle formation, (2) non-planar wrinkled structures, and (3) complex wrinkled patterns like gradient or hierarchical surfaces. These strategies have also been summarized and organized in different sections in the present review providing a broad overview of the current possibilities to modulate the interfacial properties.

Firstly, reversible stimuli-responsive wrinkled patterns were formed using smart polymers that can respond to external stimuli and actively react to environmental changes. In this section, materials that can react to stimuli were reviewed, the most common being light stimuli-responsive materials such as the one reported by Omenneto et al. [40] or Jiang et al. [43]; also, changes in environmental pH were reviewed, as in the case of Yang et al. [59], or temperature stimuli-responsive materials as in the case of Zhang and Sun et al. [60]. Other types of stimuli were also reviewed, such as changes in electric/magnetic fields or changes in environmental humidity and even reactions to mechanical stimuli such as stretching or compression. In addition, a series of applications, such as logic gates or anti-counterfeiting devices, were also mentioned and analyzed.

Secondly, non-planar wrinkled structures were reviewed, standing out several structures like solid core-shell spheres or cylinders and hollow spheres or cylinders. Interestingly, these types of structures allow taking more advantage of the surface increment that generates the wrinkling process by creating these patterns on non-planar surfaces, thus improving several characteristics that depend on the material's interaction with the environment. One of the most exciting applications is, for example, the use of core-shell or hollow spheres for drug delivery or the use of cylindrical hollow tubes that can react to temperature or NIR stimulus, thus allowing to open or close a channel depending on the external stimulus (Sun and Wu et al. [122]).

Finally, complex wrinkled patterns were also reviewed in this article, mainly focused on the formation of gradient and hierarchically ordered wrinkled patterns. The fabrication methods for this type of complex structure go from taking advantage of surface instabilities using UV, plasma, oxidation, or mechanical stretching, to more advanced fabrication methods like electrophoresis, as reported by Kato et al. [130]. These complex wrinkled patterns, together with some particular functionalities of the smart materials, allow the development of incredible applications like anti-counterfeiting devices or surfaces with controlled wettability (Chai et al. [134]).

The developments in this review about wrinkle structure formation may provide an exciting toolbox to produce surfaces adapted for a particular purpose and able to change their structure and, thus, their properties depending on the environmental conditions.

Author Contributions: M.A.S.-V.: conceptualization, data curation, writing—original draft preparation, funding acquisition. F.E.C.-I.: writing, review, and editing. D.A.P.-M.: writing, review, and editing. N.F.A.-R.: writing, review, and editing. C.A.T.-I.: review and editing. J.R.-H.: conceptualization, resources, writing, project administration, funding acquisition. C.M.G.-H.: conceptualization, resources, writing, supervision, project administration, funding acquisition. All authors have read and agreed to the published version of the manuscript.

Funding: The authors recognize the financial support from FONDECYT Grant N° 1220251 and N° 11230427. M.A. Sarabia-Vallejos acknowledges financial support from Universidad San Sebastián (USS) via project No. VRID FAI 21/17. J. Rodriguez-Hernandez acknowledges financial support from Ministerio de Ciencia, Innovación y Universidades (Project MAT2016-78437-R, FONDOS FEDER, and RTI2018-096328-B-I00). Finally, we thank the PIT FAB3D, PTI + Salud Global, and the PTI + SUSPLAST from CSIC for their support.

Institutional Review Board Statement: Not applicable.

Informed Consent Statement: Not applicable.

Data Availability Statement: Not applicable.

Conflicts of Interest: The authors declare no conflict of interest.

References

1. Yu, C.; O'Brien, K.; Zhang, Y.-H.; Yu, H.; Jiang, H. Tunable optical gratings based on buckled nanoscale thin films on transparent elastomeric substrates. *Appl. Phys. Lett.* **2010**, *96*, 041111. [CrossRef]
2. Peng, Y.; Serfass, C.M.; Kawazoe, A.; Shao, Y.; Gutierrez, K.; Hill, C.N.; Santos, V.J.; Visell, Y.; Hsiao, L.C. Elastohydrodynamic friction of robotic and human fingers on soft micropatterned substrates. *Nat. Mater.* **2021**, *20*, 1707–1711. [CrossRef] [PubMed]
3. Peng, Y.; Serfass, C.M.; Hill, C.N.; Hsiao, L.C. Bending of Soft Micropatterns in Elastohydrodynamic Lubrication Tribology. *Exp. Mech.* **2021**, *61*, 969–979. [CrossRef]
4. Schlaak, H.F.; Jungmann, M.; Matysek, M.; Lotz, P. Novel multilayer electrostatic solid state actuators with elastic dielectric (Invited Paper). In *Smart Structures and Materials 2005: Electroactive Polymer Actuators and Devices (EAPAD)*; Bar-Cohen, Y., Ed.; SPIE: Paris, France, 2005; p. 121.
5. Ren, K.; Liu, S.; Lin, M.; Wang, Y.; Zhang, Q.M. A compact electroactive polymer actuator suitable for refreshable Braille display. *Sensors Actuators A Phys.* **2008**, *143*, 335–342. [CrossRef]
6. Shivapooja, P.; Wang, Q.; Orihuela, B.; Rittschof, D.; López, G.P.; Zhao, X. Bioinspired Surfaces with Dynamic Topography for Active Control of Biofouling. *Adv. Mater.* **2013**, *25*, 1430–1434. [CrossRef]
7. Chung, J.Y.; Youngblood, J.P.; Stafford, C.M. Anisotropic wetting on tunable micro-wrinkled surfaces. *Soft Matter* **2007**, *3*, 1163–1169. [CrossRef]
8. Chan, E.P.; Smith, E.J.; Hayward, R.C.; Crosby, A.J. Surface wrinkles for smart adhesion. *Adv. Mater.* **2008**, *20*, 711–716. [CrossRef]
9. Qian, Y.; Zhang, X.; Xie, L.; Qi, D.; Chandran, B.K.; Chen, X.; Huang, W. Stretchable Organic Semiconductor Devices. *Adv. Mater.* **2016**, *28*, 9243–9265. [CrossRef]
10. Qian, Y.; Zhang, X.; Qi, D.; Xie, L.; Chandran, B.K.; Chen, X.; Huang, W. Thin-film organic semiconductor devices: From flexibility to ultraflexibility. *Sci. China Mater.* **2016**, *59*, 589–608. [CrossRef]
11. González-Henríquez, C.; Sarabia-Vallejos, M.; Rodriguez-Hernandez, J. Advances in the Fabrication of Antimicrobial Hydrogels for Biomedical Applications. *Materials* **2017**, *10*, 232. [CrossRef]
12. Shu, T.; Hu, L.; Shen, Q.; Jiang, L.; Zhang, Q.; Serpe, M.J. Stimuli-responsive polymer-based systems for diagnostic applications. *J. Mater. Chem. B* **2020**, *8*, 7042–7061. [CrossRef]
13. Saleh, T.A.; Fadillah, G.; Ciptawati, E. Smart advanced responsive materials, synthesis methods and classifications: From Lab to applications. *J. Polym. Res.* **2021**, *28*, 197. [CrossRef]
14. Jingcheng, L.; Reddy, V.S.; Jayathilaka, W.A.D.M.; Chinnappan, A.; Ramakrishna, S.; Ghosh, R. Intelligent polymers, fibers and applications. *Polymers* **2021**, *13*, 1427. [CrossRef]
15. Holman, H.; Kavarana, M.N.; Rajab, T.K. Smart materials in cardiovascular implants: Shape memory alloys and shape memory polymers. *Artif. Organs* **2021**, *45*, 454–463. [CrossRef]
16. Peponi, L.; Raquez, J.M. *How Smart Are the Polymers?* 1st ed.; Peponi, L., Raquez, J.-M., Eds.; Nova Science Publisher, Inc.: Hauppauge, NY, USA, 2018; ISBN 978-1-53613-960-0.
17. Zhou, Y.; Huang, W.M.; Kang, S.F.; Wu, X.L.; Lu, H.B.; Fu, J.; Cui, H. From 3D to 4D printing: Approaches and typical applications. *J. Mech. Sci. Technol.* **2015**, *29*, 4281–4288. [CrossRef]
18. Tibbits, S.; Cheung, K. Programmable materials for architectural assembly and automation. *Assem. Autom.* **2012**, *32*, 216–225. [CrossRef]
19. Hu, Y.Y.; Huang, W.M. Elastic and elastic-plastic analysis of multilayer thin films:Closed-form solutions. *J. Appl. Phys.* **2004**, *96*, 4154–4160. [CrossRef]
20. Huang, W.M.; Ding, Z.; Wang, C.C.; Wei, J.; Zhao, Y.; Purnawali, H. Shape memory materials. *Mater. Today* **2010**, *13*, 54–61. [CrossRef]
21. Sun, L.; Huang, W.M.; Ding, Z.; Zhao, Y.; Wang, C.C.; Purnawali, H.; Tang, C. Stimulus-responsive shape memory materials: A review. *Mater. Des.* **2012**, *33*, 577–640. [CrossRef]
22. Hornbogen, E. Comparison of Shape Memory Metals and Polymers. *Adv. Eng. Mater.* **2006**, *8*, 101–106. [CrossRef]
23. El Feninat, F.; Laroche, G.; Fiset, M.; Mantovani, D. Shape Memory Materials for Biomedical Applications. *Adv. Eng. Mater.* **2002**, *4*, 91–104. [CrossRef]
24. Gunes, I.S.; Jana, S.C. Shape Memory Polymers and Their Nanocomposites: A Review of Science and Technology of New Multifunctional Materials. *J. Nanosci. Nanotechnol.* **2008**, *8*, 1616–1637. [CrossRef] [PubMed]
25. Sarangi, M.K.; Rao, M.E.B.; Parcha, V. Smart polymers for colon targeted drug delivery systems: A review. *Int. J. Polym. Mater. Polym. Biomater.* **2020**, *70*, 1130–1166. [CrossRef]
26. Tan, L.; Davis, A.C.; Cappelleri, D.J. Smart Polymers for Microscale Machines. *Adv. Funct. Mater.* **2021**, *31*, 1–56. [CrossRef]
27. Rodríguez-Hernández, J. Wrinkled interfaces: Taking advantage of surface instabilities to pattern polymer surfaces. *Prog. Polym. Sci.* **2015**, *42*, 1–41. [CrossRef]
28. Cerda, E.A.; Ravi-Chandar, K.; Mahadevan, L. Thin films. Wrinkling of an elastic sheet under tension. *Nature* **2002**, *419*, 579–580. [CrossRef]
29. Efimenko, K.; Rackaitis, M.; Manias, E.; Vaziri, A.; Mahadevan, L.; Genzer, J. Nested self-similar wrinkling patterns in skins. *Nat. Mater.* **2005**, *4*, 293–297. [CrossRef]
30. Dimmock, R.L.; Wang, X.; Fu, Y.; El Haj, A.J.; Yang, Y. Biomedical Applications of Wrinkling Polymers. *Recent Prog. Mater.* **2020**, *2*, 1–31. [CrossRef]

31. Raj, S.S.; Mathew, R.M.; Nair, Y.; Aruna, S.T.; Vinod, T.P. Fabrication and Applications of Wrinkled Soft Substrates: An Overview. *ChemistrySelect* **2022**, *7*, e202200714. [CrossRef]
32. Li, Z.; Zhai, Y.; Wang, Y.; Wendland, G.M.; Yin, X.; Xiao, J. Harnessing Surface Wrinkling–Cracking Patterns for Tunable Optical Transmittance. *Adv. Opt. Mater.* **2017**, *5*, 1700425. [CrossRef]
33. Chen, W.; Gui, X.; Yang, L.; Zhu, H.; Tang, Z. Wrinkling of two-dimensional materials: Methods, properties and applications. *Nanoscale Horiz.* **2019**, *4*, 291–320. [CrossRef]
34. González-Henríquez, C.M.; Rodríguez-Hernández, J. *Wrinkled Polymer Surfaces*; Springer International Publishing: Cham, Switzerland, 2019; ISBN 978-3-030-05122-8.
35. Yang, S.; Khare, K.; Lin, P.C. Harnessing surface wrinkle patterns in soft matter. *Adv. Funct. Mater.* **2010**, *20*, 2550–2564. [CrossRef]
36. Deng, S.; Berry, V. Wrinkled, rippled and crumpled graphene: An overview of formation mechanism, electronic properties, and applications. *Mater. Today* **2016**, *19*, 197–212. [CrossRef]
37. Chung, J.Y.; Nolte, A.J.; Stafford, C.M. Surface wrinkling: A versatile platform for measuring thin-film properties. *Adv. Mater.* **2011**, *23*, 349–368. [CrossRef]
38. Hou, H.; Yin, J.; Jiang, X. Smart Patterned Surface with Dynamic Wrinkles. *Acc. Chem. Res.* **2019**, *52*, 1025–1035. [CrossRef]
39. Zhang, H.; Zhu, H.; Liang, X.; Liu, P.; Zhang, Q.; Zhu, S. Wrinkled smart surfaces: Enhanced switchable wettability and directional liquid transportation. *Appl. Surf. Sci.* **2020**, *513*, 145810. [CrossRef]
40. Wang, Y.; Kim, B.J.; Peng, B.; Li, W.; Wang, Y.; Li, M.; Omenetto, F.G. Controlling silk fibroin conformation for dynamic, responsive, multifunctional, micropatterned surfaces. *Proc. Natl. Acad. Sci. USA* **2019**, *116*, 21361–21368. [CrossRef]
41. Li, F.; Hou, H.; Yin, J.; Jiang, X. Multi-responsive wrinkling patterns by the photoswitchable supramolecular network. *ACS Macro Lett.* **2017**, *6*, 848–853. [CrossRef]
42. Zong, C.; Zhao, Y.; Ji, H.; Han, X.; Xie, J.; Wang, J.; Cao, Y.-P.; Jiang, S.; Lu, C. Tuning and Erasing Surface Wrinkles by Reversible Visible-Light-Induced Photoisomerization. *Angew. Chem.-Int. Ed.* **2016**, *55*, 3931–3935. [CrossRef]
43. Wang, J.; Zheng, Y.; Li, L.; Liu, E.; Zong, C.; Zhao, J.; Xie, J.; Xu, F.; König, T.A.F.; Saphiannikova, M.G.; et al. All-Optical Reversible Azo-Based Wrinkling Patterns with High Aspect Ratio and Polarization-Independent Orientation for Light-Responsive Soft Photonics. *ACS Appl. Mater. Interfaces* **2019**, *11*, 25595–25604. [CrossRef]
44. Ma, T.; Li, T.; Zhou, L.; Ma, X.; Yin, J.; Jiang, X. Dynamic wrinkling pattern exhibiting tunable fluorescence for anticounterfeiting applications. *Nat. Commun.* **2020**, *11*, 1811. [CrossRef] [PubMed]
45. Zong, C.; Azhar, U.; Zhou, C.; Wang, J.; Zhang, L.; Cao, Y.; Zhang, S.; Jiang, S.; Lu, C. Photocontrollable Wrinkle Morphology Evolution on Azo-Based Multilayers for Hierarchical Surface Micropatterns Fabrication. *Langmuir* **2019**, *35*, 2601–2609. [CrossRef] [PubMed]
46. Xu, L.; Azhar, U.; Chen, Z.; Niu, Q.; Chen, J.; Zhao, X.; Zhang, S.; Zong, C. Light-printable epoxy oligomer wrinkle-forming surface for rewritable information storage. *RSC Adv.* **2020**, *10*, 3416–3423. [CrossRef] [PubMed]
47. Li, F.; Hou, H.; Yin, J.; Jiang, X. Near-infrared light–responsive dynamic wrinkle patterns. *Sci. Adv.* **2018**, *4*, eaar5762. [CrossRef] [PubMed]
48. Xie, M.; Xu, F.; Zhang, L.; Yin, J.; Jiang, X. Reversible Surface Dual-Pattern with Simultaneously Dynamic Wrinkled Topography and Fluorescence. *ACS Macro Lett.* **2018**, *7*, 540–545. [CrossRef]
49. Hou, H.; Hu, K.; Lin, H.; Forth, J.; Zhang, W.; Russell, T.P.; Yin, J.; Jiang, X. Reversible Surface Patterning by Dynamic Crosslink Gradients: Controlling Buckling in 2D. *Adv. Mater.* **2018**, *30*, 1803463. [CrossRef]
50. Zong, C.; Zhao, Y.; Ji, H.; Xie, J.; Han, X.; Wang, J.; Cao, Y.-P.; Lu, C.; Li, H.; Jiang, S. Patterning Surfaces on Azo-Based Multilayer Films via Surface Wrinkling Combined with Visible Light Irradiation. *Macromol. Rapid Commun.* **2016**, *37*, 1288–1294. [CrossRef]
51. Xie, M.; Lin, G.; Ge, D.; Yang, L.; Zhang, L.; Yin, J.; Jiang, X. Pattern Memory Surface (PMS) with Dynamic Wrinkles for Unclonable Anticounterfeiting. *ACS Mater. Lett.* **2019**, *1*, 77–82. [CrossRef]
52. Bae, H.J.; Bae, S.; Park, C.; Han, S.; Kim, J.; Kim, L.N.; Kim, K.; Song, S.H.; Park, W.; Kwon, S. Biomimetic microfingerprints for anti-counterfeiting strategies. *Adv. Mater.* **2015**, *27*, 2083–2089. [CrossRef]
53. Zhou, L.; Hu, K.; Zhang, W.; Meng, G.; Yin, J.; Jiang, X. Regulating surface wrinkles using light. *Natl. Sci. Rev.* **2020**, *7*, 1247–1257. [CrossRef]
54. Wang, B.; Wang, J.; Han, X.; Lu, C. Light-Induced Surface Wrinkling on Azo-Based Composite Films. *Coatings* **2022**, *12*, 608. [CrossRef]
55. Bai, J.; Hu, K.; Zhang, L.; Shi, Z.; Zhang, W.; Yin, J.; Jiang, X. The Evolution of Self-Wrinkles in a Single-Layer Gradient Polymer Film Based on Viscoelasticity. *Macromolecules* **2022**, *55*, 3563–3572. [CrossRef]
56. Um, E.; Cho, Y.-K.; Jeong, J. Spontaneous Wrinkle Formation on Hydrogel Surfaces Using Photoinitiator Diffusion from Oil–Water Interface. *ACS Appl. Mater. Interfaces* **2021**, *13*, 15837–15846. [CrossRef]
57. Li, Y.; Li, X.; Guo, W.; Wu, M.; Sun, J. Spontaneous wrinkling of layer-by-layer assembled polyelectrolyte films for humidity-responsive superhydrophobicity. *Sci. China Chem.* **2016**, *59*, 1568–1573. [CrossRef]
58. Hou, J.; Li, Q.; Han, X.; Lu, C. Swelling/Deswelling-Induced Reversible Surface Wrinkling on Layer-by-Layer Multilayers. *J. Phys. Chem. B* **2014**, *118*, 14502–14509. [CrossRef]
59. Yang, J.; Hu, D.; Li, W.; Jia, Y.; Li, P. Formation mechanism of zigzag patterned P(NIPAM-co-AA)/CuS composite microspheres by in situ biomimetic mineralization for morphology modulation. *RSC Adv.* **2021**, *11*, 37904–37916. [CrossRef]

60. Zeng, S.; Li, R.; Freire, S.G.; Garbellotto, V.M.M.; Huang, E.Y.; Smith, A.T.; Hu, C.; Tait, W.R.T.; Bian, Z.; Zheng, G.; et al. Moisture-Responsive Wrinkling Surfaces with Tunable Dynamics. *Adv. Mater.* **2017**, *29*, 1700828. [CrossRef]
61. Tokudome, Y.; Kuniwaki, H.; Suzuki, K.; Carboni, D.; Poologasundarampillai, G.; Takahashi, M. Thermoresponsive Wrinkles on Hydrogels for Soft Actuators. *Adv. Mater. Interfaces* **2016**, *3*, 1500802. [CrossRef]
62. Ringsdorf, H.; Venzmer, J.; Winnik, F.M. Fluorescence studies of hydrophobically modified poly(N-isopropylacrylamides). *Macromolecules* **1991**, *24*, 1678–1686. [CrossRef]
63. Wang, Y.; Xiao, J. Programmable, reversible and repeatable wrinkling of shape memory polymer thin films on elastomeric substrates for smart adhesion. *Soft Matter* **2017**, *13*, 5317–5323. [CrossRef]
64. Jiang, B.; Liu, L.; Gao, Z.; Feng, Z.; Zheng, Y.; Wang, W. Fast Dual-Stimuli-Responsive Dynamic Surface Wrinkles with High Bistability for Smart Windows and Rewritable Optical Displays. *ACS Appl. Mater. Interfaces* **2019**, *11*, 40406–40415. [CrossRef] [PubMed]
65. Wang, Y.; Villada, A.; Zhai, Y.; Zou, Z.; Chen, Y.; Yin, X.; Xiao, J. Tunable surface wrinkling on shape memory polymers with application in smart micromirror. *Appl. Phys. Lett.* **2019**, *114*, 193701. [CrossRef]
66. Lee, D.-H.; Yun, G.-T.; Doo, G.; Yuk, S.; Guim, H.; Kim, Y.; Jung, W.-B.; Jung, H.-T.; Kim, H.-T. Hierarchical Wrinkle-Structured Catalyst Layer/Membrane Interface for Ultralow Pt-Loading Polymer Electrolyte Membrane Fuel Cells (PEMFCs). *Nano Lett.* **2022**, *22*, 1174–1182. [CrossRef] [PubMed]
67. Van den Ende, D.; Kamminga, J.-D.; Boersma, A.; Andritsch, T.; Steeneken, P.G. Voltage-Controlled Surface Wrinkling of Elastomeric Coatings. *Adv. Mater.* **2013**, *25*, 3438–3442. [CrossRef]
68. Lin, I.-T.; Choi, Y.S.; Wojcik, C.; Wang, T.; Kar-Narayan, S.; Smoukov, S.K. Electro-responsive surfaces with controllable wrinkling patterns for switchable light reflection–diffusion–grating devices. *Mater. Today* **2020**, *41*, 51–61. [CrossRef]
69. Psarra, E.; Bodelot, L.; Danas, K. Wrinkling to crinkling transitions and curvature localization in a magnetoelastic film bonded to a non-magnetic substrate. *J. Mech. Phys. Solids* **2019**, *133*, 103734. [CrossRef]
70. Bodelot, L.; Voropaieff, J.-P.; Pössinger, T. Experimental investigation of the coupled magneto-mechanical response in magnetorheological elastomers. *Exp. Mech.* **2018**, *58*, 207–221. [CrossRef]
71. Jiang, B.; Liu, L.; Gao, Z.; Wang, W. A General and Robust Strategy for Fabricating Mechanoresponsive Surface Wrinkles with Dynamic Switchable Transmittance. *Adv. Opt. Mater.* **2018**, *6*, 1800195. [CrossRef]
72. Li, Z.; Liu, Y.; Marin, M.; Yin, Y. Thickness-dependent wrinkling of PDMS films for programmable mechanochromic responses. *Nano Res.* **2020**, *13*, 1882–1888. [CrossRef]
73. Oyefusi, A.; Chen, J. Mechanical approaches to dynamic, reversible mechanochromism based on thin film interference. *Appl. Mater. Today* **2020**, *20*, 100774. [CrossRef]
74. Baek, S.; Jang, H.; Kim, S.Y.; Jeong, H.; Han, S.; Jang, Y.; Kim, D.H.; Lee, H.S. Flexible piezocapacitive sensors based on wrinkled microstructures: Toward low-cost fabrication of pressure sensors over large areas. *RSC Adv.* **2017**, *7*, 39420–39426. [CrossRef]
75. Yu, S.; Li, L.; Wang, J.; Liu, E.; Zhao, J.; Xu, F.; Cao, Y.; Lu, C. Light-Boosting Highly Sensitive Pressure Sensors Based on Bioinspired Multiscale Surface Structures. *Adv. Funct. Mater.* **2020**, *30*, 1907091. [CrossRef]
76. Ouchi, T.; Hayward, R.C. Harnessing Multiple Surface Deformation Modes for Switchable Conductivity Surfaces. *ACS Appl. Mater. Interfaces* **2020**, *12*, 10031–10038. [CrossRef]
77. Kashihara, Y.; Asoh, T.; Uyama, H. Travelling Wave Generation of Wrinkles on the Hydrogel Surfaces. *Macromol. Rapid Commun.* **2022**, *43*, 2100848. [CrossRef]
78. Wawryk, M.M.H.; Turpin, G.A.; Tabor, R.F. Surface defects on wrinkled PDMS induce droplet anisotropy. *Colloids Surfaces A Physicochem. Eng. Asp.* **2022**, *639*, 128317. [CrossRef]
79. Vandeparre, H.; Damman, P. Wrinkling of Stimuloresponsive Surfaces: Mechanical Instability Coupled to Diffusion. *Phys. Rev. Lett.* **2008**, *101*, 124301. [CrossRef]
80. Oscurato, S.L.; Formisano, F.; de Lisio, C.; D'Ischia, M.; Gesuele, F.; Maddalena, P.; Manini, P.; Migliaccio, L.; Pezzella, A. Spontaneous wrinkle emergence in nascent eumelanin thin films. *Soft Matter* **2019**, *15*, 9261–9270. [CrossRef]
81. El Haitami, A.; Bretagnol, F.; Assuid, P.; Petitet, G.; Cantournet, S.; Corté, L. Erasable and reversible wrinkling of halogenated rubber surfaces. *Langmuir* **2013**, *29*, 15664–15672. [CrossRef]
82. Zhang, Y.; Fu, Q.; Cui, Y.; Mu, R.; Jin, L.; Bao, X. Enhanced reactivity of graphene wrinkles and their function as nanosized gas inlets for reactions under graphene. *Phys. Chem. Chem. Phys.* **2013**, *15*, 19042. [CrossRef]
83. Hou, H.; Yin, J.; Jiang, X. Reversible Diels–Alder Reaction to Control Wrinkle Patterns: From Dynamic Chemistry to Dynamic Patterns. *Adv. Mater.* **2016**, *28*, 9126–9132. [CrossRef]
84. Qu, C.; Wang, S.; Liu, L.; Bai, Y.; Li, L.; Sun, F.; Hao, M.; Li, T.; Lu, Q.; Li, L.; et al. Bioinspired Flexible Volatile Organic Compounds Sensor Based on Dynamic Surface Wrinkling with Dual-Signal Response. *Small* **2019**, *15*, 1900216. [CrossRef] [PubMed]
85. Su, X.; Li, H.; Lai, X.; Chen, Z.; Zeng, X. Stimuli-responsive superhydrophobic films driven by solvent vapor for electric switch and liquid manipulation. *Chem. Eng. J.* **2020**, *394*, 124919. [CrossRef]
86. Kim, H.S.; Crosby, A.J. Solvent-responsive surface via wrinkling instability. *Adv. Mater.* **2011**, *23*, 4188–4192. [CrossRef] [PubMed]
87. Zhao, Y.; Zhu, H.; Jiang, C.; Cao, Y.; Feng, X.Q. Wrinkling pattern evolution on curved surfaces. *J. Mech. Phys. Solids* **2020**, *135*, 103798. [CrossRef]
88. Wang, T.; Yang, Y.; Fu, C.; Liu, F.; Wang, K.; Xu, F. Wrinkling and smoothing of a soft shell. *J. Mech. Phys. Solids* **2020**, *134*, 103738. [CrossRef]

89. Cutolo, A.; Pagliarulo, V.; Merola, F.; Coppola, S.; Ferraro, P.; Fraldi, M. Wrinkling prediction, formation and evolution in thin films adhering on polymeric substrata. *Mater. Des.* **2020**, *187*, 108314. [CrossRef]
90. Kundu, S.; Davis, C.S.; Long, T.; Sharma, R.; Crosby, A.J. Adhesion of nonplanar wrinkled surfaces. *J. Polym. Sci. Part B Polym. Phys.* **2011**, *49*, 179–185. [CrossRef]
91. Lin, P.-C.; Vajpayee, S.; Jagota, A.; Hui, C.-Y.; Yang, S. Mechanically tunable dry adhesive from wrinkled elastomers. *Soft Matter* **2008**, *4*, 1830. [CrossRef]
92. Li, B.; Jia, F.; Cao, Y.-P.; Feng, X.-Q.; Gao, H. Surface wrinkling patterns on a core-shell soft sphere. *Phys. Rev. Lett.* **2011**, *106*, 2–5. [CrossRef]
93. Gao, N.; Zhang, X.; Liao, S.; Jia, H.; Wang, Y. Polymer Swelling Induced Conductive Wrinkles for an Ultrasensitive Pressure Sensor. *ACS Macro Lett.* **2016**, *5*, 823–827. [CrossRef]
94. Xu, J.; Liu, Z.; Wang, J.; Liu, P.; Ahmad, M.; Zhang, Q.; Zhang, B. Preparation of core-shell C@TiO_2 composite microspheres with wrinkled morphology and its microwave absorption. *J. Colloid Interface Sci.* **2022**, *607*, 1036–1049. [CrossRef]
95. Yang, J.-L.; Zhao, S.-X.; Lu, Y.-M.; Zeng, X.-T.; Lv, W.; Cao, G.-Z. ZnS spheres wrapped by an ultrathin wrinkled carbon film as a multifunctional interlayer for long-life Li–S batteries. *J. Mater. Chem. A* **2020**, *8*, 231–241. [CrossRef]
96. Im, J.; Yoo, D.; Kim, J.; Yoon, S.; Cho, K.Y. One-Step Spontaneous Formation of Dual Wrinkling on Uniform-Sized Microparticles Induced by Surface. *Macromol. Chem. Phys.* **2017**, *218*, 1700152. [CrossRef]
97. Ziarati, A.; Badiei, A.; Luque, R.; Ouyang, W. Designer hydrogenated wrinkled yolk@shell TiO_2 architectures towards advanced visible light photocatalysts for selective alcohol oxidation. *J. Mater. Chem. A* **2018**, *6*, 8962–8968. [CrossRef]
98. Peymanfar, R.; Ahmadi, A.; Selseleh-Zakerin, E.; Ghaffari, A.; Mojtahedi, M.M.; Sharifi, A. Electromagnetic and optical characteristics of wrinkled Ni nanostructure coated on carbon microspheres. *Chem. Eng. J.* **2021**, *405*, 126985. [CrossRef]
99. Sun, X.; Li, Y. Colloidal Carbon Spheres and Their Core/Shell Structures with Noble-Metal Nanoparticles. *Angew. Chem. Int. Ed.* **2004**, *43*, 597–601. [CrossRef]
100. Cao, Y.-P.; Jia, F.; Zhao, Y.; Feng, X.-Q.; Yu, S.-W. Buckling and post-buckling of a stiff film resting on an elastic graded substrate. *Int. J. Solids Struct.* **2012**, *49*, 1656–1664. [CrossRef]
101. Cao, Y.P.; Li, B.; Feng, X.Q. Surface wrinkling and folding of core-shell soft cylinders. *Soft Matter* **2012**, *8*, 556–562. [CrossRef]
102. Wang, D.; Sheng, B.; Wu, X.; Huang, Y.; Peng, L.; Xu, B.; Li, B.; Ni, Z. Self-assembly ring wrinkles formed on the surface of polydimethylsiloxane cylinders. *Mater. Res. Express* **2019**, *6*, 095312. [CrossRef]
103. Jia, F.; Pearce, S.P.; Goriely, A. Curvature delays growth-induced wrinkling. *Phys. Rev. E* **2018**, *98*, 033003. [CrossRef]
104. Patrício, P.; Teixeira, P.I.C.; Trindade, A.C.; Godinho, M.H. Longitudinal versus polar wrinkling of core-shell fibers with anisotropic size mismatches. *Phys. Rev. E* **2014**, *89*, 012403. [CrossRef] [PubMed]
105. Guo, J.; Niu, M.; Yang, C. Highly flexible and stretchable optical strain sensing for human motion detection. *Optica* **2017**, *4*, 1285. [CrossRef]
106. Wang, A.; Zhu, Q.; Xing, Z. A functionalized chitosan wrinkled hollow sphere containing calcium ions: Efficient adsorption of sodium dodecylbenzenesulfonate (SDBS) from aqueous solutions. *J. Colloid Interface Sci.* **2019**, *555*, 203–213. [CrossRef] [PubMed]
107. Wang, A.; Zhu, Q.; Xing, Z. Design and synthesis of a calcium modified quaternized chitosan hollow sphere for efficient adsorption of SDBS. *J. Hazard. Mater.* **2019**, *369*, 342–352. [CrossRef]
108. Wang, L.; Wang, Y.; Tian, H.; Qiao, L.; Zeng, Y. Enhanced ammonia detection using wrinkled porous $CoFe_2O_4$ double-shelled spheres prepared by a thermally driven contraction process. *Sensors Actuators B Chem.* **2020**, *314*, 128085. [CrossRef]
109. Han, X.Y.; Zhao, D.L.; Meng, W.J.; Yang, H.X.; Wu, Y.Q.; Gao, R.Z.; Yang, Y.; Pu, H. Hollow tremella-like graphene sphere/SnO_2 composite for high performance Li-ion battery anodes. *Ceram. Int.* **2019**, *45*, 16244–16250. [CrossRef]
110. Mohammadi Zardkhoshoui, A.; Hosseiny Davarani, S.S. Formation of graphene-wrapped multi-shelled $NiGa_2O_4$ hollow spheres and graphene-wrapped yolk–shell $NiFe_2O_4$ hollow spheres derived from metal–organic frameworks for high-performance hybrid supercapacitors. *Nanoscale* **2020**, *12*, 1643–1656. [CrossRef]
111. Mohammadi Zardkhoshoui, A.; Hosseiny Davarani, S.S. Boosting the energy density of supercapacitors by encapsulating a multi-shelled zinc–cobalt-selenide hollow nanosphere cathode and a yolk–double shell cobalt–iron-selenide hollow nanosphere anode in a graphene network. *Nanoscale* **2020**, *12*, 12476–12489. [CrossRef]
112. Wang, Z.; Jia, H.; Cai, Y.; Li, C.; Zheng, X.; Liang, H.; Qi, J.; Cao, J.; Feng, J. Highly conductive Mn_3O_4/MnS heterostructures building multi-shelled hollow microspheres for high-performance supercapacitors. *Chem. Eng. J.* **2020**, *392*, 123890. [CrossRef]
113. Wang, S.; Xiong, Y.; Wang, X.; Liu, W.; Tian, J.; Wu, N.; Cui, H.; Wang, X. Surface-wrinkled SnO_2 hollow microspheres decorated with AuAg bimetallic nanoparticles for triethylamine detection. *Powder Technol.* **2022**, *22*, 117457. [CrossRef]
114. Li, X.; Wang, Y.; Cheng, P.; Liu, Y.; Zhang, Y.; Ma, J.; Lv, C. High-performance ethanol sensor of wrinkled microspheres by spray pyrolysis. *Sensors Actuators B Chem.* **2021**, *344*, 130309. [CrossRef]
115. Niknam, E.; Naffakh-Moosavy, H.; Moosavifard, S.E.; Afshar, M.G. Multi-shelled bimetal V-doped Co_3O_4 hollow spheres derived from metal organic framework for high performance supercapacitors. *J. Energy Storage* **2021**, *44*, 103508. [CrossRef]
116. Patrinoiu, G.; Rodriguez, J.R.; Wang, Y.; Birjega, R.; Osiceanu, P.; Musuc, A.M.; Qi, Z.; Wang, H.; Pol, V.G.; Calderon-Moreno, J.M.; et al. Versatile by design: Hollow Co_3O_4 architectures for superior lithium storage prepared by alternative green Pechini method. *Appl. Surf. Sci.* **2020**, *510*, 145431. [CrossRef]
117. Geng, J.; Taibur Rahman, M.; Panat, R.; Li, L. Self-Assembled Axisymmetric Microscale Periodic Wrinkles on Elastomer Fibers. *J. Micro Nano-Manuf.* **2017**, *5*, 021006. [CrossRef]

118. Yang, J.; Tang, J.; Guo, H.; Liu, W.; Shen, C.; Liu, J.; Qin, L. Fabrication of micron and submicron gratings by using plasma treatment on the curved polydimethylsiloxane surfaces. *Opt. Mater.* **2017**, *72*, 241–246. [CrossRef]
119. Sampathkumar, K.; Androulidakis, C.; Koukaras, E.N.; Rahova, J.; Drogowska, K.; Kalbac, M.; Vetushka, A.; Fejfar, A.; Galiotis, C.; Frank, O. Sculpturing graphene wrinkle patterns into compliant substrates. *Carbon* **2019**, *146*, 772–778. [CrossRef]
120. Zeng, S.; Li, R.; Tait, W.R.T.; Smith, A.T.; Zhang, M.; Zhu, M.; Chov, N.; Xu, G.; Zhang, D.; Sun, L. Spontaneous formation of wrinkle-driven tubular structure as a versatile platform for adaptive 3D stretchable electronics. *Mater. Horiz.* **2020**, *7*, 2368–2377. [CrossRef]
121. Wen, X.; Sun, S.; Wu, P. Dynamic wrinkling of a hydrogel-elastomer hybrid microtube enables blood vessel-like hydraulic pressure sensing and flow regulation. *Mater. Horiz.* **2020**, *7*, 2150–2157. [CrossRef]
122. Zhang, J.; Jian, Y.; Tong, J.; Deng, H.; Du, Y.; Shi, X. Hollow chitosan hydrogel tube with controllable wrinkled pattern via film-to-tube fabrication. *Carbohydr. Polym.* **2022**, *287*, 119333. [CrossRef]
123. Jung, W.-B.; Cho, S.-Y.; Yun, G.-T.; Choi, J.; Kim, Y.; Kim, M.; Kang, H.; Jung, H.-T. Hierarchical Metal Oxide Wrinkles as Responsive Chemical Sensors. *ACS Appl. Nano Mater.* **2019**, *2*, 5520–5526. [CrossRef]
124. Chang, T.-H.; Tian, Y.; Li, C.; Gu, X.; Li, K.; Yang, H.; Sanghani, P.; Lim, C.M.; Ren, H.; Chen, P.-Y. Stretchable Graphene Pressure Sensors with Shar-Pei-like Hierarchical Wrinkles for Collision-Aware Surgical Robotics. *ACS Appl. Mater. Interfaces* **2019**, *11*, 10226–10236. [CrossRef] [PubMed]
125. Yan, J.; Ma, Y.; Li, X.; Zhang, C.; Cao, M.; Chen, W.; Luo, S.; Zhu, M.; Gao, Y. Flexible and high-sensitivity piezoresistive sensor based on MXene composite with wrinkle structure. *Ceram. Int.* **2020**, *46*, 23592–23598. [CrossRef]
126. Li, K.; Wang, J.; Shao, B.; Xiao, J.; Zhou, H.; Yu, S. Wrinkling patterns of tantalum films on modulus-gradient compliant substrates. *Thin Solid Films* **2018**, *654*, 100–106. [CrossRef]
127. Yu, S.; Sun, Y.; Li, S.; Ni, Y. Harnessing fold-to-wrinkle transition and hierarchical wrinkling on soft material surfaces by regulating substrate stiffness and sputtering flux. *Soft Matter* **2018**, *14*, 6745–6755. [CrossRef] [PubMed]
128. Yu, S.; Ma, L.; Sun, Y.; Lu, C.; Zhou, H.; Ni, Y. Controlled Wrinkling Patterns in Periodic Thickness-Gradient Films on Polydimethylsiloxane Substrates. *Langmuir* **2019**, *35*, 7146–7154. [CrossRef]
129. Lee, J.S.; Hong, H.; Park, S.J.; Lee, S.J.; Kim, D.S. A simple fabrication process for stepwise gradient wrinkle pattern with spatially-controlled wavelength based on sequential oxygen plasma treatment. *Microelectron. Eng.* **2017**, *176*, 101–105. [CrossRef]
130. Kato, M.; Asoh, T.A.; Uyama, H. Electrophoretic fabrication of an active and selective wrinkle surface on hydrogels. *Chem. Commun.* **2019**, *55*, 4170–4173. [CrossRef]
131. Liguori, G.R.; Zhou, Q.; Liguori, T.T.A.; Barros, G.G.; Kühn, P.T.; Moreira, L.F.P.; van Rijn, P.; Harmsen, M.C. Directional Topography Influences Adipose Mesenchymal Stromal Cell Plasticity: Prospects for Tissue Engineering and Fibrosis. *Stem Cells Int.* **2019**, *2019*, 5387850. [CrossRef]
132. Yang, L.; Jurczak, K.M.; Ge, L.; Rijn, P. High-Throughput Screening and Hierarchical Topography-Mediated Neural Differentiation of Mesenchymal Stem Cells. *Adv. Healthc. Mater.* **2020**, *9*, 2000117. [CrossRef]
133. Shen, Y.; Wang, Y.; Luo, Z.; Wang, B. Durable, Sensitive, and Wide-Range Wearable Pressure Sensors Based on Wavy-Structured Flexible Conductive Composite Film. *Macromol. Mater. Eng.* **2020**, *305*, 2000206. [CrossRef]
134. Chai, H.; Tian, Y.; Yu, S.; Cao, B.; Peng, X.; Zhang, Z.; Liu, A.; Wu, H. Large-Range, Reversible Directional Spreading of Droplet on a Double-Gradient Wrinkled Surface Adjusted Under Mechanical Strain. *Adv. Mater. Interfaces* **2020**, *7*, 1901980. [CrossRef]
135. Park, C.; Bae, H.J.; Yoon, J.; Song, S.W.; Jeong, Y.; Kim, K.; Kwon, S.; Park, W. Gradient-Wrinkled Microparticle with Grayscale Lithography Controlling the Cross-Linking Densities for High Security Level Anti-Counterfeiting Strategies. *ACS Omega* **2021**, *6*, 2121–2126. [CrossRef]
136. Lee, W.K.; Bin Jung, W.; Nagel, S.R.; Odom, T.W. Stretchable superhydrophobicity from monolithic, three-dimensional hierarchical wrinkles. *Nano Lett.* **2016**, *16*, 3774–3779. [CrossRef]
137. Wang, D.; Cheewaruangroj, N.; Li, Y.; McHale, G.; Jiang, Y.; Wood, D.; Biggins, J.S.; Xu, B. Bin Spatially Configuring Wrinkle Pattern and Multiscale Surface Evolution with Structural Confinement. *Adv. Funct. Mater.* **2017**, *28*, 1704228. [CrossRef]
138. Li, X.; Guo, M.; Wang, F.; Shen, X.; Weng, Y.; Hu, Z. Controllable Hierarchical Surface Patterns of Supramolecular Hydrogels: Harnessing Buckling Instability by Confinement. *Chem.-Eur. J.* **2017**, *23*, 17444–17448. [CrossRef]
139. González-Henríquez, C.M.; Rodríguez-Umanzor, F.E.; Guzmán, D.; Sarabia-Vallejos, M.A.; Rodríguez-Hernández, J. Formation of responsive hierarchical wrinkled patterns on hydrogel films via multi-step methodology. *Polymer* **2019**, *179*, 121662. [CrossRef]
140. Kim, E.S.; Kim, S.H.; Lee, S.-J.; Lee, J.H.; Byeon, M.; Suh, D.H.; Choi, W.J. Facile fabrication of micro/nano-structured wrinkles by controlling elastic properties of polydimethylsiloxane substrates. *Polymer* **2021**, *212*, 123087. [CrossRef]
141. Yu, S.; Sun, Y.; Zhang, X.; Lu, C.; Zhou, H.; Ni, Y. Hierarchical wrinkles and oscillatory cracks in metal films deposited on liquid stripes. *Phys. Rev. E* **2019**, *99*, 062802. [CrossRef]
142. Zhou, H.; Sun, Y.; Jiao, Z.; Yu, S. Spontaneous hierarchical wrinkling of metal films sputter-deposited on liquid-like gel substrates. *Thin Solid Films* **2020**, *707*, 138075. [CrossRef]
143. Li, Y.; Dai, S.; John, J.; Carter, K.R. Superhydrophobic Surfaces from Hierarchically Structured Wrinkled Polymers. *ACS Appl. Mater. Interfaces* **2013**, *5*, 11066–11073. [CrossRef]
144. Wu, H.; Yu, S.; Xu, Z.; Cao, B.; Peng, X.; Zhang, Z.; Chai, G.; Liu, A. Theoretical and Experimental Study of Reversible and Stable Wetting States of a Hierarchically Wrinkled Surface Tuned by Mechanical Strain. *Langmuir* **2019**, *35*, 6870–6877. [CrossRef] [PubMed]

145. Lin, G.; Chandrasekaran, P.; Lv, C.; Zhang, Q.; Tang, Y.; Han, L.; Yin, J. Self-similar Hierarchical Wrinkles as a Potential Multifunctional Smart Window with Simultaneously Tunable Transparency, Structural Color, and Droplet Transport. *ACS Appl. Mater. Interfaces* **2017**, *9*, 26510–26517. [CrossRef] [PubMed]
146. Yang, C.; Li, L.; Zhao, J.; Wang, J.; Xie, J.; Cao, Y.; Xue, M.; Lu, C. Highly Sensitive Wearable Pressure Sensors Based on Three-Scale Nested Wrinkling Microstructures of Polypyrrole Films. *ACS Appl. Mater. Interfaces* **2018**, *10*, 25811–25818. [CrossRef] [PubMed]
147. Qi, J.; Wang, A.C.; Yang, W.; Zhang, M.; Hou, C.; Zhang, Q.; Li, Y.; Wang, H. Hydrogel-based hierarchically wrinkled stretchable nanofibrous membrane for high performance wearable triboelectric nanogenerator. *Nano Energy* **2020**, *67*, 104206. [CrossRef]
148. Wang, H.; Yan, G.; Cao, X.; Liu, Y.; Zhong, Y.; Cui, L.; Liu, J. Hierarchical $Cu(OH)_2$@MnO_2 core-shell nanorods array in situ generated on three-dimensional copper foam for high-performance supercapacitors. *J. Colloid Interface Sci.* **2020**, *563*, 394–404. [CrossRef]
149. Kato, M.; Kashihara, Y.; Asoh, T.A.; Uyama, H. Geometry Control of Wrinkle Structures Aligned on Hydrogel Surfaces. *Langmuir* **2020**, *36*, 1467–1473. [CrossRef]
150. Li, T.; Hu, K.; Ma, X.; Zhang, W.; Yin, J.; Jiang, X. Hierarchical 3D Patterns with Dynamic Wrinkles Produced by a Photocontrolled Diels–Alder Reaction on the Surface. *Adv. Mater.* **2020**, *32*, 1906712. [CrossRef]
151. Hou, H.; Li, F.; Su, Z.; Yin, J.; Jiang, X. Light-reversible hierarchical patterns by dynamic photo-dimerization induced wrinkles. *J. Mater. Chem. C* **2017**, *5*, 8765–8773. [CrossRef]
152. Zhou, L.; Ma, T.; Li, T.; Ma, X.; Yin, J.; Jiang, X. Dynamic Interpenetrating Polymer Network (IPN) Strategy for Multiresponsive Hierarchical Pattern of Reversible Wrinkle. *ACS Appl. Mater. Interfaces* **2019**, *11*, 15977–15985. [CrossRef]
153. Yang, Y.; Duan, S.; Xiao, W.; Zhao, H. Silver nanowire-based stretchable strain sensors with hierarchical wrinkled structures. *Sensors Actuators A Phys.* **2022**, *343*, 113653. [CrossRef]

polymers

Article

Stimuli-Responsive and Antibacterial Cellulose-Chitosan Hydrogels Containing Polydiacetylene Nanosheets

Edwin Shigwenya Madivoli [1,2,*], Justine Veronique Schwarte [2], Patrick Gachoki Kareru [1], Anthony Ngure Gachanja [1] and Katharina M. Fromm [2]

[1] Department of Chemistry, Jomo Kenyatta University of Agriculture and Technology, Nairobi P.O. Box 62000-00200, Kenya
[2] Department of Chemistry, University of Fribourg, 1700 Fribourg, Switzerland
* Correspondence: edwin.madivoli@jkuat.ac.ke

Abstract: Herein, we report a stimuli-responsive hydrogel with inhibitory activity against *Escherichia coli* prepared by chemical crosslinking of carboxymethyl chitosan (CMCs) and hydroxyethyl cellulose (HEC). The hydrogels were prepared by esterification of chitosan (Cs) with monochloroacetic acid to produce CMCs which were then chemically crosslinked to HEC using citric acid as the crosslinking agent. To impart a stimuli responsiveness property to the hydrogels, polydiacetylene-zinc oxide (PDA-ZnO) nanosheets were synthesized in situ during the crosslinking reaction followed by photopolymerization of the resultant composite. To achieve this, ZnO was anchored on carboxylic groups in 10,12-pentacosadiynoic acid (PCDA) layers to restrict the movement of the alkyl portion of PCDA during crosslinking CMCs and HEC hydrogels. This was followed by irradiating the composite with UV radiation to photopolymerize the PCDA to PDA within the hydrogel matrix so as to impart thermal and pH responsiveness to the hydrogel. From the results obtained, the prepared hydrogel had a pH-dependent swelling capacity as it absorbed more water in acidic media as compared to basic media. The incorporation of PDA-ZnO resulted in a thermochromic composite responsive to pH evidenced by a visible colour transition from pale purple to pale pink. Upon swelling, PDA-ZnO-CMCs-HEC hydrogels had significant inhibitory activity against *E. coli* attributed to the slow release of the ZnO nanoparticles as compared to CMCs-HEC hydrogels. In conclusion, the developed hydrogel was found to have stimuli-responsive properties and inhibitory activity against E. coli attributed to zinc nanoparticles.

Keywords: stimuli responsive; hydrogels; slow release; antimicrobial activity

1. Introduction

Biosensors play an important role in clinical diagnostics, point-of-care testing, personalized medicine, and pharmaceutical research [1]. In this area, highly sensitive detection systems that have excellent specificity are required for their ability to provide useful insights into an individual's health [2]. Over the last decade, the scientific community has focused much of its research on biosensor systems applicable to point-of-care (POC) diagnostic devices which can rapidly assess the cause of various illnesses among patients [1–5]. These diagnostic devices facilitate faster and more accurate identification of diseases, which leads to better treatment of patients [1,2]. Among these devices, hydrogels are attracting attention due to their versatility in chemical modification which makes them responsive to external stimuli such as pH [6–9], or temperature [1,10] and the prospect of encapsulating therapeutic agents such as antimicrobials, e.g., zinc and silver, containing nanoparticles within their matrix [2]. They are developed from both synthetic polymers and biopolymers, and are utilized in tissue engineering, artificial biomedical scaffolds, soft actuators, wound dressings and environmental remediation [6,11]. Chitosan [12,13], cellulose [14–16], pectin [17], alginate [18,19], and carrageenan [20,21] are the most attractive surrogates to petroleum-based starting materials due to their relative abundance and availability. Moreover, their biodegradability, biocompatibility, ease of functionalization and gelling properties make them ideal starting materials for developing hydrogels [22,23]. These features can also be enhanced through chemical and physical crosslinking of two different hydrogels to create dual crosslinked polymers [24,25]. In addition, chitosan is a polyelectrolyte that has mild swelling capacity in acidic media but its functionalization by the introduction of a carboxymethyl group and crosslinking to another high swelling capacity polymer such as hydroxyethyl cellulose will further enhance its properties. Moreover, the incorporation of conjugated polymers and other chemical moieties within their matrices further enhances their functionality as they impart the ability to respond to external stimuli [2,26,27]. Conjugated polymers such as polyaniline, polypyrrole, and polydiacetylene (PDA) enable them to function as electrochemical biosensors since the electrochemical properties of the conjugate systems are associated with visible colorimetric changes [26,28]. PDAs are especially attractive as they exhibit a blue to red color transition visible to the naked eye when they are subjected to external stimuli such as changes in temperature, pH, bacterial cells, and aromatic compounds [7,29]. They have also been used in the development of highly sensitive colorimetric probes for the detection of cholesterol [5], ammonia [30], glucose [31], microorganisms [7,32], volatile organic compounds [33], active pharmaceutical excipients, and small and large biomolecules among other compounds [28]. In this study, a hydrogel prepared from polydiacetylene-zinc oxide-carboxymethyl chitosan-hydroxyethyl cellulose (PDA-ZnO-CMCs-HEC) was evaluated for its pH responsiveness, colorimetric transitions and inhibitory activity against *E. coli*. First, chitosan (Cs) was esterified with monochloroacetic acid to obtain carboxymethyl chitosan (CMCs) which was subsequently chemically crosslinked with hydroxyethyl cellulose (HEC) using citric acid as the crosslinking agent to obtain CMCs-HEC hydrogels. To enhance the pH responsiveness of the hydrogels, PDA-ZnO was introduced into the CMCs-HEC hydrogel to obtain a colorimetric pH-responsive PDA-ZnO-CMCs-HEC hydrogel. To better understand the properties such as pH responsiveness, thermal profile, crystallinity, functional group changes upon crosslinking and antimicrobial activity of the hydrogels against *E. coli*, the hydrogels were then analyzed using a Fourier transform infrared spectrophotometer (FT-IR), powder X-ray diffraction (XRD), scanning electron microscopy (SEM), thermal gravimeter analyzer (TGA), and UV-Vis techniques as well as by antibacterial assays.

2. Materials and Methods

2.1. Materials

Chitosan (medium molecular weight: 100,000–300,000 Da), 10,12-pentacosadiynoic acid (97.0%, HPLC grade), and isopropanol (ACS reagent, ≥99.5%), glacial acetic acid (ACS reagent, ≥99.7%), sodium hydroxide (ACS reagent, ≥97.0%, pellets), and ethyl alcohol (95%) were procured from Sigma Aldrich Co. (Buchs, Switzerland).

2.2. Synthesis of CMCs-HEC Hydrogels

Carboxymethyl chitosan (CMCs) was synthesized by esterification using monochloroacetic acid in aqueous media (Scheme 1) [15,34,35].

First, 1 g chitosan (Cs) was dispersed in 20 mL 25 μM NaOH solution followed by the addition of 50 mL isopropyl alcohol and monochloroacetic acid (11–32 mmol/g), respectively. The reaction mixture was heated to 80 °C for 2 h under constant stirring, after which the product was purified by washing repeatedly with 98% ethanol and dried in an oven to constant weight [36]. This was followed by Soxhlet extraction of sodium chloride using methanol as the extracting solvent for 6 h at 70 °C [10,14,37]. To prepare CMCs-HEC hydrogels, different formulations containing HEC and CMCs in the ratios between 1:1–1:5 were prepared in petroleum ether as follows [36]. First, to a series of solutions containing 1 g CMCs in petroleum ether, between 1 and 5 g of HEC were added and stirred for 2 h at room temperature. This was followed by the addition of citric acid (1.04 mmol/g CMCs-HEC) to the solution, which was then stirred for 16 h at 60 °C for the crosslinking to occur, filtered, washed with 98% ethanol and dried to constant weight at 105 °C to obtain CMCs-HEC hydrogels [15,34–36,38]. To obtain PDA-ZnO-CMCs-HEC hydrogels, the composite was prepared in situ by crosslinking CMCs and HEC in the presence of 2.67 mmol 10,12-pentacosadiynoic acid containing 2.5 mmol ZnO followed by UV irradiation at 254 nm to form a composite (Scheme 2) [7,30,32].

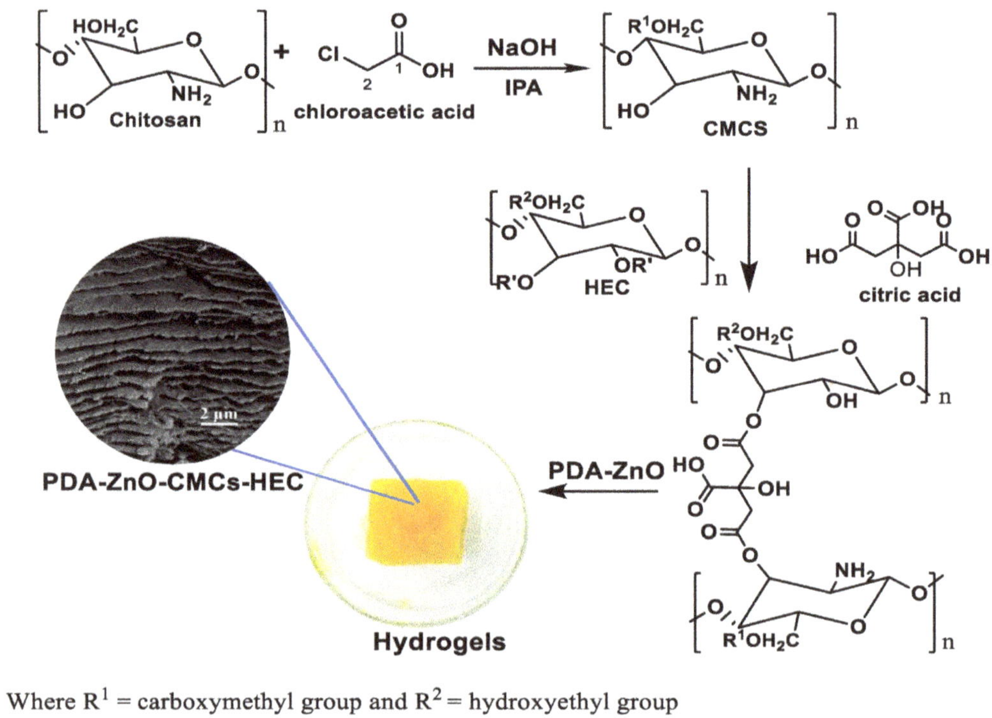

Scheme 1. Mechanism of thermochemical crosslinking of CMCs and HEC using citric acid.

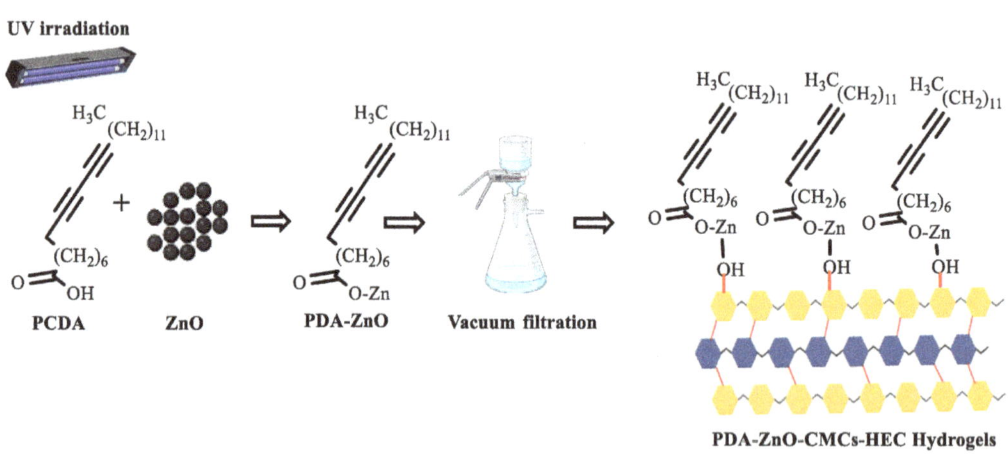

Scheme 2. Synthesis of PDA-ZnO-CMCs-HEC.

2.3. Swelling Measurements

To evaluate the swelling capacity of the hydrogels as a function of HEC:CMC ratio and pH, 100 mg of the samples were pressed using a laboratory hand press (International Crystal Laboratories, Garfield, NJ, USA) to obtain 7 mm discs. To evaluate the effect of CMC:HEC ratio on the swelling capacity of the hydrogels, the discs were prepared from samples containing HEC-CMC crosslinked in the ratios between 1:1 and 1:5 *g/g*. To study the effect of pH of the swelling media on the swelling capacity, 7 mm disc of HEC:CMC (1:4 *g/g*) were then immersed in ultrapure water whose pH was adjusted to 3, 5, 7, 11, and 14 at 20 °C [27,38,39]. The swelling capacity was then determined in triplicate by removing the discs from the swelling media at 1 min intervals, blotting to remove excess water and recording the weight gained. The degree of swelling was then calculated using the following equation and reported as mean ± SD:

$$\text{Swelling capacity} = \frac{M_t - M_i}{M_i} \times 100 \tag{1}$$

where M_i and M_t are the weights of the hydrogels before and after swelling, respectively [38].

2.4. Colorimetric Response to pH Change

The colorimetric response to changes in pH was evaluated by preparing different solutions of PDA-ZnO-CMCs-HEC at 3.3 g/L in distilled water basified with 1 M NaOH to adjust the pH to 7, 8, 10, 12, and 14, respectively. These solutions were then analyzed with a fluorimeter using excitation wavelengths 320 and 365 nm, and measuring from 360 to 620 nm and 400 to 700 nm, respectively. The solutions were then diluted to 400 µL in 2.0 mL of pH-adjusted distilled water and their absorbance was measured using a UV-Vis spectrophotometer from 800 to 200 nm [7,28,29,40].

2.5. Characterization of PDA-ZnO-CMCs-HEC

The IR spectra of the PDA-ZnO-CMCs-HEC hydrogels were acquired in the frequency range of 4000–400 cm^{-1} using a Bruker Tensor II FT-IR spectrophotometer model (Bruker, Ettlingen, Germany) [41,42]. UV measurements, fluorescence measurements, and bacterial optical density measurements were measured using Perkin Elmer Lambda 40 UV-Vis Spectrophotometer (Perkin Elmer, Waltham, MA, USA), Perkin Elmer Luminescence spectrometer LS50B (Perkin Elmer, Waltham, MA, USA) and Spark 10 M multimode microplate reader, (Tecan Austria GmbH, Grodig, Austria), respectively. The X-ray diffractograms were obtained using a Bruker D8 Advance Diffractometer (Bruker, Ettlingen, Germany) with a copper tube operating at a voltage and current of 40 kV and 40 mA. The samples were irradiated with a monochromatic CuKα radiation of 0.1542 nm and the diffractograms were acquired between 2θ values of 5°–90° at 0.05° intervals with a measurement time of 1 s per 2θ intervals. The thermal profile of the hydrogels was evaluated using a Mettler Toledo TGA/DSC (Mettler-Toledo GmbH, Greifensee, Switzerland) [41,43]. Approximately 5 mg samples were weighed into 40 µL aluminium crucibles which were then heated from 25–500 °C at 10 °C/min and cooled to 25 °C. Morphological analysis of the composite hydrogels was observed using a Tescan Mira3 LM FE scanning electron microscope (Tescan, Brno-Kohoutovice, Czech Republic) operating under an accelerating voltage of 3 kV. The samples were sputter coated with 4 nm gold before analysis to avoid charging using AGB7340 Agar Sputter Coater (Agar Scientific, Essex, UK) [42,43]. The elemental mapping of the hydrogel was obtained with a Thermo Fischer SEM FEIXL30SFEG equipped with an Oxford Aztec advanced system equipped with an X-MAX 150 mm^2 Silicon Drift Detector. The EDX mapping was obtained with an accelerating voltage of 15 kV and an acquisition time of 2 h.

2.6. Zinc Ion Release Experiments

To determine the concentration of Zn^{2+} ions diffusing from the hydrogels [44], 1 g of the composite containing 0.6 mmol ZnO was immersed in 300 mL distilled water to allow swelling of the hydrogels [45]. Aliquots of 15 mL were subsequently drawn from the solution at 30 min intervals, acidified with 1 mL concentrated nitric acid and the concentration of the solution was measured using a Perkin Elmer ICP-OES Optima7000 DV (Perkin Elmer, Waltham, MA, USA) [46].

3. Killing Curve Assay

To test the minimum inhibitory concentration (MIC) and minimum bacterial concentrations (MBC), a colony of *Escherichia coli* 25,922 was picked up from a Muller Hinton (MH) agar plate and incubated in 5 mL of MH broth overnight at 37 °C and 180 rpm. Then, a 24 g/L solution of PDA-ZnO-CMCs-HEC in H_2O was prepared and agitated at 10 rpm overnight [45,46]. A 96 well-plate was filled with different volumes of the hydrogel solutions, and total volumes were completed to 100 uL with MH broth. Bacteria preculture was diluted to 2.0×10^4 cfu/mL MH broth [45,46] after which 100 uL of this culture was inoculated in each well except sterility controls where 100 uL of MH broth was added. The plate was incubated in a humidity cassette at 37 °C and stirred at 72 rpm with absorbance monitoring for 23 h. MIC was then determined as the first concentration for which the absorbance did not increase during the incubation. Then serials of dilutions were performed for all wells, and 100 µL were spread on MH agar plates. They were incubated at 37 °C for 20 h, and the bacterial colonies were counted. MBC was then determined as the first concentration for which no colony grew at all. The same procedure was applied for the negative control [45,46].

4. Statistical Analysis

The results obtained in this study were analyzed using Originlab statistical software package (V. 22, OriginLab Corp, Northampton, MA, USA) and reported as their mean ± standard deviations while the least significance difference test (ANOVA, LSD test $p \leq 0.05$) was used to evaluate the influence of pH and CMC/HEC ratio on the swelling capacity of the hydrogels.

5. Results and Discussions

5.1. Swelling Behaviour of CMC-HEC Hydrogels

The swelling capacities of the CMCs-HEC hydrogels (Scheme 1) are depicted in Figure 1.

As can be observed from Figure 1, while CMCs had a higher swelling capacity as compared to Cs, chemical crosslinking of CMCs with HEC using citric acid resulted in a subsequent increase in the swelling capacity of the resultant hydrogels [36]. Higher swelling capacities were observed when the amount of HEC present in the hydrogels was increased from 0 to 5 g with the highest swelling capacity observed when the ratio of CMCs-HEC was 1:4 *g/g*, respectively. It is worth noting also that maximum water uptake was attained after 20 min in which 1 g of the hydrogel absorbed up to 7 g of water (≈700 %) to attain equilibrium (Figure 2). This fluid uptake is relatively higher as compared to carboxymethyl cellulose (CMC)-hydroxyethyl cellulose (HEC) hydrogels crosslinked with fumaric acid which exhibited a 400% swelling capacity [15,35]. The swelling capacity is dependent on the behaviour of individual components making up the hydrogels, the degree of crosslinking and the type of crosslinker used in the synthesis. For instance, CMC and HEC have been reported to have lower swelling capacities when individually crosslinked with citric acid as compared to when they are crosslinked together [35,36,47,48]. In HEC, the decrease in the swelling capacity was attributed to a lower potential for citric acid adsorption leading to a lower degree of crosslinking brought about by fewer OH groups [36,49,50]. While in CMC the observed lower swelling capacity of 200% was attributed to an increase in electrostatic repulsion between the charged macromolecules of the polyelectrolyte chains that limit water uptake [10,36,51]. As such, the increase in the swelling capacity in CMCs-HEC can be associated with the formation of intermolecular rather than intramolecular crosslinks when HEC is added to the citric acid solution containing CMCs [36]. Moreover, higher swelling capacities are normally observed when the ratio of the constituent polymers favours a higher degree of crosslinks between individual polymers and the crosslinking agent increases hydrogen bond interactions [36,50]. For instance, when crosslinked with divinyl sulfone, CMC and HEC have been reported to have a swelling capacity of 200% as compared to citric acid crosslinked CMC-HEC which had an equilibrium swelling capacity of 900% [52].

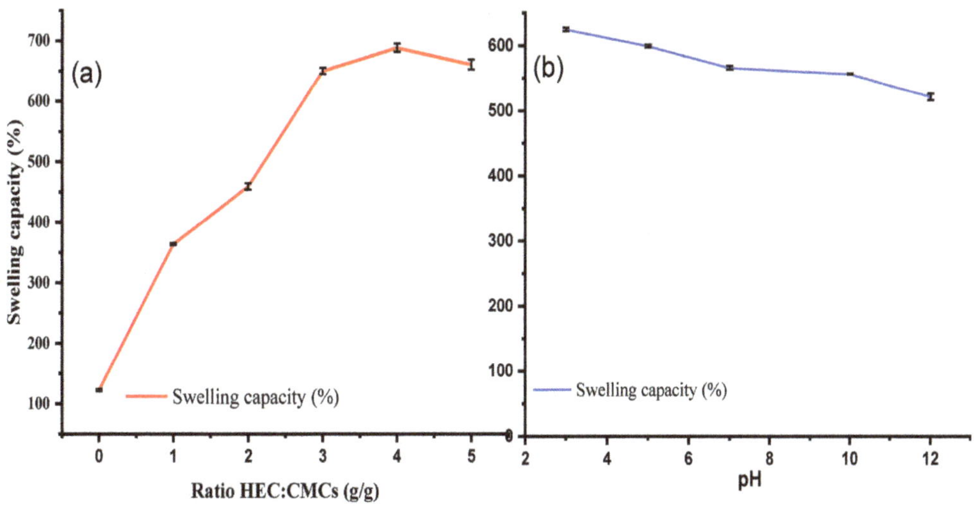

Figure 1. Swelling capacity as a function of (**a**) HEC content and (**b**) pH of the swelling medium.

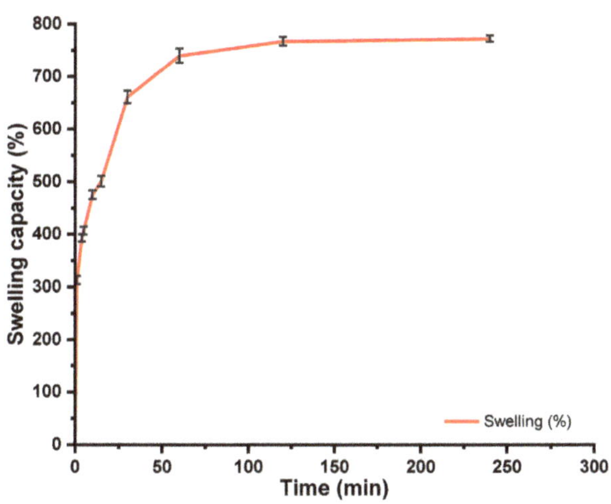

Figure 2. Swelling capacity as a function of time.

The effect of pH on the swelling capacity of the hydrogels was evaluated at pH 3, 5, 7, 10, and 12 (Figure 1b). At low pH values, the swelling capacity of the hydrogels was higher but it gradually decreased as the pH of the swelling media became more basic. The highest swelling capacity of 625% (6g of water per gram of hydrogels) was observed when the pH of the swelling media was 3 while the lowest swelling capacity was observed at pH 12. This phenomenon was attributed to the strong interactions between the two polymers and citric acid, including hydrogen bonding between –COOH and –OH groups and electrostatic interactions between $-NH_3^+$ and $-COO^-$ groups. Indeed, as the pH of the swelling media increased, hydrogen bond interactions between $-NH_2$ groups and –OH groups and repulsions in the polymer chains resulted in a decline in the swelling capacity. Moreover, it should be noted that using citric acid as the crosslinker leads to a formation of an ester bond (C–O–C) between OH groups present in the two polymers and COOH groups present in citric acid [52]. In the presence of OH^- ions, this bond is cleaved forming carboxylic acid salts thereby reducing hydrogen bond interactions between water molecules and the hydrogels. However, at low pH, the protonation of amino groups in chitosan, a water-soluble cationic polyelectrolyte, led to repulsion in the polymer chains thus dissociation of intramolecular hydrogen bond interactions allowing more water uptake into the gel network. Moreover, displacement of Na^+ present in COONa of carboxymethyl chitosan enhances hydrogen bond formation hence the observed increase in swelling capacity at low pH values. Thus, the ease of solution uptake into the hydrogel at low pH values is attributed to the protonated amine groups in chitosan while the "charge screening effect" of Na^+ that occurs at higher pH values accounts for the decrease in swelling capacity in a basic medium [38]. In addition, at higher pH values, the electronegative oxygen atom of the water molecule is electrostatically attracted to the positive charge of the metal ions having a larger hydration sphere thereby deswelling the hydrogel. The pH value of normal skin is usually below 5 but once the surface is damaged, the underlying tissue with a pH value of 7.4 is normally exposed. During the early stages of wound healing where the pH is slightly acidic, the PDA-CMC-HEC hydrogels can accelerate cell infiltration and proliferation and facilitate oxygen osmosis thereby aiding wound healing [39,53]. The potential applications of such a hydrogel include releasing antimicrobial agents such as metallic nanoparticles during the initial wound healing phase, which will reduce the extent of inflammation in the inflammation stage and avoid bacterial overgrowth in the fibroblast proliferation phase [54–56].

5.2. IR Results of CMCs, CMCs-HEC Hydrogels

FTIR spectroscopy was used to follow the introduction of the carboxymethyl group in Cs (Figure 3) and the subsequent crosslinking of HEC and CMCs with citric acid monomers (Figure 4).

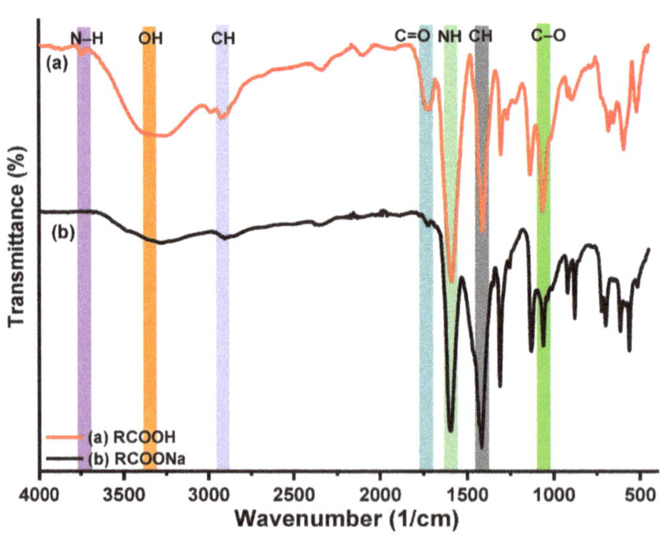

Figure 3. IR spectra of (a) RCOOH, (b) RCOONa.

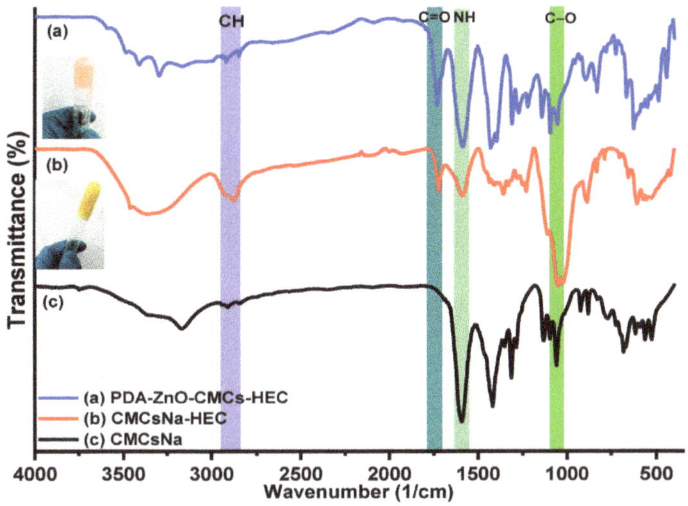

Figure 4. IR spectra of (a) PDA-ZnO-CMCs-HEC, (b) CMCs-HEC, and (c) CMCsNa. Inset CMCs-HEC hydrogels and PDA-ZnO-CMCs-HEC hydrogels.

The introduction of the carboxymethyl group on CS to form CMCs was manifested by a weak band associated with C=O bonds at 1727 cm^{-1} (Figure 3b). Initially, this band was not observed in both CS and CMCs but ion exchange of CMCs in acetic acid (Figure 3a) resulted in the conversion of RCOONa groups to RCOOH observed by the characteristic C=O vibrational band (Figure 3b) [52]. Characteristic bands of chitosan were observed at 3740 cm^{-1} attributed to NH stretching, OH stretching between 3000–3500 cm^{-1}, CH stretching at 2900–2879 cm^{-1}, NH bending at 1599 cm^{-1}, CH bending at 1401 cm^{-1} and C–O stretching at 1030 cm^{-1} [35,38,47,57]. The band at 1628 cm^{-1} assigned to the NH_2 group overlapped with the asymmetrical stretching vibration of COONa groups. Treatment of CMCs with acetic acid converted the COONa group to the COOH group evidenced by the formation of a strong characteristic band of C=O at 1738 cm^{-1} present in the free protonated carboxylic groups (Figure S2) [49,57]. Note that, the substitution of carboxymethyl groups on the oxygen in the C_6 position imparts a pH-dependent anionic character to the molecule [49,52]. From Figure 4, the chemical crosslinking of CMCs with HEC using citric acid as the crosslinking agent led to the formation of a sharp band associated with C=O vibration at 1737 cm^{-1} [57]. This band was attributed to the ester linkage formed when the hydroxyl groups of CMCs and HEC reacted with the carboxyl group of citric acid. On the other hand, the intensity of the NH peak observed at 1599 cm^{-1} decreased as the amount of HEC was increased during the crosslinking reaction (Figure S3) [35,38,52]. PDA nanosheets synthesized in the absence of the hydrogels were also characterized by IR (Figure S4). The vibrational bands originating at IR ν_{max} (cm^{-1}): 2918−2847 (C−H stretch), 1690 (C=O stretch), and 722 (C−H bend) were observed [7]. In the presence of CMCs-HEC, a doublet was observed at 2922 cm^{-1} and 2852 cm^{-1} (C−H stretch), while the bands at 1419 cm^{-1} in CS shifted to 1422 cm^{-1} with a shoulder at 1394 cm^{-1}.

5.3. XRD Results of CMCS-HEC Composite

The X-ray diffractograms of CMCs-HEC and PDA-ZnO-CMCs-HEC hydrogels are depicted in Figure 5.

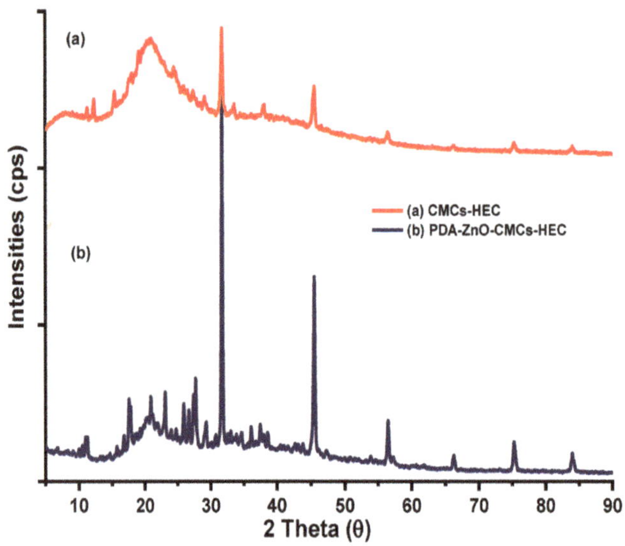

Figure 5. X-ray diffractograms of (a) CMCs-HEC and (b) PDA-ZnO-CMCs-HEC hydrogels.

From Figure 5, it can be observed that there were no remarkable changes in the diffractograms after chemical crosslinking of CMCs and HEC with citric acid. Typically, chitosan exhibits broad peaks centred at about 2θ = 18°and 32° attributed to the presence of OH and NH_2 groups which form stronger interactions that impart some degree of crystallinity. On the other hand, being a precursor of cellulose, HEC exhibits diffraction peaks that are typically observed in cellulose nanofibers at 2θ angles of 16°, 22°, and° which have been attributed to the diffraction planes of (101) and (002) characteristic of cellulose I (Figure S5) [43,58]. Chemical crosslinking CMCs and HEC with citric acid resulted in one peak centred at 2θ = 21 being observed with sharp peaks at 2θ values of 31°, 37°, 45°, 56°, and 75°. The broad peak was attributed to the overlap of broad peaks of Cs and HEC while the sharp peaks were attributed to citric acid used as a crosslinking agent in the hydrogels (Figure S5). The incorporation of PDA-ZnO within the hydrogels was linked to the sharp peaks observed at 2θ values of 32°, 36°, 48°, 57°, 68°, 75°, and 85° associated with (100), (101), (102), (110), (103), and (113) crystal planes of crystalline ZnO nanoparticles which had an average crystallite size of 18.22 ± 4.7 nm calculated from the Scherrer equation [7,59].

5.4. Thermal Profile of Cs, CMCS, and Hydrogels

The thermal profiles of Cs, CMCs, and the prepared hydrogels were investigated using TGA/DTGA and the results are depicted in Figures 6 and 7.

Both CMCs and Cs had two degradation stages in which, the first degradation stage accounted for water loss in CMCs and the hydrogels, which was slower than for Cs particles (Figures 6 and 7). The second and third mass loss in the DTGA thermograms of the hydrogels was observed in the range between 175–300 °C. The onset degradation temperature and ash content for Cs was higher as compared to that of CMCs and the hydrogels. The first degradation stage observed below 100 °C was attributed to the evaporation of water molecules adsorbed on the surface of the Cs, CMCs, and hydrogels [15,35]. For Cs and CMCs, the second degradation stage that is linked to the breaking down of the anhydroglucose units was observed to occur at 286 and 259 °C, respectively [60]. As for the hydrogels, this degradation stage was observed to occur at 273 °C since a second degradation stage associated with the degradation of the crosslinking agent was observed at 175 °C. In this study, the CMCs-HEC hydrogel possessed lower thermal stability than the corresponding precursors as CMCs and HEC have maximum degradation temperatures of 284 and 280 °C, respectively [15,35].

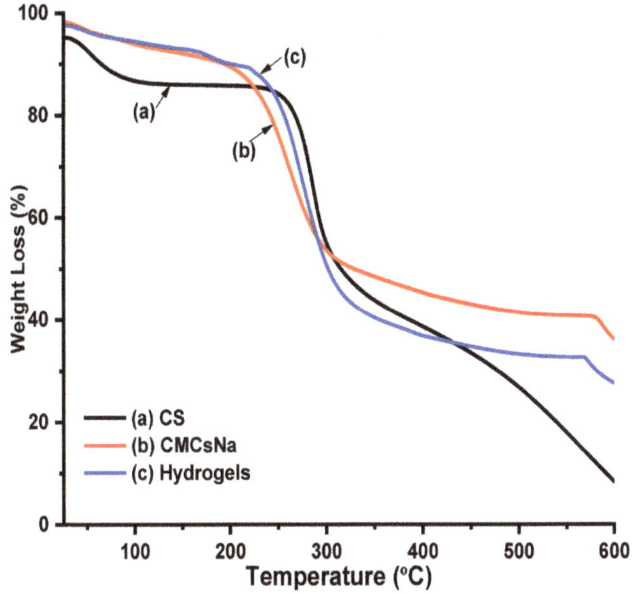

Figure 6. TGA thermograms of (a) Cs, (b) CMCsNa, and (c) hydrogels.

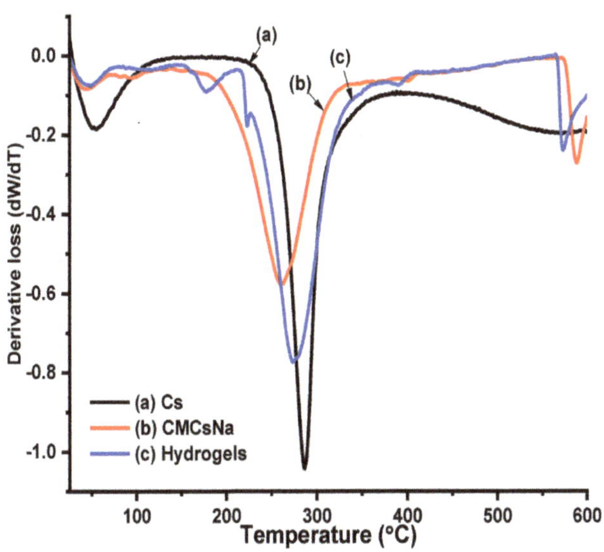

Figure 7. DTGA thermograms of (a) Cs, (b) CMCsNa, and (c) hydrogels.

5.5. UV-Vis Spectra of Polydiacetylene

Figure 8 depicts the UV-Vis spectra of the photopolymerization reaction that converted PDCA to PDA and the subsequent thermochromic shift.

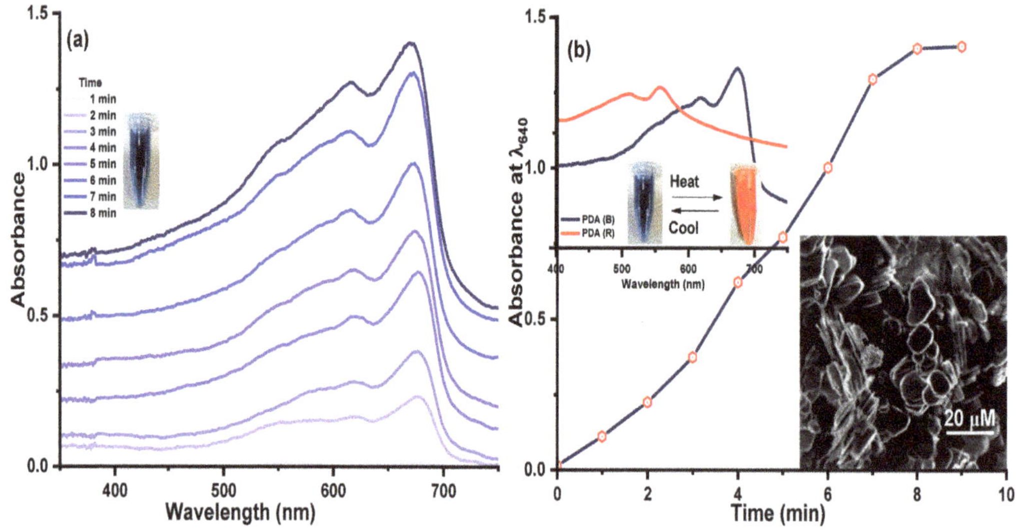

Figure 8. Absorption spectra of (**a**) PDA-ZnO, (**b**) plot of A_{670nm} vs. time. Inset thermochromic shift at 65 °C and PDA nanosheets micrograph.

As illustrated in Figure 8, the photopolymerization of 10,12-pentacosadiynoic acid (PCDA) upon irradiation with UV radiation occurred as a function of time. The plot shows a sharp increase in the absorbance at the onset of the photopolymerization which starts to level off after irradiation for 6 min, an indication of the formation of PDA nanosheets (Figure 8b inset). The colour of the solution, which changed from clear colourless to dark blue, was also another indicator of the complete polymerization, but upon exposure of this solution to higher temperatures, it changed to red (Figure 8b inset) [7,33,59,61]. Interaction of the monomers upon radiation resulted in the formation of a peak centred at 650 nm that increased with longer photopolymerization time as the nanosheets are formed (Figure 8a). However, due to the presence of an extended π-electron system composed of alternating carbon multiple bonds, this peak can shift to higher or lower wavelengths when the sheets are exposed to external stimuli such as thermal, chemical, and mechanical. At higher temperatures, however, the hypsochromic shift observed at 550 nm when blue PDA nanosheets change to red PDA nanosheets is a result of the conjugation effect which endows the chromatic properties to the material giving rise to the observed shift in λ_{max} (Figure 8b) [2].

5.6. Colorimetric Response of the Hydrogels to pH

Figures 9 and 10 depict the fluorescence and UV spectra of the hydrogel solutions as a function of pH.

From Figure 9, it was observed that PDA-ZnO-CMCs-HEC was only slightly fluorescent, and no major changes were observed when the pH was varied. Nevertheless, the hydrogel depicted negative solvatochromism when in contact with basic media which is illustrated in the UV absorbance spectra (Figure 10). The observed hypsochromic shift upon changes in pH between pH 5 (max at 364 nm) and pH 14 (max at 364 nm) was also visually observed as the hydrogel changed its colour from pale purple to pale pink (Figure 10 inset). This colorimetric response was also evident when PDA-ZnO was exposed to pH changes, though in this case, it resulted in a hypsochromic shift on exposure to basic pH [7,28,29,40]. Similar observations have been made in the literature as various composite materials containing PDA showed colorimetric transition when they were exposed to basic pH [7,28,29,40]. This transition has been reported to be important as it can be used as a means to detect the onset of infection in wounds as, during infection, the pH around the wound area has been reported to be basic [7,28,29,40].

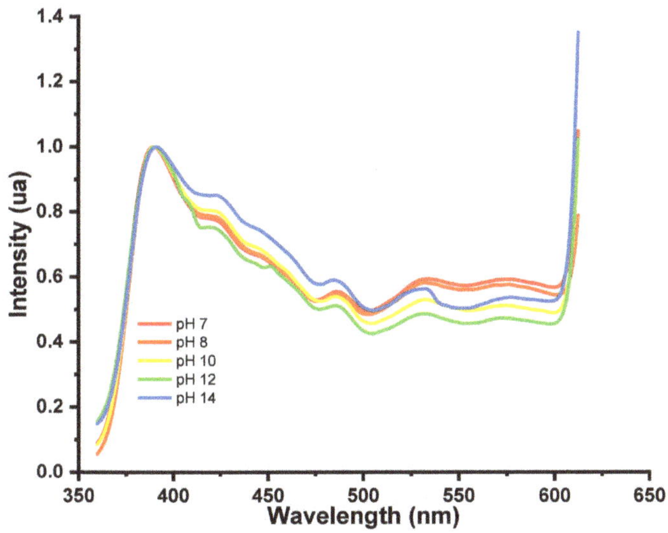

Figure 9. Fluorescence spectra of PDA-ZnO-CMCs-HEC as a function of pH.

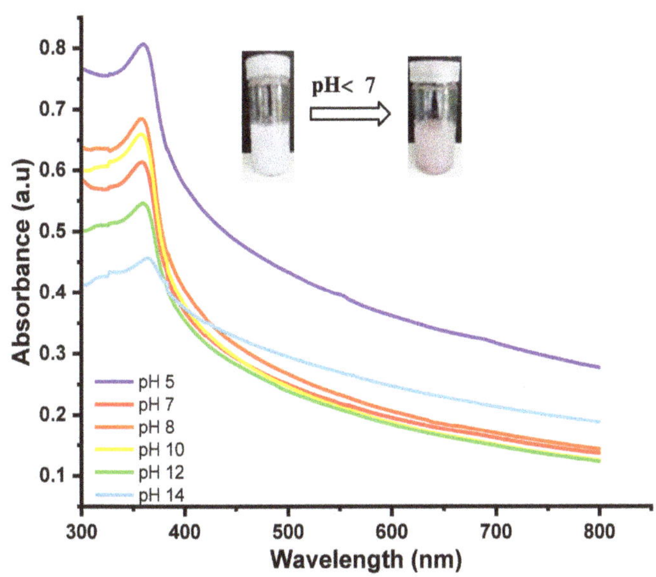

Figure 10. Absorbance spectra of PDA-ZnO-CMCs-HEC as a function of pH. Inset colour transition at higher pH.

5.7. SEM Micrographs

The nanostructures formed during the formation of PDA-ZnO-CMCs-HEC were studied using SEM and the results are depicted in Figures 11 and 12.

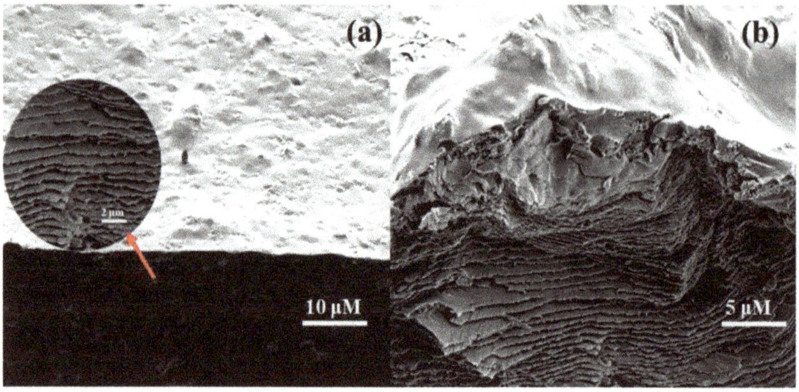

Figure 11. SEM micrographs of hydrogels (**a**) surface and (**b**) cross-section. Inset hydrogel cross-section.

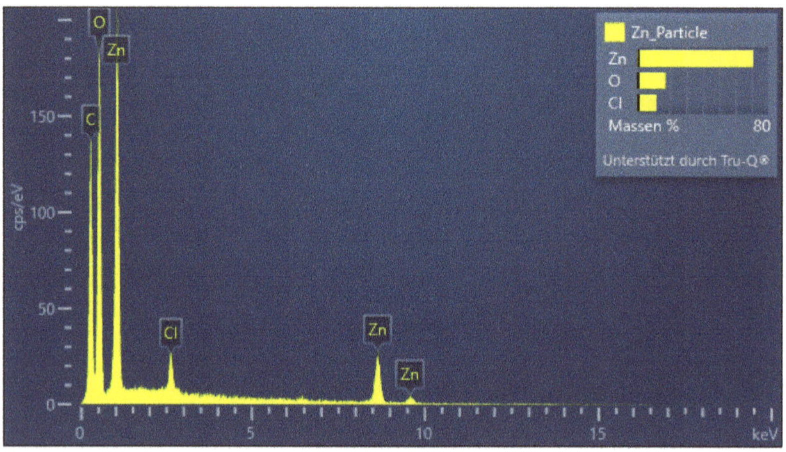

Figure 12. EDS spectra showing the elemental composition of the composite films.

In this study, the presence of amine groups and a plethora of OH functional groups in the hydrogel network are instrumental in the structuring and stabilization of the PDA assemblies within the hydrogel network. The nanocomposite comprised a mixture of nanorods, nanosheets and spherical hollow rings which was attributed to the composition making up its matrix (Figures S6–S8) [62]. Before chemical crosslinking, HEC appeared as thick fibres while CMCs appeared as sponge-like amorphous materials (Figure S7) but after chemical crosslinking and in situ synthesis of PDA-ZnO, the composition of the resultant hydrogel changed (Figure S8). ZnO nanoclusters which appeared as rods of different sizes and lengths within the hydrogel network were confirmed by EDS spectra (Figure 12) and elemental mapping of the composite (Figure S9). These nanorods were also observed when the hydrogels were used to prepare composite films through dissolution and solvent casting in PET moulds. In this case, while the surface of the resultant dried hydrogel films was smooth, observation of its cross-section revealed that it comprised multiple layers stacked together in a repeating pattern (Figure 11) similar to multi-layered hydrogels reported in the literature. These highly porous and interconnected multiple layers are a result of strong ionic and electrostatic interactions within the hydrogel network which have been shown to be responsible for the formation of interlayer spaces in a composite hydrogel [63,64]. The formation and growth of each layer are related to the diffusion of the crosslinker with the interlayer spaces fine-tuned by changing the degree of crosslinking between layers [63–65]. For the case of PDA-ZnO-CMCs-HEC, the interaction between zinc ions and the hydroxyl and amine functional groups found in CMCs-HEC induced a chelation effect which lead to the formation of multi-layered hydrogels [63–65]. In addition, polyanions such as Ca^{2+}, Al^{3+}, Cu^{2+} and other polyvalent inorganic cations have been used as crosslinkers to prepare multi-layered hydrogels with arbitrary shapes, including onion-like, tubular, and star-like from combinations of different polymers such as alginate, carboxymethyl cellulose, chitosan, and agar, among others [63–65].

5.8. Kinetics of Zn^{2+} Release from the Nanocomposites

Figure 13 depicts the concentration of zinc ions released from the hydrogels over a period of 120 h as determined by ICP-OES measurements.

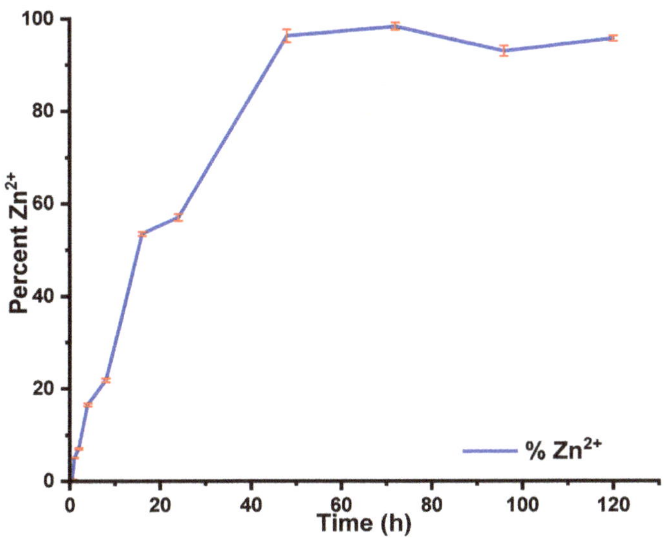

Figure 13. Percent Zn^{2+} ions diffusing from the hydrogels per unit time (h).

From Figure 13, it can be observed that the percent concentration of zinc ions released increases over time with the highest amount released being observed after 72 h before it gradually became constant. From the experimental data obtained in this study, it was observed that the amount of zinc ions released was initially low during the first few hours after the hydrogels were immersed in the absorbing media but it gradually increased with time. After 1 h, the concentration of Zn^{2+} ions in solution was found to be 6.74 ppm but it gradually increased to 121 ppm (93% Zn^{2+}) after 96 h (Figure S10) remaining constant afterwards [46,66].

Similar observations were made when AgNPs were embedded within chitosan-PEG hydrogels and titanium dioxide nano-capsules in which there was a gradual increase in the concentration of AgNPs being released over several days [46,66]. It is worth mentioning that diffusion of Zn^{2+} ions loaded into a polymer matrix usually implies water penetration into the matrix, hydration, swelling, diffusion of the dissolved substance and or erosion of the gelatinous layer. The amount of encapsulated moiety released from the matrix is dependent on the loading efficiency, the solution pH, and the nature of the encapsulated substance and polymer used [46,66].

5.9. Antimicrobial Assays

The inhibitory activity of the hydrogel was evaluated by killing curve assay and the results are depicted in Figure 14.

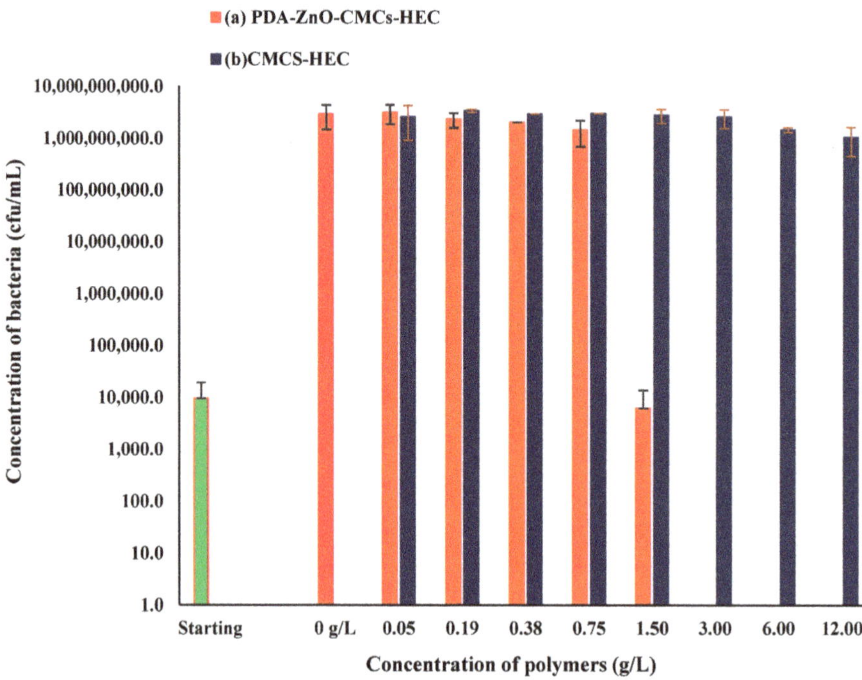

Figure 14. Minimum bacterial concentration (MBC) of alive *E. coli* 25,922 isolates after 23 h in the presence of (a) PDA-ZnO-CMCs-HEC and (b) CMCs-HEC hydrogels.

As can be observed in Figure 14, the bacterial concentration at the start was 1.0×10^{-4} cfu/mL but PDA-ZnO-CMCs-HEC was able to inhibit the growth of *E. coli* as compared to CMCs-HEC that was used as a control (Figures 15 and 16). The composite hydrogels were able to absorb liquid from the culture media which necessitated the flow of Zn^{2+} ions from the hydrogels into the surrounding medium thereby preventing the growth of bacteria [66]. As for the case of CMCs-HEC, the low inhibition activity observed was a result of the absence of ZnO nanoparticles which were present in PDA-ZnO-CMCs-HEC [66]. The PDA-ZnO-CMCs-HEC scaffolds showed excellent antibacterial activity toward clinical *E. coli* compared with CMCs-HEC scaffolds in vivo attributed to the release of Zn^{2+} ions from the composite hydrogels. From kinetic measurements (Figure 13), it was observed that after 24 h, the amount of zinc ion released from the composite hydrogels into swelling media was 74 ppm which was linked to the antibacterial activity being observed during bacterial studies (Figures 14 and 15) as the composite hydrogel without ZnO that did not inhibit bacterial growth (Figure 16). This strain of *E. coli* has an MBC (Figure 14) and an MIC (Figure 15) of 1.5 g/L for PDA-ZnO-CMCs-HEC as compared to CMCs-HEC which did not exhibit any inhibitory activity against the bacteria. It has been shown that the incorporation of metal nanoparticles such as ZnO and AgNPs in a nanocomposite enhances their inhibitory activity thereby preventing the onset of infections. This inhibitory activity is associated with the gradual and sustained release of the antimicrobial ZnO nanoparticles which interact with the bacterial cell causing cell membrane disruption, the infiltration of cell components and, ultimately, bacterial cell lysis (Figure 17) [45,46,66]. Moreover, chitosan alone and chitosan functionalized with cyclohexanone and 2-N-methylpyrrolidone have been shown to have a minimum inhibitory concentration of 50 g/mL for *E. coli* [39].

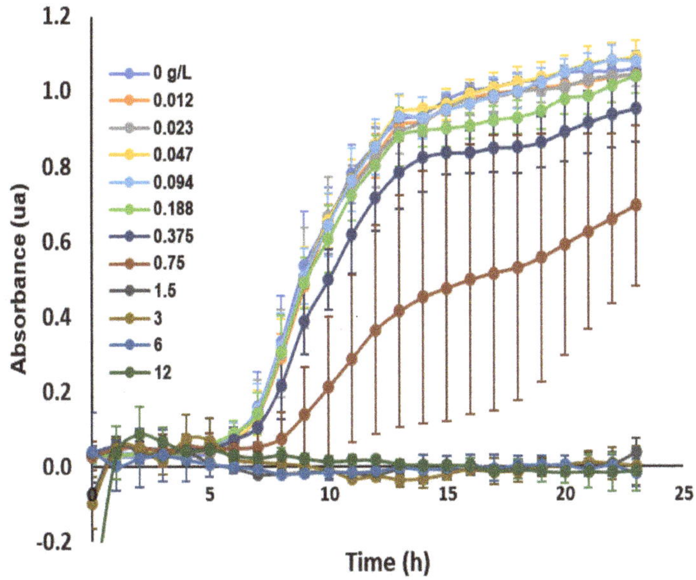

Figure 15. Growing bacteria in the presence of (MIC) of PDA-ZnO-CMCs-HEC.

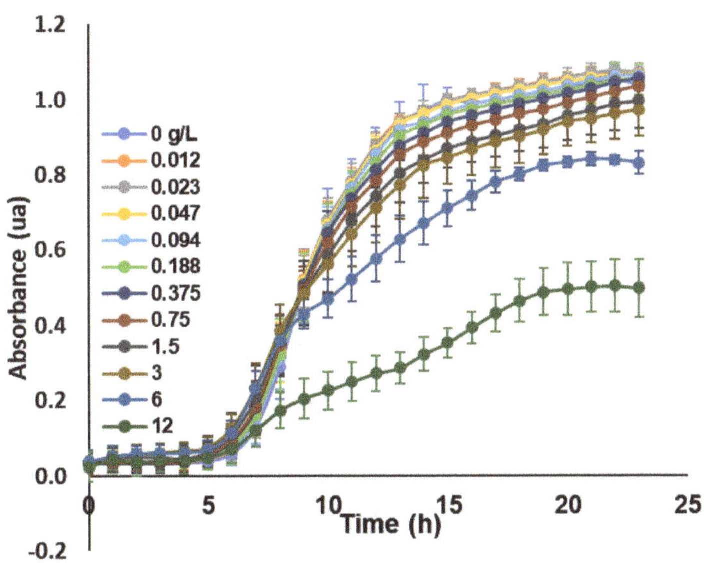

Figure 16. Growing of bacteria in the presence of CMCs-HEC (negative control).

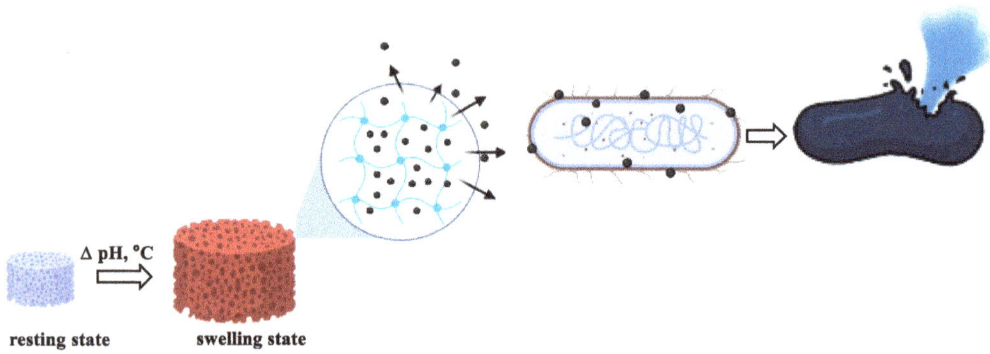

Figure 17. Mechanism of antimicrobial activity of hydrogels against *E. coli.*

6. Conclusions

We rationally combined PDA-ZnO nanosheets as a stimuli-responsive matrix into CMCs-HEC hydrogel to obtain a pH- and thermal-responsive PDA-ZnO-CMCs-HEC hydrogel with inhibitory activity against *E. coli.* The swelling capacity of the hydrogels was pH dependent as it had a higher swelling capacity in acidic pH as compared to basic pH, though the presence of PDA-ZnO within the hydrogel imparted a colorimetric response triggered by a change in pH from acidic to basic pH. This response resulted in a colour transition from purple to pale pink visible to the naked eye and as such, it is hypothesized that this property can be utilized as a sensing mechanism to indicate the onset of infection in a wound. Moreover, the incorporation of ZnO within the hydrogel ensured that the composite had inhibitory activity against *E. coli* as compared to the composite without nanoparticles.

Supplementary Materials: The following supporting information can be downloaded at: https://www.mdpi.com/article/10.3390/polym15051062/s1, Figure S1: IR spectra of (a) chitosan (CS) and (b) carboxymethyl chitosan (RCOOH); Figure S2: IR spectra of hydrogels prepared by crosslinking different ratios of HEC, Figure S3: IR spectra of (a) PDA and (b) PDA-ZnO NPs; Figure S4: X-ray diffractogram of CS, HEC and CMCs-HEC; Figure S5: X-ray diffractogram of citric acid; Figure S6: SEM Micrographs of (a) HEC and (b) CMCsNa; Figure S7: SEM Micrographs of PDA nanosheets on PDA-CMCs-ZnO-HEC; Figure S8: SEM micrographs of ZnO nanorods on PDA-ZnO-CMCs-HEC; Figure S9: Elemental mapping showing (a) carbon (b) oxygen (c) zinc (d) chlorine; Figure S10: Concentration of Zn^{2+} ions diffusing from hydrogels per unit time.

Author Contributions: Conceptualization, E.S.M., P.G.K. and A.N.G.; methodology, E.S.M., J.V.S.; validation, E.S.M. and J.V.S.; formal analysis, E.S.M. and J.V.S.; investigation, E.S.M. and J.V.S.; resources, K.M.F.; data curation, E.S.M.; writing—original draft preparation, E.S.M.; writing—review and editing, K.M.F., P.G.K. and A.N.G.; supervision, K.M.F. and P.G.K.; project administration, E.S.M.; funding acquisition, K.M.F. and E.S.M. All authors have read and agreed to the published version of the manuscript.

Funding: The authors acknowledge financial support from Swiss National Science Foundation (SNSF) awarded to Justine V. Schwarte grant number 178827.

Data Availability Statement: The data presented in this study are openly available in Zenodo at 10.5281/zenodo.7500538., reference number 7500539.

Acknowledgments: The authors take this opportunity to acknowledge the financial support of the Swiss National Science Foundation (SNSF) under grant number IZSEZ0_200290 awarded to Edwin Madivoli to undertake a scientific exchange at the University of Fribourg, Switzerland in the group of Katharina Fromm. The authors also take this opportunity to acknowledge Jozef Adamcik, Adolf Merkel Institute, University of Fribourg for his invaluable insights on SEM analysis and the Department of Chemistry, Jomo Kenyatta University of Agriculture and technology for their support.

Conflicts of Interest: The authors declare no conflict of interest.

References

1. Lee, H.; Lee, J.; Lee, S.G.; Doyle, P.S. Hydrogel-Based Colorimetric Assay for Multiplexed MicroRNA Detection in a Microfluidic Device. *Anal. Chem.* **2020**, *92*, 5750–5755. [CrossRef] [PubMed]
2. Lebègue, E.; Farre, C.; Jose, C.; Saulnier, J.; Lagarde, F.; Chevalier, Y.; Chaix, C.; Jaffrezic-Renault, N. Responsive polydiacetylene vesicles for biosensing microorganisms. *Sensors* **2018**, *18*, 599. [CrossRef] [PubMed]
3. Al Sulaiman, D.; Shapiro, S.J.; Gomez-Marquez, J.; Doyle, P.S. High-Resolution Patterning of Hydrogel Sensing Motifs within Fibrous Substrates for Sensitive and Multiplexed Detection of Biomarkers. *ACS Sens.* **2021**, *6*, 203–211. [CrossRef] [PubMed]
4. Shapiro, S.J.; Dendukuri, D.; Doyle, P.S. Design of Hydrogel Particle Morphology for Rapid Bioassays. *Anal. Chem.* **2018**, *90*, 13572–13579. [CrossRef] [PubMed]
5. Ning, Y.; Lu, F.; Liu, Y.; Yang, S.; Wang, F.; Ji, X.; He, Z. Glow-type chemiluminescent hydrogels for point-of-care testing (POCT) of cholesterol. *Analyst* **2021**, *146*, 4775–4780. [CrossRef] [PubMed]
6. Xu, C.; Cao, L.; Bilal, M.; Cao, C.; Zhao, P.; Zhang, H.; Huang, Q. Multifunctional manganese-based carboxymethyl chitosan hydrogels for pH-triggered pesticide release and enhanced fungicidal activity. *Carbohydr. Polym.* **2021**, *262*, 117933. [CrossRef]
7. Yapor, J.P.; Alharby, A.; Gentry-Weeks, C.; Reynolds, M.M.; Alam, A.K.M.M.; Li, Y.V. Polydiacetylene Nanofiber Composites as a Colorimetric Sensor Responding to Escherichia coli and pH. *ACS Omega* **2017**, *2*, 7334–7342. [CrossRef]
8. Suhail, M.; Wu, P.C.; Minhas, M.U. Development and characterization of pH-sensitive chondroitin sulfate-co-poly(acrylic acid) hydrogels for controlled release of diclofenac sodium. *J. Saudi Chem. Soc.* **2021**, *25*, 101212. [CrossRef]
9. Ciolacu, D.E.; Nicu, R.; Ciolacu, F. Cellulose-based hydrogels as sustained drug-delivery systems. *Materials* **2020**, *13*, 5270. [CrossRef]
10. Cao, J.; Wu, P.; Cheng, Q.; He, C.; Chen, Y.; Zhou, J. Ultrafast Fabrication of Self-Healing and Injectable Carboxymethyl Chitosan Hydrogel Dressing for Wound Healing. *ACS Appl. Mater. Interfaces* **2021**, *13*, 24095–24105. [CrossRef]
11. Chandran, R.; Mohd Tohit, E.R.; Stanslas, J.; Tuan Mahmood, T.M.; Salim, N. Factors influencing the swelling behaviour of polymethyl vinyl ether-co-maleic acid hydrogels crosslinked by polyethylene glycol. *J. Drug Deliv. Sci. Technol.* **2022**, *68*, 103080. [CrossRef]
12. Alvarez-Lorenzo, C.; Blanco-Fernandez, B.; Puga, A.M.; Concheiro, A. Crosslinked ionic polysaccharides for stimuli-sensitive drug delivery. *Adv. Drug Deliv. Rev.* **2013**, *65*, 1148–1171. [CrossRef]
13. Wang, T.; Chen, L.; Shen, T.; Wu, D. Preparation and properties of a novel thermo-sensitive hydrogel based on chitosan/hydroxypropyl methylcellulose/glycerol. *Int. J. Biol. Macromol.* **2016**, *93*, 775–782. [CrossRef] [PubMed]
14. Lin, P.; Liu, L.; He, G.; Zhang, T.; Yang, M.; Cai, J.; Fan, L.; Tao, S. Preparation and properties of carboxymethyl chitosan/oxidized hydroxyethyl cellulose hydrogel. *Int. J. Biol. Macromol.* **2020**, *162*, 1692–1698. [CrossRef] [PubMed]
15. Akar, E.; Altinişik, A.; Seki, Y. Preparation of pH- and ionic-strength responsive biodegradable fumaric acid crosslinked carboxymethyl cellulose. *Carbohydr. Polym.* **2012**, *90*, 1634–1641. [CrossRef]
16. Lin, G.; Chen, X.; Zhou, H.; Zhou, X.; Xu, H.; Chen, H. Elaboration of a feather keratin/carboxymethyl cellulose complex exhibiting pH sensitivity for sustained pesticide release. *J. Appl. Polym. Sci.* **2019**, *136*, 47160. [CrossRef]
17. Singh, P.; Baisthakur, P.; Yemul, O.S. Synthesis, characterization and application of crosslinked alginate as green packaging material. *Heliyon* **2020**, *6*, e03026. [CrossRef]
18. Zhao, D.; Liu, Y.; Liu, B.; Chen, Z.; Nian, G.; Qu, S.; Yang, W. 3D Printing Method for Tough Multifunctional Particle-Based Double-Network Hydrogels. *ACS Appl. Mater. Interfaces* **2021**, *13*, 13714–13723. [CrossRef]
19. Fatimi, A.; Okoro, O.V.; Podstawczyk, D.; Siminska-Stanny, J.; Shavandi, A. Natural Hydrogel-Based Bio-Inks for 3D Bioprinting in Tissue Engineering: A Review. *Gels* **2022**, *8*, 179. [CrossRef]
20. Tsirigotis-Maniecka, M. Alginate-, Carboxymethyl Cellulose-, and κCarrageenan-Based Microparticles as Storage Vehicles for Cranberry Extract. *Molecules* **2020**, *25*, 3998. [CrossRef]
21. Tavakoli, S.; Kharaziha, M.; Nemati, S.; Kalateh, A. Nanocomposite hydrogel based on carrageenan-coated starch/cellulose nanofibers as a hemorrhage control material. *Carbohydr. Polym.* **2021**, *251*, 117013. [CrossRef]
22. Barrett-Catton, E.; Ross, M.L.; Asuri, P. Multifunctional hydrogel nanocomposites for biomedical applications. *Polymers* **2021**, *13*, 856. [CrossRef]
23. Mokhtari, H.; Tavakoli, S.; Safarpour, F.; Kharaziha, M.; Bakhsheshi-Rad, H.R.; Ramakrishna, S.; Berto, F. Recent advances in chemically-modified and hybrid carrageenan-based platforms for drug delivery, wound healing, and tissue engineering. *Polymers* **2021**, *13*, 1744. [CrossRef] [PubMed]
24. Mo, F.; Wang, Z.; Jiang, R.; Gai, W.; Li, Q.; Lv, S.; Zhi, C. Energy-dissipative dual-crosslinked hydrogels for dynamically super-tough sensors. *Sci. China Mater.* **2021**, *64*, 2764–2776. [CrossRef]
25. Chu, W.; Nie, M.; Ke, X.; Luo, J.; Li, J. Recent Advances in Injectable Dual Crosslinking Hydrogels for Biomedical Applications. *Macromol. Biosci.* **2021**, *21*, 2100109. [CrossRef] [PubMed]
26. Pérez-Luna, V.H.; González-Reynoso, O. Encapsulation of biological agents in hydrogels for therapeutic applications. *Gels* **2018**, *4*, 61. [CrossRef]
27. Budianto, E.; Amalia, A. Swelling behavior and mechanical properties of Chitosan-Poly(N-vinyl-pyrrolidone) hydrogels. *J. Polym. Eng.* **2020**, *40*, 551–560. [CrossRef]

28. Huang, Q.; Wu, W.; Ai, K.; Liu, J. Highly Sensitive Polydiacetylene Ensembles for Biosensing and Bioimaging. *Front. Chem.* **2020**, *8*, 565782. [CrossRef]
29. Saenjaiban, A.; Singtisan, T.; Suppakul, P.; Jantanasakulwong, K.; Punyodom, W.; Rachtanapun, P. Novel color change film as a time–temperature indicator using polydiacetylene/silver nanoparticles embedded in carboxymethyl cellulose. *Polymers* **2020**, *12*, 2306. [CrossRef]
30. Tang, J.; Weston, M.; Kuchel, R.P.; Lisi, F.; Liang, K.; Chandrawati, R. Fabrication of polydiacetylene particles using a solvent injection method. *Mater. Adv.* **2020**, *1*, 1745–1752. [CrossRef]
31. Sawayama, J.; Okitsu, T.; Nakamata, A.; Kawahara, Y.; Takeuchi, S. Hydrogel Glucose Sensor with In Vivo Stable Fluorescence Intensity Relying on Antioxidant Enzymes for Continuous Glucose Monitoring. *iScience* **2020**, *23*, 101243. [CrossRef]
32. Kim, C.; Lee, K. Polydiacetylene (PDA) Liposome-Based Immunosensor for the Detection of Exosomes. *Biomacromolecules* **2019**, *20*, 3392–3398. [CrossRef] [PubMed]
33. Weston, M.; Tjandra, A.D.; Chandrawati, R. Tuning chromatic response, sensitivity, and specificity of polydiacetylene-based sensors. *Polym. Chem.* **2020**, *11*, 166–183. [CrossRef]
34. Kimani, P.K.; Kareru, P.G.; Madivoli, S.E.; Kairigo, P.K.; Maina, E.G. Comparative Study of Carboxymethyl Cellulose Synthesis from Selected Kenyan Biomass. *Chem. Sci. Int. J.* **2016**, *17*, 1–8. [CrossRef]
35. Seki, Y.; Altinisik, A.; Demircioğlu, B.; Tetik, C. Carboxymethylcellulose (CMC)-hydroxyethylcellulose (HEC) based hydrogels: Synthesis and characterization. *Cellulose* **2014**, *21*, 1689–1698. [CrossRef]
36. Durpekova, S.; Filatova, K.; Cisar, J.; Ronzova, A.; Kutalkova, E.; Sedlarik, V. A Novel Hydrogel Based on Renewable Materials for Agricultural Application. *Int. J. Polym. Sci.* **2020**, *2020*, 8363418. [CrossRef]
37. Muslim, T.; Rahman, M.H.; Begum, H.A.; Rahman, M.A. Chitosan and Carboxymethyl Chitosan from Fish Scales of Labeo rohita. *Dhaka Univ. J. Sci.* **2013**, *61*, 145–148. [CrossRef]
38. Wen, X.; Bao, D.; Chen, M.; Zhang, A.; Liu, C.; Sun, R. Preparation of CMC/HEC Crosslinked Hydrogels for Drug Delivery. *BioResources* **2015**, *10*, 8339–8351. [CrossRef]
39. Hassan, M.A.; Tamer, T.M.; Omer, A.M.; Baset, W.M.A.; Abbas, E.; Mohy-Eldin, M.S. Therapeutic potential of two formulated novel chitosan derivatives with prominent antimicrobial activities against virulent microorganisms and safe profiles toward fibroblast cells. *Int. J. Pharm.* **2023**, *634*, 122649. [CrossRef]
40. Li, Q.; Ren, S.; Peng, Y.; Lv, Y.; Wang, W.; Wang, Z.; Gao, Z. A Colorimetric Strip for Rapid Detection and Real-Time Monitoring of Histamine in Fish Based on Self-Assembled Polydiacetylene Vesicles. *Anal. Chem.* **2020**, *92*, 1611–1617. [CrossRef]
41. Madivoli, E.S.; Kareru, P.G.; Gachanja, A.N.; Mugo, S.M.; Makhanu, D.S. Synthesis and characterization of dialdehyde cellulose nanofibers from O. sativa husks. *SN Appl. Sci.* **2019**, *1*, 723. [CrossRef]
42. Kian, L.K.; Jawaid, M.; Ariffin, H.; Alothman, O.Y. Isolation and characterization of microcrystalline cellulose from roselle fibers. *Int. J. Biol. Macromol.* **2017**, *103*, 931–940. [CrossRef] [PubMed]
43. Otenda, B.V.; Kareru, P.G.; Madivoli, E.S.; Salim, A.M.; Gichuki, J.; Wanakai, S.I. Starch-Hibiscus-Cellulose Nanofibrils Composite Films as a Model Antimicrobial Food Packaging Material. *J. Nat. Fibers* **2022**, *19*, 12371–12384. [CrossRef]
44. Lin, C.C.; Lee, M.H.; Chi, M.H.; Chen, C.J.; Lin, H.Y. Preparation of Zinc Oxide Nanoparticles Containing Spray and Barrier Films for Potential Photoprotection on Wound Healing. *ACS Omega* **2019**, *4*, 1801–1809. [CrossRef]
45. Priebe, M.; Widmer, J.; Suhartha Löwa, N.; Abram, S.L.; Mottas, I.; Woischnig, A.K.; Brunetto, P.S.; Khanna, N.; Bourquin, C.; Fromm, K.M. Antimicrobial silver-filled silica nanorattles with low immunotoxicity in dendritic cells. *Nanomed. Nanotechnol. Biol. Med.* **2017**, *13*, 11–22. [CrossRef] [PubMed]
46. Hérault, N.; Wagner, J.; Abram, S.L.; Widmer, J.; Horvath, L.; Vanhecke, D.; Bourquin, C.; Fromm, K.M. Silver-containing titanium dioxide nanocapsules for combating multidrug-resistant bacteria. *Int. J. Nanomed.* **2020**, *15*, 1267–1281. [CrossRef] [PubMed]
47. Mali, K.K.; Dhawale, S.C.; Dias, R.J.; Dhane, N.S.; Ghorpade, V.S. Citric acid crosslinked carboxymethyl cellulose-based composite hydrogel films for drug delivery. *Indian J. Pharm. Sci.* **2018**, *80*, 657–667. [CrossRef]
48. Chen, Y.-C.; Ho, H.-O.; Liu, D.-Z.; Siow, W.-S.; Sheu, M.-T. Swelling/Floating Capability and Drug Release Characterizations of Gastroretentive Drug Delivery System Based on a Combination of Hydroxyethyl Cellulose and Sodium Carboxymethyl Cellulose. *PLoS ONE* **2015**, *10*, e0116914. [CrossRef] [PubMed]
49. Barbucci, R.; Magnani, A.; Consumi, M. Swelling Behavior of Carboxymethylcellulose Hydrogels in Relation to Cross-Linking, pH, and Charge Density. *Macromolecules* **2000**, *33*, 7475–7480. [CrossRef]
50. Bashir, S.; Hina, M.; Iqbal, J. Fundamental concepts of hydrogels: Synthesis, properties, and their applications. *Polymers* **2020**, *12*, 2702. [CrossRef]
51. Astrini, N.; Anah, L.; Haryono, A. Crosslinking Parameter on the Preparation of Cellulose Based Hydrogel with Divynilsulfone. *Procedia Chem.* **2012**, *4*, 275–281. [CrossRef]
52. Raucci, M.G.; Alvarez-Perez, M.A.; Demitri, C.; Giugliano, D.; De Benedictis, V.; Sannino, A.; Ambrosio, L. Effect of citric acid crosslinking cellulose-based hydrogels on osteogenic differentiation. *J. Biomed. Mater. Res. Part A* **2015**, *103*, 2045–2056. [CrossRef] [PubMed]
53. Sim, P.; Strudwick, X.L.; Song, Y.; Cowin, A.J.; Garg, S. Influence of Acidic pH on Wound Healing In Vivo: A Novel Perspective for Wound Treatment. *Int. J. Mol. Sci.* **2022**, *23*, 13655. [CrossRef] [PubMed]

54. Jing, X.; Feng, P.; Chen, Z.; Xie, Z.; Li, H.; Peng, X.F.; Mi, H.Y.; Liu, Y. Highly Stretchable, Self-Healable, Freezing-Tolerant, and Transparent Polyacrylic Acid/Nanochitin Composite Hydrogel for Self-Powered Multifunctional Sensors. *ACS Sustain. Chem. Eng.* **2021**, *9*, 9209–9220. [CrossRef]
55. Cao, J.; You, J.; Zhang, L.; Zhou, J. Homogeneous synthesis and characterization of chitosan ethers prepared in aqueous alkali/urea solutions. *Carbohydr. Polym.* **2018**, *185*, 138–144. [CrossRef]
56. Hao, Y.; Zhao, W.; Zhang, H.; Zheng, W.; Zhou, Q. Carboxymethyl chitosan-based hydrogels containing fibroblast growth factors for triggering diabetic wound healing. *Carbohydr. Polym.* **2022**, *287*, 119336. [CrossRef]
57. Socrates, G. *Infrared and Raman Characteristic Group Frequencies*, 3rd ed.; John Wiley & Sons: Hoboken, NJ, USA, 2001; ISBN 978-0-470-09307-8.
58. Madivoli, E.S.; Kareru, P.G.; Gichuki, J.; Elbagoury, M.M. Cellulose nanofibrils and silver nanoparticles enhances the mechanical and antimicrobial properties of polyvinyl alcohol nanocomposite film. *Sci. Rep.* **2022**, *12*, 19005. [CrossRef]
59. Kim, J.; Moon, B.S.; Hwang, E.; Shaban, S.; Lee, W.; Pyun, D.G.; Lee, D.H.; Kim, D.H. Solid-state colorimetric polydiacetylene liposome biosensor sensitized by gold nanoparticles. *Analyst* **2021**, *146*, 1682–1688. [CrossRef]
60. Abreu, F.D.; Campana-Filho, S.P. Preparation and characterization of carboxymethylchitosan. *Polimeros* **2005**, *15*, 79–83. [CrossRef]
61. Tian, X.; Wu, S.; Zhang, Q.; Zou, G. Colorimetric sensor for fine differentiation of organic solvents based on only one kind of polydiacetylene coated on polymer optical fiber. *IEEE Sens. J.* **2012**, *12*, 1946–1949. [CrossRef]
62. Jang, S.; Son, S.U.; Kang, B.; Lim, J.; Seo, S.B.; Kang, T.; Jung, J.; Seo, S.; Lim, E.-K. Polydiacetylene-Based Hydrogel Beads as Colorimetric Sensors for the Detection of Biogenic Amines in Spoiled Food. *SSRN Electron. J.* **2022**. [CrossRef]
63. Liu, G.; Ding, Z.; Yuan, Q.; Xie, H.; Gu, Z. Multi-Layered Hydrogels for Biomedical Applications. *Front. Chem.* **2018**, *6*, 1–10. [CrossRef] [PubMed]
64. Jin, L.; Xu, J.; Xue, Y.; Zhang, X.; Feng, M.; Wang, C.; Yao, W.; Wang, J.; He, M. Research progress in the multilayer hydrogels. *Gels* **2021**, *7*, 172. [CrossRef] [PubMed]
65. Nie, J.; Wang, Z.; Hu, Q. Chitosan Hydrogel Structure Modulated by Metal Ions. *Sci. Rep.* **2016**, *6*, 36005. [CrossRef]
66. Masood, N.; Ahmed, R.; Tariq, M.; Ahmed, Z.; Masoud, M.S.; Ali, I.; Asghar, R.; Andleeb, A.; Hasan, A. Silver nanoparticle impregnated chitosan-PEG hydrogel enhances wound healing in diabetes induced rabbits. *Int. J. Pharm.* **2019**, *559*, 23–36. [CrossRef]

polymers MDPI

Article

Role of Bovine Serum Albumin Addition in Micellization and Gel Formation of Poloxamer 407

Namon Hirun [1,*], Pakorn Kraisit [1] and Siriwat Soontaranon [2]

[1] Thammasat University Research Unit in Smart Materials and Innovative Technology for Pharmaceutical Applications (SMIT-Pharm), Faculty of Pharmacy, Thammasat University, Pathumthani 12120, Thailand; pakorn54@tu.ac.th

[2] Synchrotron Light Research Institute (Public Organization), Nakhon Ratchasima 30000, Thailand; siriwat@slri.or.th

* Correspondence: namon.hirun@yahoo.co.th or namon.hi@tu.ac.th

Abstract: The combination of the thermoresponsive polymer and protein has demonstrated great promise in its applications in drug delivery and tissue engineering fields. This study described the impact of bovine serum albumin (BSA) on the micellization and sol–gel transition behaviors of poloxamer 407 (PX). The micellization of aqueous PX solutions with and without BSA was examined using isothermal titration calorimetry. In the calorimetric titration curves, the pre-micellar region, the transition concentration region, and the post-micellar region were observed. The presence of BSA had no noticeable impact on critical micellization concentration, but the inclusion of BSA caused the pre-micellar region to expand. In addition to studying the self-organization of PX at a particular temperature, the temperature-induced micellization and gelation of PX were also explored using differential scanning calorimetry and rheology. The incorporation of BSA had no discernible effect on critical micellization temperature (CMT), but it did affect gelation temperature (T_{gel}) and gel integrity of PX-based systems. The response surface approach illustrated the linear relation between the compositions and the CMT. The major factor affecting the CMT of the mixtures was the concentration of PX. The alteration of the T_{gel} and the gel integrity were discovered to be a consequence of the intricate interaction between PX and BSA. BSA mitigated the inter-micellar entanglements. Hence, the addition of BSA demonstrated a modulating influence on T_{gel} and a softening effect on gel integrity. Understanding the influence of serum albumin on the self-assembly and gelation of PX will enable the creation of thermoresponsive drug delivery and tissue engineering systems with controlled gelation temperatures and gel strength.

Keywords: poloxamer 407; bovine serum albumin; thermoresponsive gelation; micellization

1. Introduction

Over the past few years, the self-assembly of amphiphilic copolymers has increased attention. It has been considered for various applications, such as templates for nanomaterial synthesis, vehicles for drug encapsulation, and matrices for drug delivery and tissue engineering [1–3]. Among the several amphiphilic copolymers, poloxamers are becoming more prevalent in the biochemical and pharmaceutical fields due to their minimal immunological response and low toxicity [3–5]. The triblock structure of poloxamers consists of a central hydrophobic poly(propylene oxide) (PPO) block and terminal hydrophilic poly(ethylene oxide) (PEO) blocks. The amphiphilicity of poloxamers depends on the molecular weight and the number and proportion of PEO and PPO blocks [3]. The self-assembly of the unimers in an aqueous environment varies according to polymer concentration and temperature [6]. Moreover, the addition of cosolvent and some additives refines the self-assembly of poloxamers in water [4,5,7]. The worsening interaction between the hydrophobic block of the polymer chain and water with a rise in temperature triggers heat-induced self-assembly in aqueous poloxamer solutions, resulting in micelle formation [5]. The formation of the polymeric micelles and further micelle organization relate to the alteration of physicochemical behaviors such as rheological characteristics, encapsulation performance, and thermogelation ability [5,7]. Since the underlying self-assembly behavior governs the structural features and applications, several efforts have been made to tune, characterize, and comprehend the additive impact on the phase transitions of poloxamers [5].

Poloxamer 407 (PX), also known as pluronic F127, is one particular type of amphiphilic copolymer belonging to a family of poloxamers. The self-organization of PX unimers in aqueous solutions has become an emerging topic with fundamental as well as technological aspects that arise from their capacity to form micellar liquids and physical soft gels [5,8]. At a certain temperature, most of the PX unimers form micelles when the polymer concentration reaches a critical micellization concentration (CMC). The CMC value of PX in water at 30 °C has been reported to be in the range of 0.04 to 0.1 % *w*/*v* [9–11]. Apart from the concentration-dependent self-organization behavior, the aqueous solutions of PX become micellar liquids with increasing temperature. By raising the PX concentration above the CMC or by elevating the temperature to a critical micellization temperature (CMT), the process of self-organization can be triggered. Furthermore, PX solutions show thermoresponsive gelation, changing from the low-viscosity micellar liquid to the viscoelastic gel upon heating as a result of the development of an orderly packed inter-micellar structure. At concentrations greater than 15% *w*/*v*, aqueous PX solutions display the thermoresponsive gelation feature [12]. The temperature at which the change from liquid to gel happens is known as a gelation temperature (T_{gel}) [13]. Although the thermoresponsive gelation makes aqueous PX solution an appealing thermoresponsive system for drug delivery and tissue engineering, the low T_{gel} of plain PX solution is a drawback and limits its potential performance [5,12,14]. To avoid the usage of new synthetic polymer derivatives with potential toxicities, the search for suitable small-molecular or macromolecular additives is a promising strategy for tailoring the T_{gel} of the aqueous PX solution [5,15]. In addition, understanding the impact of additives on the temperature-dependent characteristics of PX-based solutions is of great importance for the development of the thermoresponsive system [15]. The various types of poloxamers have their own specific molecular weights and proportions of PEO and PPO blocks [16]. The hydrophilic–hydrophobic balance of each poloxamer is associated with the PEO/PPO ratio in the chain structure [17–19]. Poloxamer self-assembly in water is more likely to occur when the polymer possesses more hydrophobic PPO [17]. The PEO/PPO ratio in PX is lower than that in poloxamer 188 [20]. At temperatures below 100°C, PX at concentrations greater than 15% can form gel, while poloxamer 188 at concentrations below 40% cannot form gel [21]. In addition, it has been reported that poloxamer mixtures rich in hydrophobic PPO have low T_{gel}, whereas solutions rich in hydrophilic PEO have a high T_{gel} [20,22]. The micellization and gelation of PX-based matrices can be altered by combining PX with another type of poloxamer containing different lengths of the hydrophobic and/or hydrophilic blocks. The thermal and rheological properties of systems containing PX and poloxamer 188 have previously been assessed in order to alter their phase transition tendencies [13,22]. Moreover, it has been noted that combining PX with polymeric additives such as poly(acrylic acid) and poly(acrylic acid) derivatives might alter the T_{gel} of a gelling vehicle based on PX [23–25]. Early studies were particularly concerned with the impact of polymeric additives on the self-assembly and gelation of PX. In addition to synthetic polymers, proteins can interact with amphiphilic copolymers, altering their self-assembly properties [26].

Bovine serum albumin (BSA) is a hydrophilic globular protein utilized as a nutrient in cell culture. It has a wide range of biochemical and pharmaceutical applications because it is biocompatible, affordable, and readily available [27–29]. It has been reported that the presence of albumin that simulated the abundant protein in inflamed synovia influenced the phase transition of PX [30]. Furthermore, Neacsu et al. discovered that human serum albumin (HSA), which has a high sequence homology as well as a high 3D structure similarity to BSA [31], affected the liquid-to-gel transition of PX [32]. Recently, the opportunities of developing thermoresponsive formulations for the delivery of BSA-based therapeutics have been explored and suggested for use as an injectable platform [33]. Understanding the influence of additives on the self-assembly of PX is necessary for the rationale design of prospective thermoresponsive gels. More information is required to comprehend how BSA affects the micellization and gelation of PX.

In this study, the impact of BSA on the micellar properties and gelation of PX was examined. Using isothermal titration calorimetry (ITC), the aqueous associated colloids from PX and the mixtures were characterized, and the CMC was determined. Further, response surface methodology was applied to rationally design the experiments and statistically examine the relationship between the composition and the thermoresponsive micellization and gelation characteristics. To comprehend the phenomena of temperature-induced micellization and thermogelation of PX in the presence of BSA, differential scanning calorimetry (DSC) and rheological investigations were conducted.

2. Materials and Methods

2.1. Materials and Sample Preparation

PX was purchased from Sigma-Aldrich (Saint Louis, MO, USA). BSA was supplied by Fisher BioReagents (Thermo Fisher Scientific, Waltham, MA, USA). Ultrapure water was obtained from a Simplicity® water purification system (Millipore, Molsheim, France) and used as the solvent for all samples.

The plain PX solutions were prepared using the cold method [34], which involved dispersing the desired quantity of PX in two-thirds of the required cold water while stirring to obtain a homogenous dispersion. The sample's volume was then adjusted to the desired total volume by adding the remaining cold water. To prepare the mixtures of BSA and PX, the required quantity of BSA was dissolved in the PX solution before adjusting the volume of each mixture to the final volume. In order to prepare BSA solutions, the necessary amount of BSA was dissolved in water.

2.2. ITC Experiments

In order to ensure that the polymer concentration in the sample cell varies to span the micellization process as the titration progresses, the concentration of the amphiphilic polymer placed into the injection syringe as titrant should be higher than the CMC [35]. In addition, the heat change in each titration should not exceed the measuring capacity or be too small compared to background noise [35]. According to our preliminary study, 1.5% w/v PX was suitable for the ITC experiment. For studying the micellization of plain PX, the sample cell was initially loaded with water while 1.5% w/v PX was loaded into the syringe. Then the PX solution was successively titrated into water.

To examine the impact of BSA on PX micellization, the mixture of PX and BSA was continuously titrated into water that contained the same BSA composition as in the titrant. In this manner, the concentration of BSA remains constant during the titration procedure, and any heat change resulting from BSA dilution can be ruled out [6]. The mixtures of PX and BSA used as titrants were 0.75% and 1.5% w/v of BSA in 1.5% w/v PX.

The ITC experiments were conducted at 30 °C using a MicroCal PEAQ-ITC microcalorimeter (Malvern Panalytical, Malvern, UK). An initial small injection of 0.4 µL was followed by 38 injections of 1 µL of the titrant. Spacing was 120 s, and stirring speed was set to 750 rpm. All experiments were performed in triplicate. Data were processed and analyzed using MicroCal PEAQ-ITC Analysis Software and MATLAB 2018a (Mathworks, Natick, MA, USA). The CMC value was estimated from the maximum of the first derivative of the calorimetric titration curve, as described previously [35,36].

2.3. Response Surface Methodology

Response surface methodology with a 3-level factorial design was adopted to study how the factors affect responses including the CMT, the T_{gel}, and the elastic modulus at 37 °C ($G'_{37°C}$), which reflected gel strength [15]. The variables used for 3-level factorial design are represented in Table 1. The factors were the concentrations of PX and BSA in the thermoresponsive formulations. The actual values for low, medium, and high levels of PX were 16, 18, and 20% *w*/*v*, respectively. The concentration of 16% *w*/*v* was chosen as the low level since it is fractionally greater than the minimum concentration that forms gel as reported in the literature [12]. According to Table 1, the chosen highest concentration of BSA was 1.5% *w*/*v* which covered the BSA concentration generally used for biomedical and pharmaceutical applications [14,37,38]. The sample notation and description are listed in Table 2. With Design-Expert® software (version 13; Stat-Ease Inc., Minneapolis, MN, USA), thirteen runs for three levels, two factors, and five center points were produced as shown in Table 3. The relations between the factors and the experimental results of the responses were established. Analysis of variance (ANOVA) was used to figure out whether the regression model was statistically significant. Statistical significance was regarded as a *p*-value of less than 0.05.

Table 1. Variables used for 3-level factorial design.

Factors	Actual Level (Coded Level)		
	Low	Medium	High
X_1: PX concentration (% *w*/*v*)	16 (−1)	18 (0)	20 (+1)
X_2: BSA concentration (% *w*/*v*)	0 (−1)	0.75 (0)	1.5 (+1)
Responses			
Y_1: CMT (°C)			
Y_2: T_{gel} (°C)			
Y_3: $G'_{37°C}$ (Pa)			

Table 2. List of sample notation and description.

Sample Notation	Description of Sample
16PX	16% *w*/*v* PX
16PX0.75BSA	The mixture containing 16% *w*/*v* PX and 0.75% *w*/*v* BSA
16PX1.5BSA	The mixture containing 16% *w*/*v* PX and 1.5% *w*/*v* BSA
18PX	18% *w*/*v* PX
18PX0.75BSA	The mixture containing 18% *w*/*v* PX and 0.75% *w*/*v* BSA
18PX1.5BSA	The mixture containing 18% *w*/*v* PX and 1.5% *w*/*v* BSA
20PX	20% *w*/*v* PX
20PX0.75BSA	The mixture containing 20% *w*/*v* PX and 0.75% *w*/*v* BSA
20PX1.5BSA	The mixture containing 20% *w*/*v* PX and 1.5% *w*/*v* BSA

2.4. DSC Experiments

The DSC investigations were conducted using a Mettler Toledo DSC 3+ model (Mettler-Toledo, Viroflay, France). Samples were placed into aluminum pans with a pinhole in the lid. Temperature scans from 5 to 45 °C under nitrogen purge were conducted at a heating rate of 1 °C/min. STARe Evaluation Software was used for data analysis. All measurements were in triplicate.

2.5. Rheological Experiments

A HAAKE MARS 40 rheometer (ThermoFisher Scientific, Bremen, Germany) was used to evaluate the rheological characteristics. By utilizing a 60 mm parallel plate geometry with a 0.5 mm gap, the dynamic moduli—the elastic modulus (G′) and viscous modulus (G′′)—were evaluated through temperature ramps at a heating rate of 1 °C/min from 5 to 45 °C. At a frequency of 1 Hz, the oscillation temperature ramp tests were conducted within the linear viscoelastic region. All measurements were in triplicate.

Table 3. The 3-level factorial design and response data.

Run	Sample Notation	Factors		Responses *		
		X_1	X_2	Y_1	Y_2	Y_3
1	16PX0.75BSA	16	0.75	18.10	30.50	2375.67
2	18PX0.75BSA	18	0.75	16.85	26.66	13,493.30
3	20PX	20	0	15.93	23.43	19,253.30
4	18PX0.75BSA	18	0.75	16.86	26.96	12,910.00
5	18PX0.75BSA	18	0.75	16.89	26.87	11,713.30
6	18PX1.5BSA	18	1.5	16.96	28.01	11,140.00
7	20PX0.75BSA	20	0.75	15.81	23.96	17,900.00
8	16PX1.5BSA	16	1.5	18.03	31.79	758.30
9	18PX0.75BSA	18	0.75	16.79	26.76	13,286.70
10	16PX	16	0	18.29	29.00	9331.33
11	18PX0.75BSA	18	0.75	17.02	26.83	12,693.30
12	18PX	18	0	17.10	25.43	15,120.00
13	20PX1.5BSA	20	1.5	15.48	24.44	18,743.30

* Mean value of triplicate experiments.

2.6. Statistical Analysis

ANOVA together with *post hoc* test was used to assess whether there were significant differences between the groups. A statistical difference was considered significant for p-value < 0.05.

3. Results and Discussion

3.1. ITC Measurements on the Self-Organization of PX into Micelles

ITC analysis is a precise method for monitoring the self-organization of the amphiphilic molecules and for determining CMC [35,39]. This method enables the observation of micellization in real time without the addition of dye molecules or measurements of surface tension [39]. Small amounts of a concentrated polymeric surfactant solution are titrated into the sample cell containing dispersion medium. This causes demicellization with every injection until the polymer concentration reaches CMC, which can be evaluated from the calorimetric titration data.

3.1.1. PX Self-Organization in Water

Figure 1a represents the raw ITC thermogram of aqueous micellar PX solution titrated into water. Each injection caused a negative heat flow that deviated from the baseline, suggesting that the injection of micellar PX solution into water produced the exothermic heat flow effect. The concentration of the PX solution in the syringe was much higher than the expected CMC of PX; therefore, it may be inferred that the polymer solution in the syringe consisted of micellar colloids due to the self-organization. After the first injection of micellar solution from the syringe into the dispersion medium, the concentration of PX in the sample cell was far below CMC owing to the dilution of the injected PX solution. The dilution of the injected PX solution inside the sample cell destabilized the micellar colloids and triggered the micelle dissociation into unimers. This particular phenomenon is described as demicellization [35]. Therefore, the first few injections resulted in a large heat release (Figure 1a), which was attributed to the micelle dilution, the dissociation of the injected micelles into unimers, and the unimer dilution [35]. With subsequent injections, the magnitude of the observed heat release decreased. Eventually, the heat flow became almost constant for the last few injections.

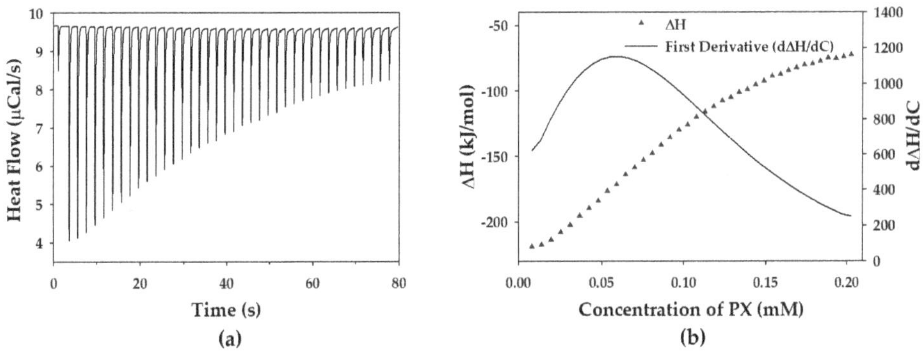

Figure 1. (**a**) Raw ITC thermogram of 1.5% *w*/*v* PX titrated into water at 30 °C. (**b**) Normalized titration curve of 1.5% *w*/*v* PX titrated into water at 30 °C. The first derivative of the titration curve is represented by a solid line.

The normalized titration curve was obtained by integrating the heat effect over time and subsequent normalization per amount of polymer. According to Figure 1b, the normalized titration curve of aqueous micellar PX solution titrated into water has a sigmoidal shape, which can be divided into three regions: a pre-micellar region, a transition concentration region, and a post-micellar region [35,39,40]. In the pre-micellar region, the concentration of PX in the sample cell was much below the CMC, and the observed enthalpy change was associated with the breakup of the added micelles as well as the dilution of unimers. As more of the concentrated PX was injected, the concentration of PX in the sample cell increased to the transition concentration region. The majority of injected micelles remained in the micellar form, with only a small portion dissociating into unimers in this region. The micelles gradually depress dissociating in this concentration range, as evidenced by the progressive decline in the enthalpy change magnitude. The PX concentration in the sample cell reaches the CMC in the transition concentration region. The differential curve (Figure 1b) was obtained by calculating the first derivative of the titration profile, and the CMC value was taken to be the concentration at which the differential curve reached its maximum. The CMC value of PX determined in this study was 0.057 ± 0.003 mM (0.072 ± 0.004 % *w*/*v*) which is in line with the previously reported range of CMC values [10,11,41]. In the post-micellar area, a plateau was reached on the titration curve. The predominant process occurring in the post-micellar region is the dilution of the PX micelles when the micelles no longer dissociate into unimers. The small heat release in this region reflected the dilution enthalpy of micelles.

3.1.2. Effect of BSA on PX Self-Organization

The normalized titration curves of PX in the absence and presence of BSA are shown in Figure 2. The titration curves of PX in the presence of BSA exhibited a negative enthalpy change, indicating that the demicellization of PX in the presence of BSA remained the exothermic attribute. While having the sigmoidal shape, the concentration range of the pre-micellar region of PX in the presence of BSA was greater than that of plain PX. Using the concentration at the intersection of the extrapolated initial and linear ascent lines (inset of Figure 2) [42], the end of the pre-micellar region was estimated. The concentrations at the end of the pre-micellar region for plain PX, PX containing 0.75% *w*/*v* BSA, and PX containing 1.5% *w*/*v* BSA were 0.017 ± 0.001, 0.028 ± 0.000, and 0.026 ± 0.002 mM, respectively. Both samples containing BSA had significantly higher concentrations at the end of the pre-micellar region than plain PX ($p < 0.05$). The CMC of PX containing 0.75% *w*/*v* BSA, and PX containing 1.5% *w*/*v* BSA were 0.054 ± 0.003 and 0.052 ± 0.003 mM, respectively. For the CMC values, there was no statistical difference among plain PX, PX containing 0.75% *w*/*v* BSA, and PX containing 1.5% *w*/*v* BSA. The interaction between BSA and PX unimers may play a role in the shift of the transition from the pre-micellar region to the transition concentration region. It has been suggested that the hydrophobic interaction between PPO residues of PX unimers and BSA induced the PPO globule of PEO-PPO-PEO block copolymers to expand in water [43]. The corresponding interaction might retard the association between PPO blocks at low concentrations of PX. Therefore, the presence of BSA extended the concentration range of the pre-micellar region.

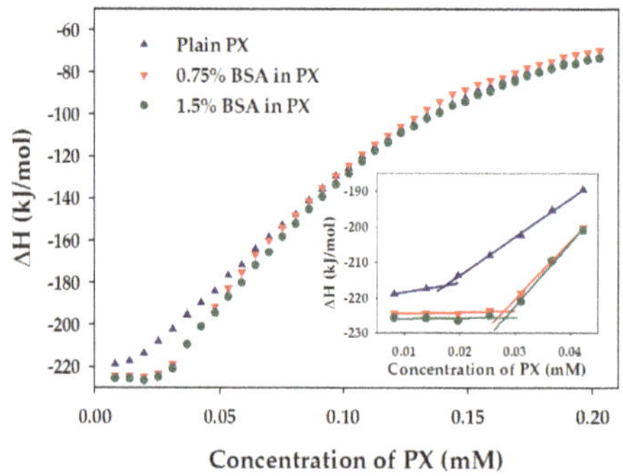

Figure 2. Normalized titration curve of 1.5% w/v PX (plain PX) titrated into water and the mixtures of 1.5% w/v PX and BSA titrated into water with the same BSA content at 30 °C. The BSA concentration for each mixture is shown in the legend.

3.2. *Investigation on the Temperature-Induced Micellization and Gelation*

The response surface methodology based on the three-level factorial design is a practical approach that uses a reasonable number of experiments and statistical analysis to establish a model correlating the casual factors with studied responses [44]. The potential effect of BSA on the temperature-induced micellization and gelation of PX was explored using the response surface approach. The DSC experiments were performed to monitor the temperature-induced micellization, and the thermoresponsive viscoelasticity of the samples was evaluated to describe the temperature-induced gelation.

The three-level factorial design was implemented to statistically identify and quantify the impact of two factors, the concentrations of PX (X_1) and BSA (X_2), on the responses including the CMT (Y_1), the T_{gel} (Y_2) and the $G'_{37°C}$ (Y_3). The three-level factorial design and response data are shown in Table 3. The sample notation is also listed in Table 3.

3.2.1. DSC Thermograms and Temperature Ramp Rheograms

The temperature-dependent micellization of PX in the presence and absence of BSA was investigated using DSC. The DSC thermograms for plain PX samples—16PX, 18PX, and 20PX—are depicted in Figure 3. The temperature-induced dehydration of PX molecules led to the association of the hydrophobic PPO blocks, resulting in the formation of PX micelles composed of PPO cores and PEO coronas [5,13]. A broad endotherm observed in the DSC thermogram of PX could be attributed to the micellization [13,22]. In response to heating, the onset of the endothermic signal corresponded to the initiation of micelle formation, and the end of the endotherm reflected the completion of the micellization process. In general, the CMT is defined as the temperature at which the endothermic signal reaches its peak [5,45]. Evidence for this temperature-induced micellization was provided by the endothermic traces for all plain PX as shown in Figure 3. The CMT as indicated by arrows in Figure 3 was slightly shifted to lower temperature with increasing PX concentration, indicating the concentration-facilitated self-association of amphiphilic co-polymers as previously described [46,47]. Figure 4 displays the DSC traces for the temperature-induced micellization of PX in the presence of BSA. The thermal characteristic of PX micellization was not noticeably altered by the addition of BSA.

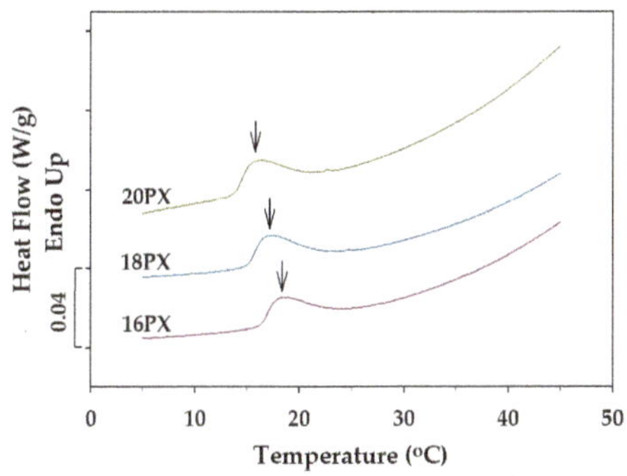

Figure 3. DSC thermograms of aqueous PX at various concentrations (16PX, 18PX, and 20PX). Arrows denote the critical micellization temperature.

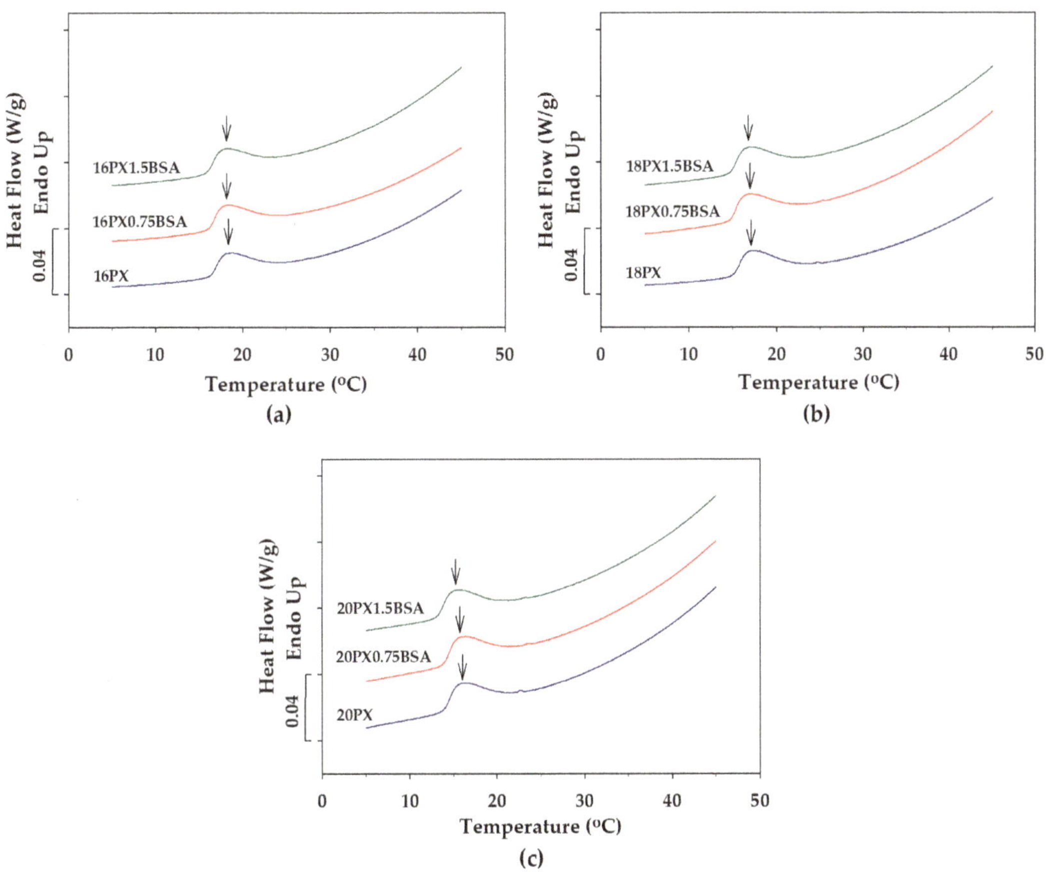

Figure 4. DSC thermograms of (**a**) 16PX and mixtures containing 16PX and various concentrations of BSA (16PX0.75BSA and 16PX1.5BSA), (**b**) 18PX and mixtures containing 18PX and various concentrations of BSA (18PX0.75BSA and 18PX1.5BSA), and (**c**) 20PX and mixtures containing 20PX and various concentrations of BSA (20PX0.75BSA and 20PX1.5BSA). Arrows denote the critical micellization temperature.

Temperature ramp rheometry was performed to examine how BSA affected the thermoresponsive gel formation of PX. The variation of the dynamic moduli with temperature for plain PX and the mixtures of PX and BSA is depicted in Figures 5 and 6. All samples exhibited thermogelation behavior. At low temperatures, the values of G′ were below those of G″ for all plain PX (Figure 5). This reflected the liquid stage of the sample. Both dynamic moduli increased with increasing temperature, and the values of G′ became superior to those of G″ at high temperatures. The predominant G′ at high temperatures revealed the gel characteristic of all samples. The liquid-to-gel transition took place in the temperature range at which the abrupt increase in G′ was detected, and the temperature at which the crossover of G′ and G″ was observed was denoted as the T_{gel} [13,22]. From Figure 5, increasing PX concentration caused the crossover points of G′ and G″ to shift to a lower temperature. This indicated that the thermoresponsive gelation was facilitated by raising the concentration of PX. On the contrary, the presence of BSA resulted in a shift of the crossover points of G′ and G″ to a higher temperature (Figure 6). The gelation of PX could be attributed to the inter-micellar entanglements caused by raising the temperature [48]. The high concentration of PX produced a large number of micelles available for the formation of the gel structure; therefore, the increase in the PX concentration facilitated the temperature-induced gelation. The increase in gelation temperature with the incorporation of BSA could be associated with the steric hindrance of serum albumin. The relationship between the micellization and gelation characteristics, and the concentration of each component was further determined using response surface methodology based on the three-level factorial design.

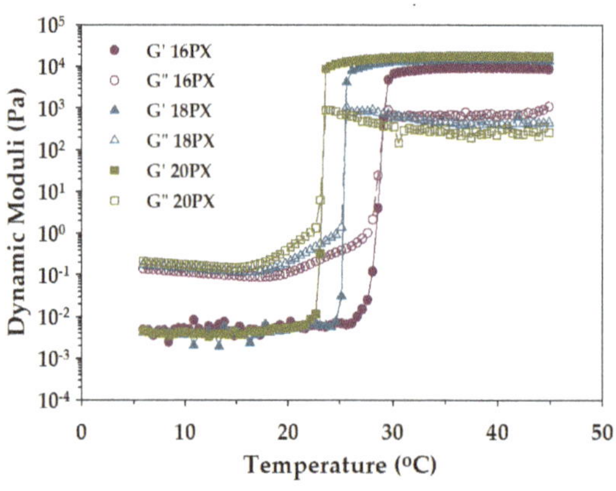

Figure 5. Temperature-dependent dynamic moduli of aqueous PX at various concentrations (16PX, 18PX, and 20PX).

3.2.2. Experimental Data and Response Surface Modeling of Three-Level Factorial Design

The experimental data presented in Table 3 were fitted to linear, two-factor interaction (2FI), and quadratic models to obtain the regression equation. For checking model adequacy, the comparative values of sequential *p*-value, lack of fit *p*-value, correlation coefficient (R^2), adjusted R^2, and predicted R^2 are summarized in Table 4. When the sequential *p*-value is lower than 0.05 and a non-significant lack of fit (p-value > 0.05) is obtained, the model is considered to be adequate [49]. In addition, the feasibility of the model is determined by the values of R^2, adjusted R^2, and predicted R^2. The adjusted R^2 and the predicted R^2 should be close to the same or different by no more than 0.2 [50,51]. All R^2 values of the satisfactory model should fall within the range of 0.7 to 1.0, and the adjusted R^2 should be close to 1.0 [50,51]. Accordingly, the response Y_1 was found to follow the linear model, while the quadratic models were selected to describe the effects of the factors on the responses Y_2 and Y_3.

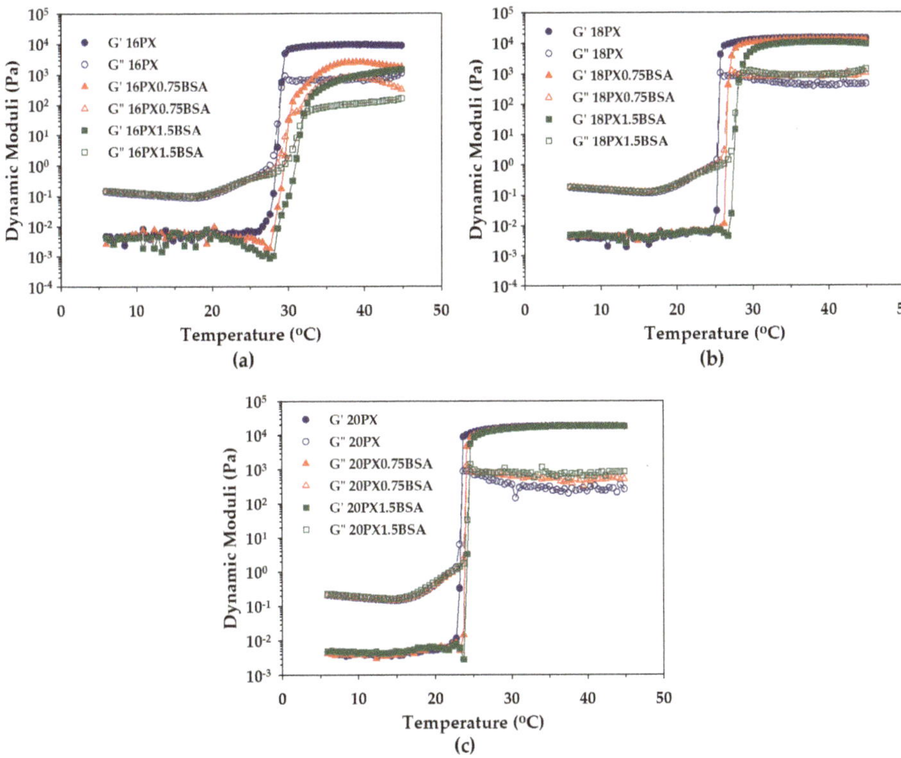

Figure 6. Temperature-dependent dynamic moduli of (**a**) 16PX and mixtures containing 16PX and various concentrations of BSA (16PX0.75BSA and 16PX1.5BSA), (**b**) 18PX and mixtures containing 18PX and various concentrations of BSA (18PX0.75BSA and 18PX1.5BSA), and (**c**) 20PX and mixtures containing 20PX and various concentrations of BSA (20PX0.75BSA and 20PX1.5BSA).

Table 4. Statistic comparison of regression models for responses Y_1, Y_2, and Y_3.

Models	Sequential p-Value	Lack of Fit p-Value	R^2	Adjusted R^2	Predicted R^2	Remarks
Response Y_1						
Linear model	<0.0001	0.3571	0.9892	0.9871	0.9817	Suggested
2FI model	0.3574	0.3403	0.9903	0.9870	0.9793	
Quadratic model	0.5725	0.2507	0.9917	0.9858	0.9516	
Response Y_2						
Linear model	<0.0001	0.0064	0.9786	0.9743	0.9483	
2FI model	0.0126	0.0197	0.9897	0.9862	0.9639	
Quadratic model	0.0119	0.1040	0.9971	0.9950	0.9766	Suggested
Response Y_3						
Linear model	<0.0001	0.0207	0.9161	0.8994	0.8994	
2FI model	0.0125	0.0618	0.9596	0.9461	0.9461	
Quadratic model	0.0150	0.2819	0.9878	0.9791	0.9791	Suggested

The relationship between the CMT and the concentrations of PX and BSA was represented by the following linear model in terms of coded factors.

$$Y_1 = \mathbf{16.93} - \mathbf{1.20X_1} - 0.1417X_2 \quad (1)$$

whereas a negative sign of a factor denotes an inverse relationship between the factor and the response, a positive sign of a factor signifies a positive effect of the factor on the response. In an equation, any term with statistical significance is bolded. As illustrated in Figure 7a, the plot of experimental response against predicted response shows an adequate correlation and behaves with a uniform distribution of data points around the 45° line. A high correlation between predicted and actual experimental data indicated that the model was reliable [52]. According to the response surface plot in Figure 7b, the CMT was mainly dependent on the concentration of PX. As the concentration of PX increased, the CMT decreased in a linear fashion. Although the concentration of BSA also had a negative effect on the CMT, the effect of BSA concentration on the response Y_1 was minor and insignificant. The insignificant influence of BSA on the CMT found in this study is in accordance with a study by Perinelli et al., which investigated the effect of BSA on the CMT of PX at low PX concentrations (2.5% and 5% w/w) in phosphate buffer and found no notable effect on the CMT value of PX [30].

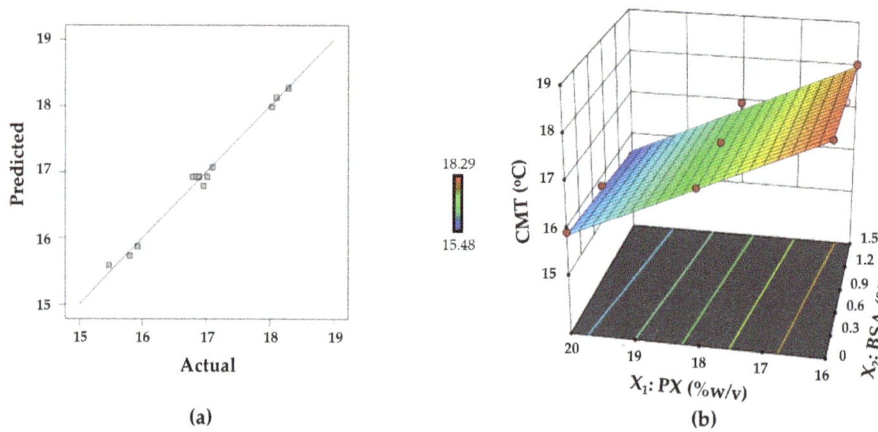

Figure 7. (**a**) Linear plot between predicted and actual response values for the critical micellization temperature, and (**b**) response surface plot representing the effect of PX and BSA on the critical micellization temperature.

The relationship between the T_{gel} and the concentrations of PX and BSA was represented by the following quadratic model in terms of coded factors.

$$Y_2 = \mathbf{26.81} - \mathbf{3.24X_1} + \mathbf{1.06X_2} - \mathbf{0.4450X_1X_2} + \mathbf{0.4290X_1^2} - 0.0810X_2^2 \quad (2)$$

The coefficient of a linear term (X_1 or X_2) represents the effect of a particular factor, whereas the coefficients of an interactive term (X_1X_2) and a quadratic term (X_1^2 or X_2^2) describe the interaction between two factors and quadratic effect, respectively. In Equation (2) X_1, X_2, X_1X_2, and X_1^2 were significant model terms, as highlighted in bold. When the experimental response is plotted against the predicted response (Figure 8a), the data points are close to the straight line. This reflected an adequate agreement between the experimental data and the predicted responses. According to Equation (2), both PX and BSA concentrations significantly impacted on the T_{gel}. The increase in the PX concentration lowered T_{gel}, as revealed by the negative coefficient of its linear term. This is consistent with the gelling behavior of PX reported by Liu et al. [48]. With the increase in PX concentration, the polymer solution gains in density and volume fraction of micelles. As a result, the distance between micelles was reduced, favoring the temperature-induced formation of a micellar network [48]. Hence, the increase in PX concentration gave rise to the reduction of T_{gel}. The positive coefficient for the linear term of BSA, on the other hand, showed that increasing the concentration of BSA shifted T_{gel} to higher values. As a result, these two factors have opposing effects. However, the curvature depicted on the response surface plot (Figure 8b) indicates the complexity of the interaction between the components in the mixtures. The influence of one component may vary depending on the amount of another in the mixture. At the lower level of PX, BSA had a more noticeable effect on T_{gel} increase. The steric hindrance caused by the additives may be responsible for the increase in T_{gel} of the PX-based mixtures [15,30]. The presence of BSA could produce steric hindrance and hinder the temperature-induced organization of PX micelles into the inter-micellar packing, causing a positive effect on T_{gel}. As discussed previously, the density and volume fraction of PX micelles depended on the concentration of PX. The micellar volume fraction of PX at low concentration may be low compared with that of PX at high concentration. Therefore, the addition of BSA to PX at a low concentration had a more pronounced effect on T_{gel}.

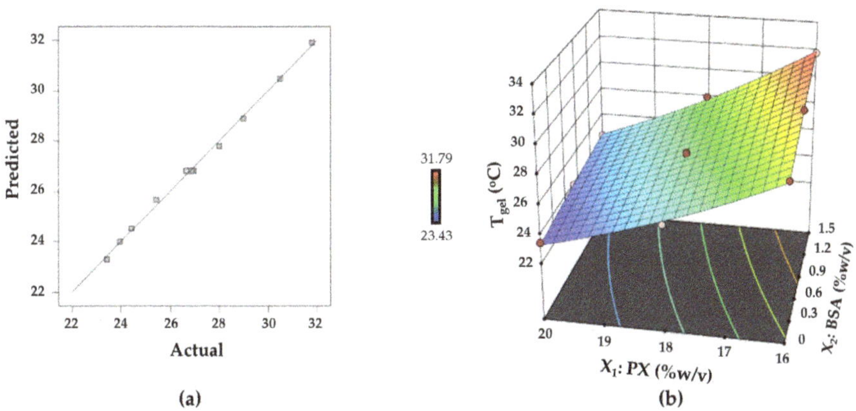

Figure 8. (**a**) Linear plot between predicted and actual response values for the gelation temperature, and (**b**) response surface plot representing the effect of PX and BSA on the gelation temperature.

The relationship between the $G'_{37°C}$ and the concentrations of PX and BSA was represented by the following quadratic model in terms of coded factors.

$$Y_3 = \mathbf{12602.36 + 7238.56X_1 - 2177.17X_2 + 2015.76X_1X_2 - 1922.08X_1^2} + 1070.08X_2^2 \quad (3)$$

X_1, X_2, X_1X_2, and X_1^2 were significant terms for the storage modulus at 37 °C. Negative coefficients of X_2 and X_1^2 indicated the inverse relationship for the $G'_{37°C}$ while other terms showed a positive effect on the $G'_{37°C}$. The adequate agreement between the experimental data and the predicted responses is demonstrated in Figure 9a. The response surface plot is depicted in Figure 9b to further highlight the relationship between these factors and the response. As can be seen, increasing the concentration of PX led to a significant increase in the $G'_{37°C}$. For PX-based gels, the increment of the elastic characteristic is associated with the increase in inter-micellar entanglements, which promote the deformation resistance of the gels [53]. The increase in PX concentration produced more entanglements in the gel structure, enhancing the gel strength. The surface response of the $G'_{37°C}$ tended to tilt down with increasing BSA concentration. In addition, the complexity of the interplay between the components was shown by the curvature of the response surface plot. At the lower level of PX, the effect of BSA on the $G'_{37°C}$ is more prominent. The reduction in elasticity of gels has been reported for the mixtures of PX and poly(acrylic acid) and the mixtures of PX and HSA [23,32]. This softening effect could be explained by the attachment of the macromolecules to the micelles, which in turn limited the movement of the micelles and obstructed inter-micellar entanglements.

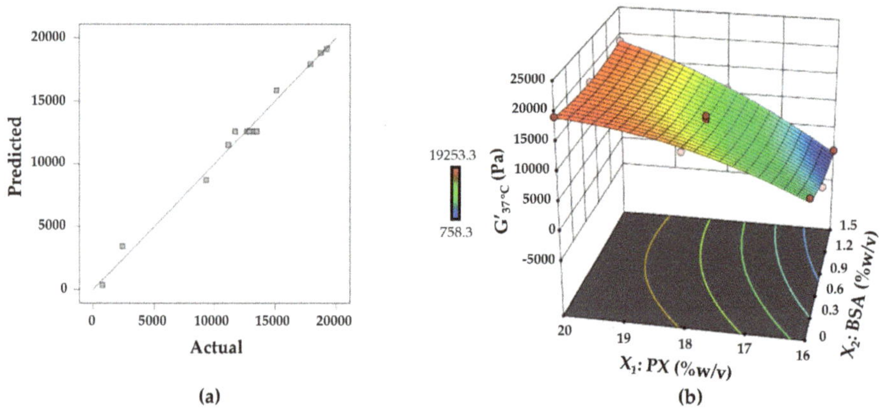

Figure 9. (**a**) Linear plot between predicted and actual response values for the elastic modulus at 37 °C, and (**b**) response surface plot representing the effect of PX and BSA on the elastic modulus at 37 °C.

4. Conclusions

The micellization of PX in the presence and absence of BSA was investigated using ITC. The pre-micellar region, the transition concentration region, and the post-micellar region were observed in the calorimetric titration curves. The presence of BSA did not show a discernible effect on the CMC of PX. However, the addition of BSA caused the extension of the pre-micellar region. This could be attributed to the hydrophobic interaction between PPO residues of PX unimers and BSA. In addition to the self-organization of PX at a certain temperature, the temperature-dependent micellization and gelation of PX were also investigated. The presence of BSA did not show a significant effect on the CMT but altered the T_{gel} as well as the gel integrity of PX-based solutions. Response surface methodology represented the linear relationship between the composition and the CMT. The change in the CMT of the mixtures was mainly attributed to the concentration of PX. The complex interplay between PX and BSA was found for the T_{gel} and the $G'_{37°C}$. BSA interfered in the inter-micellar entanglements into gel structure. Accordingly, the presence of BSA showed the modulation effect on the T_{gel} and the softening effect on the gel integrity. Understanding the impact of serum albumin on the self-assembly and gelling characteristics of PX would allow the development of thermoresponsive systems with controlled gelation temperatures and viscoelasticity for drug delivery and tissue engineering.

Author Contributions: Conceptualization, N.H.; data curation, N.H., P.K. and S.S.; funding acquisition, N.H.; investigation, N.H. and P.K.; methodology, N.H.; project administration, N.H.; supervision, N.H.; writing—original draft, N.H.; writing—review and editing, N.H. and P.K. All authors have read and agreed to the published version of the manuscript.

Funding: This work was supported by Thammasat University Research Unit in Smart Materials and Innovative Technology for Pharmaceutical Applications (SMIT-Pharm) (Project ID 6305016).

Institutional Review Board Statement: Not applicable.

Data Availability Statement: Not applicable.

Acknowledgments: This work was supported by Thammasat University Research Unit in Smart Materials and Innovative Technology for Pharmaceutical Applications (SMIT-Pharm) (Project ID 6305016). Supaporn Santhan, a research assistant, is acknowledged for her help in laboratory work.

Conflicts of Interest: The authors declare no conflict of interest.

References

1. Zhao, J.; Huang, W.; Si, P.; Ulstrup, J.; Diao, F.; Zhang, J. General syntheses of nanotubes induced by block copolymer self-assembly. *Macromol. Rapid Commun.* **2018**, *39*, 1800125. [CrossRef] [PubMed]
2. Wei, D.; Ge, L.; Guo, R. Effect of hydrophilically modified ibuprofen on thermoresponsive gelation of pluronic copolymer. *Colloids Surf. A Physicochem. Eng. Asp.* **2018**, *553*, 1–10. [CrossRef]
3. Chang, D.; Du, C.; Liu, J.; Sun, W.; Su, Y.; Zang, D.; Liu, T. Effect of hydrophobic modification of block copolymers on the self-assembly, drug encapsulation and release behavior. *J. Mol. Liq.* **2022**, *368*, 120635. [CrossRef]
4. Ashraf, U.; Chat, O.A.; Maswal, M.; Jabeen, S.; Dar, A.A. An investigation of pluronic P123–sodium cholate mixed system: Micellization, gelation and encapsulation behavior. *RSC Adv.* **2015**, *5*, 83608–83618. [CrossRef]
5. White, J.M.; Calabrese, M.A. Impact of small molecule and reverse poloxamer addition on the micellization and gelation mechanisms of poloxamer hydrogels. *Colloids Surf. A Physicochem. Eng. Asp.* **2022**, *638*, 128246. [CrossRef]
6. He, Z.; Ma, Y.; Alexandridis, P. Comparison of ionic liquid and salt effects on the thermodynamics of amphiphile micellization in water. *Colloids Surf. A Physicochem. Eng. Asp.* **2018**, *559*, 159–168. [CrossRef]
7. Djekic, L.; Čalija, B.; Medarević, Đ. Gelation behavior, drug solubilization capacity and release kinetics of poloxamer 407 aqueous solutions: The combined effect of copolymer, cosolvent and hydrophobic drug. *J. Mol. Liq.* **2020**, *303*, 112639. [CrossRef]
8. Ramburrun, P.; Khan, R.A.; Choonara, Y.E. Design, preparation, and functionalization of nanobiomaterials for enhanced efficacy in current and future biomedical applications. *Nanotechnol. Rev.* **2022**, *11*, 1802–1826. [CrossRef]
9. Alexandridis, P.; Holzwarth, J.F.; Hatton, T.A. Micellization of poly(ethylene oxide)-poly(propylene oxide)-poly(ethylene oxide) triblock copolymers in aqueous solutions: Thermodynamics of copolymer association. *Macromolecules* **1994**, *27*, 2414–2425. [CrossRef]
10. Li, Y.; Xu, R.; Couderc, S.; Bloor, M.; Wyn-Jones, E.; Holzwarth, J.F. Binding of sodium dodecyl sulfate (SDS) to the ABA block copolymer pluronic F127 (EO97PO69EO97): F127 aggregation induced by SDS. *Langmuir* **2001**, *17*, 183–188. [CrossRef]
11. Sharma, R.K.; Shaikh, S.; Ray, D.; Aswal, V.K. Binary mixed micellar systems of PEO-PPO-PEO block copolymers for lamotrigine solubilization: A comparative study with hydrophobic and hydrophilic copolymer. *J. Polym. Res.* **2018**, *25*, 73. [CrossRef]
12. Haddow, P.; McAuley, W.J.; Kirton, S.B.; Cook, M.T. Poly(N-isopropyl acrylamide)–poly(ethylene glycol)–poly(N-isopropyl acrylamide) as a thermoreversible gelator for topical administration. *Mater. Adv.* **2020**, *1*, 371–386. [CrossRef]
13. Al Khateb, K.; Ozhmukhametova, E.K.; Mussin, M.N.; Seilkhanov, S.K.; Rakhypbekov, T.K.; Lau, W.M.; Khutoryanskiy, V.V. In situ gelling systems based on pluronic F127/pluronic F68 formulations for ocular drug delivery. *Int. J. Pharm.* **2016**, *502*, 70–79. [CrossRef] [PubMed]
14. Chang, E.I.; Galvez, M.G.; Glotzbach, J.P.; Hamou, C.D.; El-ftesi, S.; Rappleye, C.T.; Sommer, K.-M.; Rajadas, J.; Abilez, O.J.; Fuller, G.G.; et al. Vascular anastomosis using controlled phase transitions in poloxamer gels. *Nat. Med.* **2011**, *17*, 1147–1152. [CrossRef]
15. Abou-Shamat, M.A.; Stair, J.L.; Kirton, S.B.; Calvo-Castro, J.; Cook, M.T. A Design-of-Experiments approach to developing thermoresponsive gelators from complex polymer mixtures. *Mol. Syst. Des. Eng.* **2020**, *5*, 1538–1546. [CrossRef]
16. Carvalho, G.C.; Araujo, V.H.S.; Fonseca-Santos, B.; de Araújo, J.T.C.; de Souza, M.P.C.; Duarte, J.L.; Chorilli, M. Highlights in poloxamer-based drug delivery systems as strategy at local application for vaginal infections. *Int. J. Pharm.* **2021**, *602*, 120635. [CrossRef]
17. Cui, N.; Dai, C.-Y.; Mao, X.; Lv, X.; Gu, Y.; Lee, E.-S.; Jiang, H.-B.; Sun, Y. Poloxamer-based scaffolds for tissue engineering applications: A review. *Gels* **2022**, *8*, 360. [CrossRef]
18. Pratap-Singh, A.; Guo, Y.; Lara Ochoa, S.; Fathordoobady, F.; Singh, A. Optimal ultrasonication process time remains constant for a specific nanoemulsion size reduction system. *Sci. Rep.* **2021**, *11*, 9241. [CrossRef]
19. Sepulveda, A.F.; Kumpgdee-Vollrath, M.; Franco, M.K.K.D.; Yokaichiya, F.; de Araujo, D.R. Supramolecular structure organization and rheological properties modulate the performance of hyaluronic acid-loaded thermosensitive hydrogels as drug-delivery systems. *J. Colloid Interface Sci.* **2023**, *630*, 328–340. [CrossRef]
20. Ibrahim, E.-S.; Ismail, S.; Fetih, G.; Shaaban, O.; Hassanein, K.; Abdellah, N. Development and characterization of thermosensitive pluronic-based metronidazole gelling formulations for vaginal application. *Acta Pharm.* **2012**, *62*, 59–70. [CrossRef]
21. Xia, H.; Jin, H.; Cheng, Y.; Cheng, Z.; Xu, Y. The controlled release and anti-inflammatory activity of a tetramethylpyrazine-loaded thermosensitive poloxamer hydrogel. *Pharm. Res.* **2019**, *36*, 52. [CrossRef] [PubMed]
22. Hirun, N.; Kraisit, P.; Tantishaiyakul, V. Thermosensitive polymer blend composed of poloxamer 407, poloxamer 188 and polycarbophil for the use as mucoadhesive in situ gel. *Polymers* **2022**, *14*, 1836. [CrossRef] [PubMed]
23. Pragatheeswaran, A.M.; Chen, S.B. The influence of poly(acrylic acid) on micellization and gelation characteristics of aqueous pluronic F127 copolymer system. *Colloid Polym. Sci.* **2016**, *294*, 107–117. [CrossRef]
24. De Souza Ferreira, S.B.; Da Silva, J.B.; Borghi-Pangoni, F.B.; Junqueira, M.V.; Bruschi, M.L. Linear correlation between rheological, mechanical and mucoadhesive properties of polycarbophil polymer blends for biomedical applications. *J. Mech. Behav. Biomed. Mater.* **2017**, *68*, 265–275. [CrossRef] [PubMed]
25. Hirun, N.; Tantishaiyakul, V.; Sangfai, T.; Boonlai, W.; Soontaranon, S.; Rugmai, S. The effect of poly(acrylic acid) on temperature-dependent behaviors and structural evolution of poloxamer 407. *Polym. Int.* **2021**, *70*, 1282–1289. [CrossRef]
26. Jain, A.; Kishore, N. Micellar properties of pluronics in combination with cationic surfactant and interaction with lysozyme: Thermodynamic evaluation. *J. Mol. Liq.* **2022**, *359*, 119303. [CrossRef]

27. Li, D.; Lu, X.; Lin, H.; Ren, F.; Leng, Y. Chitosan/bovine serum albumin co-micropatterns on functionalized titanium surfaces and their effects on osteoblasts. *J. Mater. Sci. Mater. Med.* **2013**, *24*, 489–502. [CrossRef]
28. Sabaa, M.W.; Hanna, D.H.; Abu Elella, M.H.; Mohamed, R.R. Encapsulation of bovine serum albumin within novel xanthan gum based hydrogel for protein delivery. *Mater. Sci. Eng. C Mater. Biol. Appl.* **2019**, *94*, 1044–1055. [CrossRef]
29. Yuan, H.; Zheng, X.; Liu, W.; Zhang, H.; Shao, J.; Yao, J.; Mao, C.; Hui, J.; Fan, D. A novel bovine serum albumin and sodium alginate hydrogel scaffold doped with hydroxyapatite nanowires for cartilage defects repair. *Colloids Surf. B Biointerfaces* **2020**, *192*, 111041. [CrossRef]
30. Perinelli, D.R.; Bonacucina, G.; Pucciarelli, S.; Cespi, M.; Casettari, L.; Polzonetti, V.; Carpi, F.M.; Palmieri, G.F. Could albumin affect the self-assembling properties of a block co-polymer system and drug release? An in-vitro study. *Pharm. Res.* **2015**, *32*, 1094–1104. [CrossRef]
31. Negrea, E.; Oancea, P.; Leonties, A.; Ana Maria, U.; Avram, S.; Raducan, A. Spectroscopic studies on binding of ibuprofen and drotaverine with bovine serum albumin. *J. Photochem. Photobiol. A Chem.* **2023**, *438*, 114512. [CrossRef]
32. Neacsu, M.V.; Matei, I.; Micutz, M.; Staicu, T.; Precupas, A.; Popa, V.T.; Salifoglou, A.; Ionita, G. Interaction between albumin and pluronic F127 block copolymer revealed by global and local physicochemical profiling. *J. Phys. Chem. B* **2016**, *120*, 4258–4267. [CrossRef] [PubMed]
33. Patel, N.; Ji, N.; Wang, Y.; Li, X.; Langley, N.; Tan, C. Subcutaneous delivery of albumin: Impact of thermosensitive hydrogels. *AAPS PharmSciTech* **2021**, *22*, 120. [CrossRef] [PubMed]
34. Abdelnabi, D.M.; Abdallah, M.H.; Elghamry, H.A. Buspirone hydrochloride loaded in situ nanovesicular gel as an anxiolytic nasal drug delivery system: In vitro and animal studies. *AAPS PharmSciTech* **2019**, *20*, 134. [CrossRef] [PubMed]
35. He, Z.; Alexandridis, P. Micellization thermodynamics of pluronic P123 (EO20PO70EO20) amphiphilic block copolymer in aqueous ethylammonium nitrate (EAN) solutions. *Polymers* **2018**, *10*, 32. [CrossRef]
36. Shimizu, S.; Pires, P.A.R.; Loh, W.; El Seoud, O.A. Thermodynamics of micellization of cationic surfactants in aqueous solutions: Consequences of the presence of the 2-acylaminoethyl moiety in the surfactant head group. *Colloid Polym. Sci.* **2004**, *282*, 1026–1032. [CrossRef]
37. Xing, L.; Sun, J.; Tan, H.; Yuan, G.; Li, J.; Jia, Y.; Xiong, D.; Chen, G.; Lai, J.; Ling, Z.; et al. Covalently polysaccharide-based alginate/chitosan hydrogel embedded alginate microspheres for BSA encapsulation and soft tissue engineering. *Int. J. Biol. Macromol.* **2019**, *127*, 340–348. [CrossRef]
38. Kahya, N.; Erim, F.B. Surfactant modified alginate composite gels for controlled release of protein drug. *Carbohydr. Polym.* **2019**, *224*, 115165. [CrossRef]
39. Smith, O.E.P.; Waters, L.J.; Small, W.; Mellor, S. CMC determination using isothermal titration calorimetry for five industrially significant non-ionic surfactants. *Colloids Surf. B Biointerfaces* **2022**, *211*, 112320. [CrossRef]
40. Loh, W.; Brinatti, C.; Tam, K.C. Use of isothermal titration calorimetry to study surfactant aggregation in colloidal systems. *Biochim. Biophys. Acta Gen. Subj.* **2016**, *1860*, 999–1016. [CrossRef]
41. Bouchemal, K.; Agnely, F.; Koffi, A.; Ponchel, G. A concise analysis of the effect of temperature and propanediol-1, 2 on pluronic F127 micellization using isothermal titration microcalorimetry. *J. Colloid Interface Sci.* **2009**, *338*, 169–176. [CrossRef]
42. Olofsson, G.; Loh, W. On the use of titration calorimetry to study the association of surfactants in aqueous solutions. *J. Braz. Chem. Soc.* **2009**, *20*, 577–593. [CrossRef]
43. Guo, C.; Wang, J.; Liang, X.; Zheng, L.; Liu, H. Effect of bovine serum albumin on the micellization of poly(ethylene oxide)-poly(propylene oxide)-poly(ethylene oxide) block copolymers in aqueous solutions by fluorescence spectroscopy. *Sci. China B Chem.* **2006**, *49*, 541–549. [CrossRef]
44. Sherif, A.Y.; Mahrous, G.M.; Alanazi, F.K. Novel in-situ gel for intravesical administration of ketorolac. *Saudi Pharm. J.* **2018**, *26*, 845–851. [CrossRef] [PubMed]
45. Boonlai, W.; Tantishaiyakul, V.; Hirun, N.; Phaisan, S.; Uma, T. The effect of the preservative methylparaben on the thermoresponsive gelation behavior of aqueous solutions of poloxamer 407. *J. Mol. Liq.* **2017**, *240*, 622–629. [CrossRef]
46. Scherlund, M.; Brodin, A.; Malmsten, M. Micellization and gelation in block copolymer systems containing local anesthetics. *Int. J. Pharm.* **2000**, *211*, 37–49. [CrossRef] [PubMed]
47. Boonlai, W.; Tantishaiyakul, V.; Hirun, N.; Sangfai, T.; Suknuntha, K. Thermosensitive poloxamer 407/poly(acrylic acid) hydrogels with potential application as injectable drug delivery system. *AAPS PharmSciTech* **2018**, *19*, 2103–2117. [CrossRef] [PubMed]
48. Liu, S.; Li, L. Multiple phase transition and scaling law for poly(ethylene oxide)–poly(propylene oxide)–poly(ethylene oxide) triblock copolymer in aqueous solution. *ACS Appl. Mater. Interfaces* **2015**, *7*, 2688–2697. [CrossRef] [PubMed]
49. Vera Candioti, L.; De Zan, M.M.; Cámara, M.S.; Goicoechea, H.C. Experimental design and multiple response optimization. Using the desirability function in analytical methods development. *Talanta* **2014**, *124*, 123–138.
50. Thite, V.S.; Nerurkar, A.S.; Baxi, N.N. Optimization of concurrent production of xylanolytic and pectinolytic enzymes by Bacillus safensis M35 and Bacillus altitudinis J208 using agro-industrial biomass through response surface methodology. *Sci. Rep.* **2020**, *10*, 3824. [CrossRef]
51. Breig, S.J.M.; Luti, K.J.K. Response surface methodology: A review on its applications and challenges in microbial cultures. *Mater. Today Proc.* **2021**, *42*, 2277–2284. [CrossRef]

52. Singh, S.; Parhi, R.; Garg, A. Formulation of topical bioadhesive gel of aceclofenac using 3-level factorial design. *Iran. J. Pharm. Res.* **2011**, *10*, 435–445. [PubMed]
53. Junqueira, M.V.; Borghi-Pangoni, F.B.; Ferreira, S.B.d.S.; Bruschi, M.L. Evaluation of the methylene blue addition in binary polymeric systems composed by poloxamer 407 and Carbopol 934P using quality by design: Rheological, textural, and mucoadhesive analysis. *Drug Dev. Ind. Pharm.* **2016**, *42*, 2009–2019. [CrossRef] [PubMed]

polymers

MDPI

Article

Synthesis and Comparative Study of Polyether-*b*-polybutadiene-*b*-polyether Triblock Copolymers for Use as Polyurethanes

Pengzhi Bi [1,2], Xiuzhong Zhu [1,2,3,*], Jinbang Han [1,2], Li Tian [1,2] and Wanbin Zhang [3]

1 State Key Laboratory of Biobased Material and Green Papermaking, Key Laboratory of Pulp and Paper Science & Technology of Ministry of Education, Faculty of Light Industry, Shandong Academy of Sciences, Qilu University of Technology, Jinan 250353, China; bipengzhi@163.com (P.B.); han1819583@163.com (J.H.); tianli20220222@163.com (L.T.)

2 Key Laboratory for Green Leather Manufacture Technology of China National Light Industry Council, Faculty of Light Industry, Shandong Academy of Sciences, Qilu University of Technology, Jinan 250353, China

3 Key Laboratory of Auxiliary Chemistry and Technology for Chemical Industry Ministry of Education, Shaanxi Collaborative Innovation Center of Industrial Auxiliary Chemistry and Technology, Shaanxi University of Science and Technology, Xi'an 710021, China; zhangwanbin@sust.edu.cn

* Correspondence: zhuxiuzhong@qlu.edu.cn

Abstract: In this paper, the effects of HTPBs with different main-chain microstructures on their triblock copolymers and polyurethane properties were investigated. Three polyether-modified HTPB triblock copolymers were successfully synthesized via a cationic ring-opening copolymerization reaction using three HTPBs with different microstructures prepared via three different polymerization methods as the macromolecular chain transfer agents and tetrahydrofuran (THF) and propylene oxide (PO) as the copolymerization monomers. Finally, the corresponding polyurethane elastomers were prepared using the three triblock copolymers as soft segments and toluene diisocyanate (TDI) as hard segments. The results of an analysis of the triblock copolymers showed that the triblock copolymers had lower viscosity and glass transition temperature (T_g) values as the HTPB 1,2 structure content decreased, although the effect on the thermal decomposition temperature was not significant. An analysis of the polyurethane elastomers revealed that as the content of the 1,2 structure in HTPB increased, its corresponding polyurethane elastomers showed a gradual increase in breaking strength and a gradual decrease in elongation at break. In addition, PU-1 had stronger crystallization properties compared to PU-2 and PU-3. However, the differences in the microstructures of the HTPBs did not seem to have much effect on the surface properties of the polyurethane elastomers.

Keywords: hydroxyl-terminated polybutadiene; polyether modified; triblock copolymer; comparative study

1. Introduction

Hydroxyl-terminated polybutadiene (HTPB) is a commonly used polyurethane matrix resin in defense and civil applications [1–6]. A large number of unsaturated double bonds in the backbone give the polymer excellent low-temperature resistance and flexibility [7]. However, HTPB is non-polar and, therefore, poorly compatible with polar components. As such, there have been numerous reports on modification strategies to increase the polarity of HTPB, such as main-chain epoxidation, side-group polarization, and polyester or polyether block modification [8–14]. Among these strategies, polyether block modification of HTPB, which refers to the preparation of triblock copolymers, is one of the more feasible routes for HTPB polarization and has been widely investigated [15–17]. To date, research on polyether-modified HTPB triblock copolymers has mainly focused on the effect of the type of polyether chain segment on the performance of the resulting HTPB-based copolymers;

however, the effects of the HTPB microstructure, which ultimately determines the triblock copolymer performance, have not been reported systematically.

It is well known that there are three different isomeric forms of HTPB: cis-1,4-; trans-1,4-; and 1,2-polybutadiene [18]. The different isomers have significantly different properties. For example, the glass transition temperature (T_g) of pure cis-1,4 polybutadiene can be as low as −106 °C, while HTPB with a high trans-1,4 content has a glass transition temperature on the order of −15 °C, good crystallinity, and properties characteristic of a plastic [19,20]. Compared with the 1,4 isomers, increasing the 1,2-polybutadiene content increases the viscosity of HTPB. It can be seen that the different microstructures of HTPB have a great influence on its own properties, and the mechanism of its influence may be explored through a reasonable molecular simulation [21]. At present, there are three main methods for the polymerization of HTPB, namely, free radical polymerization [22,23], anionic polymerization [24–26], and ring-opening metathesis polymerization (ROMP) [27,28]. In a previous study, our group reported in detail the differences in the structures and properties of HTPB prepared via the three polymerization methods. It was found that the contents of cis-1,4; trans-1,4; and 1,2 units within the main-chain structures of HTPB prepared via the three polymerization methods were different, which led to different glass transition temperatures and viscosities for the synthesized HTPBs [29]. As matrix resins, the differences in HTPB microstructure should directly affect the processing and application of the materials. Likewise, the different microstructures of the HTPB chain segments in the polyether blocks of the triblock copolymers will certainly affect the use of the materials, but unfortunately, there are few related studies of these effects.

Based on the motivation discussed above, in this study, three polyether-modified HTPB triblock copolymers were designed, synthesized, and then used to prepare thermoplastic polyurethane elastomers (Figure 1). To investigate the effects of the different polybutadiene isomers on the performance of the polyether-modified HTPB triblock copolymers, three triblock copolymers with the same polyether blocks and different HTPB chain segments were prepared via free radical polymerization, anionic polymerization, and ring-opening metathesis polymerization. The HTPBs prepared via free radical polymerization, anionic polymerization, and ring-opening metathesis polymerization are referred to as FHTPB, AHTPB, and RHTPB, and the corresponding triblock copolymers are referred to as TBC_{FHTPB}, TBC_{AHTPB}, and TBC_{RHTPB}, respectively. The molecular weights of AHTPB and RHTPB synthesized via anionic and ring-opening metathesis polymerization, respectively, were similar to those of the commercial FHTPB. Then, the three HTPBs were used as macromolecular chain transfer agents and copolymerized with THF and PO via a cationic ring-opening reaction to obtain the corresponding triblock copolymers. The molecular structures of the three triblock copolymers were characterized using FT-IR, NMR, and SEC. Subsequently, the viscosity and the thermal and processing properties of the synthesized triblock copolymers were studied using rotational viscometry, differential scanning thermal calorimetry (DSC), and thermogravimetric analysis (TGA), respectively, and compared. Finally, the TBC_{FHTPB}, TBC_{AHTPB}, and TBC_{RHTPB} were reacted with toluene diisocyanate (TDI) to obtain thermoplastic elastomers (named PU-1, PU-2, and PU-3, respectively), and the effects of different HTPB microstructures on the breakage strength, elongation at break, and surface properties of the corresponding elastomers were investigated via tensile testing, XRD, XPS, and contact angle measurements. We hope that the comparative study of molecular structures and properties presented here is a useful reference for the application of polyether-modified HTPB block copolymers.

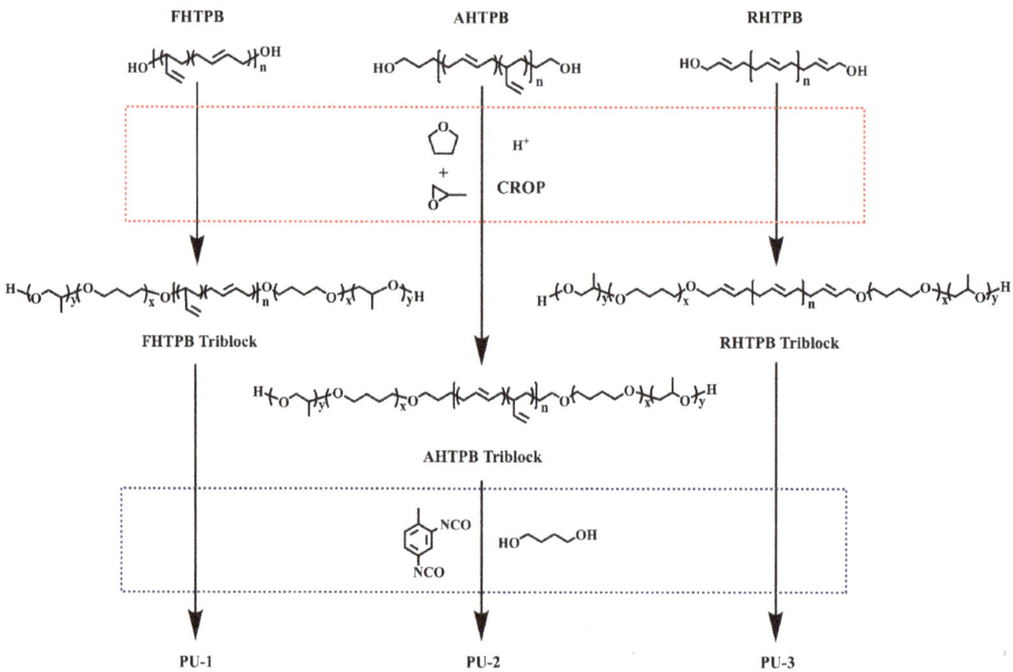

Figure 1. Synthesis routes of three HTPB triblock copolymers and corresponding polyurethane elastomers.

2. Experimental Section

2.1. Materials

FHTPB (M_n = 3400 g mol^{-1}, M_w/M_n = 1.67) was kindly supplied by Qilong Chemical Corp. (Shanghai, China). 3-tert-butyldimethysilyloxy-1-propyllithium (tBDMSOPrLi) was made in the laboratory. 1,3-butadiene (15% solution in cyclohexane) was purchased from TCI (Shanghai) Development Co., Ltd., Tokyo, Japan. Cyclohexane (Cy, used immediately after drying and distilling anhydrous calcium chloride prior to use) was purchased from Sinopharm Chemical Reagent Co., Ltd., Shanghai, China. N,N-dimethylformamide (DMF) was purchased from Sinopharm Chemical Reagent Co., Ltd., Shanghai, China. Tetrabutylammonium fluoride (TBAF) was purchased from Shanghai Aladdin Biochemical Technology Co., Ltd., Shanghai, China. Methanol (MeOH) was purchased from Sinopharm Chemical Reagent Co., Ltd., Shanghai, China. Ethylene Oxide (EO) was purchased from Sinopharm Chemical Reagent Co., Ltd., Shanghai, China. Tetrahydrofuran (THF, 99.9% chromatographic purity) was purchased from Concord Technology (Tianjin) Co., Ltd., Tianjin, China. Sodium metal wire and benzophenone were added under nitrogen atmosphere before use and refluxed until the system was blue. The resulting compound was distilled, collected, and used immediately. Sodium (Na, analytical purity) was purchased from Sinopharm Chemical Reagent Co., Ltd., Shanghai, China. Cis-2-butene-1,4-diol was purchased from Shanghai Aladdin Biochemical Technology Co., Ltd., Shanghai, China. 1,5-cyclooctadiene (COD) was purchased from TCI (Shanghai) Development Co., Ltd., Tokyo, Japan. Grubbs Catalyst, 2nd Generation was purchased from Sun Chemical Technology (Shanghai) Co., Ltd., Shanghai, China. Toluene (99.5% analytical purity) was purchased from Sinopharm Chemical Reagent Co., Ltd., Shanghai, China. Sodium metal wire and benzophenone were added under nitrogen atmosphere before use and refluxed until the system was blue. The resulting compound was collected by distillation and used immediately. Propylene oxide (PO, 99.0% analytically pure, using a type 3A molecular sieve to remove water, and stored in an airtight container) was purchased from TCI (Shanghai)

Development Co., Ltd., Tokyo, Japan. Boron trifluoride diethyl etherate ($BF_3 \cdot OEt_2$, 98% analytically pure) was purchased from Sun Chemical Technology (Shanghai) Co., Ltd., Shanghai, China. Dichloromethane (DCM, 99.5% analytical purity) was purchased from Tianjin Fuyu Fine Chemical Co., Ltd., Tianjin, China. Toluene diisocyanate (TDI, 98% chromatographically pure) was purchased from Shanghai Macklin Biochemical Co., Ltd., Shanghai, China. Dibutyltin dilaurate (DBTDL) was purchased from Shanghai Aladdin Biochemical Technology Co., Ltd., Shanghai, China.

2.2. *Synthesis of HTPBs*

2.2.1. Synthesis of AHTPB via Living Anionic Polymerization

AHTPB was synthesized according to previous references, and the preparation process was divided into two main steps. The polymerization of 1,3-butadiene was first initiated with the monolithium initiator tBDMSOPrLi in a cyclohexane solution to obtain a mono-terminal hydroxy polybutadiene with a protecting group at the end group (tBDMSO-PB-OH). Then, the protective group was removed in a THF solution of TBAF to obtain hydroxyl-terminated polybutadiene. The preparation process was as follows: A certain amount of a 1,3-butadiene solution (15% cyclohexane solution) was added to a Schlenk tube under argon atmosphere, followed by 5 mL (2.50 mmol) of a tBDMSOPrLi solution (0.5 mol/L), and the reaction was stirred at 50 °C in an oil bath. After 5 h, 1 mL of EO was added to the reaction system as a terminator, and the reaction was continued for 2 h under argon atmosphere. Finally, 2 mL of methanol was added to the reaction system to stop the reaction, and the reaction solution was added dropwise to the methanol and left to precipitate to obtain a clear and viscous mono-terminal hydroxy polybutadiene. The reaction was stirred at room temperature for 24 h. Finally, the crude product obtained via the evaporation of THF was dissolved in DCM and extracted with water to remove impurities. Then, the final product, AHTPB, was obtained via anhydrous magnesium sulfate removal, filtration, and distillation under reduced pressure.

^{1}H NMR ($CDCl_3$, δ): 5.66–5.51 (-CH-C*H*=CH_2), 5.40 (-CH_2-C*H*=C*H*-CH_2-), 5.09–4.86 (-CH-CH=CH_2), 3.68 (-CH_2-OH), 2.06 (-CH_2-CH=CH-CH_2-), 1.55–1.17 (-CH-CH_2-).

2.2.2. Synthesis of RHTPB via ROMP

RHTPB was prepared using cis-1,4-butenediol as the chain transfer agent and 1,5-cyclooctadiene as the reaction monomer. The preparation process was as follows: COD (20 g, 5.71 mmol) and 1,4-butenediol (503.10 mg, 5.71 mmol) were added to a Schlenk tube. Then, 9 mL of toluene was added as a solvent and mixed well. The reaction system was chilled for three cycles, a Grubbs II catalyst (16.6 mg dissolved in 1 mL of toluene) was added during the second cycle with nitrogen gas, and the reaction was stirred in an oil bath at 40 °C for 12 h. After the reaction, the reaction solution was added dropwise to cold methanol and placed in a refrigerator to stand, and the crude product was obtained by removing the upper layer of methanol after it was completely layered. Finally, the crude product was dissolved using DCM, and RHTPB was obtained by distillation under reduced pressure.

^{1}H NMR ($CDCl_3$, δ): 5.78–5.52 (-CH=C*H*-CH_2-OH), 5.40 (-CH_2-C*H*=C*H*-CH_2-), 4.13 (-CH=CH-CH_2-OH), 2.31–1.81 (-CH_2-CH=CH-CH_2-).

2.3. *Synthesis of HTPB Triblock Copolymers*

2.3.1. Synthesis of FHTPB Triblock Copolymers (TBC_{FHTPB})

The polymerization reaction was carried out at a low temperature using a cationic ring-opening polymerization technique with THF and PO as the copolymerization monomers, FHTPB as the macromolecular initiator, $BF_3 \cdot OEt_2$ as the Lewis acid catalyst, and toluene as the solvent. The preparation process was as follows: First, azeotropic HTPB (30 g, 8.57 mmol) was placed in a round-bottom flask, 65 mL of toluene was added as a solvent, and the system was transferred to a low-temperature thermostatic reaction bath at 0 °C after the complete dissolution of FHTPB. THF (34.72 mL, 428.50 mmol) was added one time after 15 min using a syringe. After 15 min, the catalyst $BF_3 \cdot OEt_2$ (2.16 mL, 17.14 mmol) was added once using a syringe. After 30 min, the copolymer PO (3.00 mL, 42.85 mmol) was added drop by drop using a syringe, and the drop acceleration was controlled so that the PO was added uniformly over 40 min. After the drop addition, the reaction continued. The reaction was terminated by adding distilled water to the reaction system and left to stand. After stratification, the aqueous phase of the lower layer was aspirated out with a syringe, and the upper layer was distilled under reduced pressure to remove toluene and the small amount of remaining water. The crude product was obtained via distillation under reduced pressure until the system no longer bubbled. The crude product was dissolved with dichloromethane, poured into a beaker, washed with an equal volume of distilled water until the product was washed to neutral, and left to stand and stratify. The upper layer of water was removed, and the small amount of remaining water was removed by adding anhydrous magnesium sulfate and filtering to remove the magnesium sulfate. The filtrate was distilled under reduced pressure until no longer bubbling to obtain a colorless or light yellow product.

^{1}H NMR ($CDCl_3$, δ): 5.55 (-CH-C*H*=CH_2), 5.47–5.26 (-CH_2-C*H*=C*H*-CH_2-), 5.02–4.87 (-CH-CH=CH_2), 4.21–3.81 (-C*H*-CH=CH_2), 3.72–3.13 (-CH_2-O-), 2.04 (-CH_2-CH=CH-), 1.70–1.53 (-CH_2-), 1.52–1.18 (-CH-CH_2-), 1.11 (-CH_3).

2.3.2. Synthesis of AHTPB Triblock Copolymers (TBC_{AHTPB})

TBC_{AHTPB} was prepared via the cationic ring-opening polymerization of AHTPB with THF and PO, and its synthesis and post-processing methods were similar to those of TBC_{FHTPB}.

^{1}H NMR ($CDCl_3$, δ): 5.66–5.50 (-CH-C*H*=CH_2), 5.38 (-CH_2-C*H*=C*H*-CH_2-), 5.05–4.85 (-CH-CH=CH_2), 3.74–3.15 (-C*H*-CH=CH_2, -CH_2-O-), 2.05 (-CH_2-CH=CH-), 1.75–1.51 (-CH_2-), 1.50–1.18 (-CH-CH_2-), 1.13 (-CH_3).

2.3.3. Synthesis of RHTPB Triblock Copolymers (TBC_{RHTPB})

TBC_{RHTPB} was prepared via cationic ring-opening polymerization between RHTPB, THF, and PO, and its synthesis and post-processing methods were similar to those of TBC_{FHTPB}.

^{1}H NMR ($CDCl_3$, δ): 5.76–5.51 (-C*H*=C*H*-CH_2-O-), 5.50–5.13 (-CH_2-C*H*=C*H*-CH_2-), 4.21–3.81 (-CH=CH-CH_2-O-), 3.62–3.24 (-CH_2-O-), 2.16–1.95 (-CH_2-CH=CH-CH_2-), 1.62 (-CH_2-), 1.11 (-CH_3).

2.4. Synthesis of Polyurethane Elastomer using HTPB Triblock Copolymers

TBC_{FHTPB}, TBC_{AHTPB}, and TBC_{RHTPB} were used as soft segments to synthesize PU-1, PU-2, and PU-3, respectively. All the polyurethane elastomers were prepared using a similar strategy. Take the synthesis of PU-1 as an example: TBC_{FHTPB} (2.53 g, 0.50 mmol) was placed in a round-bottom flask and 20 mL of toluene was added as a solvent. Then, TDI (0.28 g, 1.50 mmol) was added, and the reaction system was placed in an oil bath at 60 °C and stirred. After mixing well, 2 drops of a dibutyltin dilaurate catalyst were added, and the stirring of the reaction in the oil bath continued. After 6 h of stirring, BDO (90.4 mg, 1.00 mmol) was added to the reaction system, and the temperature of the oil bath was raised to 80 °C. After 12 h, the reaction was stopped, and the reaction solution was concentrated and poured into a polytetrafluoroethylene mold, which was cured to form a film to obtain the polyurethane elastomer PU-1.

2.5. Structural Characterization of HTPBs and HTPB Triblock Copolymers

2.5.1. FT-IR Test

The FT-IR spectra of all samples were tested using a Nicolet iS10 FT-IR instrument (Nicolet Instrument Corporation, Waltham, MA, USA). The samples were prepared via the coating method, and the CH_2Cl_2 solution of the samples was uniformly coated on KBr, dried, and tested. The scanning area was from 4000 cm^{-1} to 500 cm^{-1}.

The contents of the 1,4 unit and 1,2 unit microstructures in the HTPB backbones were calculated based on the FT-IR spectra using the following equations:

$$Cc\% = \frac{D_{724cm^{-1}} / Kc}{D_{724cm^{-1}} / Kc + D_{911cm^{-1}} / Kv + D_{966cm^{-1}} / Kt} \times 100\%$$

$$Cv\% = \frac{D_{724cm^{-1}} / Kv}{D_{724cm^{-1}} / Kc + D_{911cm^{-1}} / Kv + D_{966cm^{-1}} / Kt} \times 100\%$$

$$Ct\% = \frac{D_{724cm^{-1}} / Kt}{D_{724cm^{-1}} / Kc + D_{911cm^{-1}} / Kv + D_{966cm^{-1}} / Kt} \times 100\%$$

where $C_c\%$, $C_v\%$, and $C_t\%$ are the contents of the cis-1,4 unit; 1,2 unit; and trans-1,4 unit of the microstructures in the HTPB backbone. The D values are the absorbance values of the three peaks (724 cm^{-1}, 911 cm^{-1}, and 967 cm^{-1}) corresponding to the three different units of the HTPB backbone. The K values are the absorbance coefficients (K_c: 31.4 mol^{-1} cm^{-1}, K_v: 151.0 mol^{-1} cm^{-1}, and K_t: 117.0 mol^{-1} cm^{-1}) of the three peaks corresponding to the three different units contained in the HTPBs [24].

2.5.2. NMR Test

The ^{1}H NMR spectra and ^{13}C NMR spectra of all samples were tested using a Bruker Avance-400 NMR instrument (Bruker Corporation, Berlin, Germany). Deuterated chloroform ($CDCl_3$) was used as a solvent, and tetramethylsilane (TMS) was used as an internal reference.

2.5.3. SEC-MALLS Test

The number-average molecular weight and polydispersity indexes of all samples were determined using a DAWN EOS-type gel permeation chromatography-multi-angle laser light scattering coupler (SEC-MALLS, Wyatt, Milford, MA, USA). The chromatographic column type was 5 µm 500 Å MZ-Gel SDplus 103 Å (300 × 8.0 mm). HPLC-grade THF was used as an eluent, the flow rate was 0.5 mL min-1, and the temperature was 25 °C.

2.6. Comparative Study of the Properties of the Triblock Copolymers and Polyurethane Elastomers

2.6.1. DSC Test

The glass transition, crystalline behavior, and melting behavior of all samples were studied via DSC measurements (TA DSC250, TA Instruments, New Castle, DE, USA). The test was performed in the presence of a N_2 atmosphere in a temperature range from −120 to 100 °C with a ramp rate of 10 °C/min. The temperature was increased to 150 °C to eliminate the thermal history of the samples before the test. Later, the temperature was reduced to −120 °C and held for 10 min.

2.6.2. TGA Test

The thermal stability of all samples was studied via TGA (PerkinElmer, STA-8000, Waltham, MA, USA). The test was performed in the presence of a N_2 atmosphere in a temperature range from 25 to 100 °C with a ramp rate of 10 °C/min.

2.6.3. Viscosity Test

The viscosities of all samples were determined using a CAP2000+ Brookfield viscometer (Brookfield Corporation, New York, NY, USA). The viscometer rotated at 5.0 rpm with a temperature range of 20 to 65 °C and a temperature-controlled accuracy of ±0.2 °C.

2.6.4. Tensile Test

Stress–strain curves were measured using a universal testing machine (High-speed machine VWN-20) (Kunshan Keruite Testing Instruments Corporation, Kunshan, China) at room temperature with a tensile speed of 100 mm/min.

2.6.5. XRD Test

Measurements were made at 40 kV and 40 mA at room temperature with a measurement speed of 10 (°)/min and a scanning angle range of 10° to 80°.

2.6.6. XPS Test

The elemental contents of the PU surfaces were measured via X-ray photoelectron spectroscopy (XPS) using an Axis Ultra Kratos instrument (Kratos Analytical Ltd., Wharfside, Manchester, UK).

2.6.7. Contact Angle Test

The wetting properties of PU films were studied using water contact angle measurements.

3. Results and Discussion

The synthetic routes of the polyether-modified HTPB triblock copolymers is shown in Figure 1. The synthetic strategy consisted of two main steps. First, we prepared three triblock copolymers (TBC_{FHTPB}, TBC_{AHTPB}, and TBC_{RHTPB}) via cationic ring-opening polymerization using FHTPB, AHTPB, and RHTPB as macroinitiators, respectively. Second, the corresponding polyurethane elastomers, named PU-1, PU-2, and PU-3, were prepared using the HTPB triblock copolymers as the soft segment, TDI as the hard segment, and butanediol (BDO) as the chain extender.

3.1. Synthesis and Characterization of TBC_{FHTPB}

The ^{1}H NMR spectrum of TBC_{FHTPB} is shown in Figure 2d. There were peaks assigned to the methylene groups attached to the ether bond in TBC_{FHTPB} (δ = 3.72–3.13), the methylene groups in THF (δ = 1.62), and the methyl groups in PO (δ = 1.11), indicating the successful ring opening of THF with PO and confirming the successful synthesis of TBC_{FHTPB}. Compared to the ^{13}C NMR spectrum of pure FHTPB (Figure 2e), the spectrum for TBC_{FHTPB} had peaks assigned to the α-C and β-C in the copolyether segment at 70 ppm and 26 ppm, respectively, which further confirmed the successful synthesis of the desired triblock copolymer (as shown in Figure 2f and Table 1). The FT-IR spectrum of TBC_{FHTPB} is shown in Figure 2a, and vibrational absorption peaks from the double bonds (C=C) in the trans-1,4 isomer; cis-1,4 isomer; and 1,2 isomer were seen at 966 cm^{-1}, 724 cm^{-1}, and 911 cm^{-1}, respectively. Compared to the FT-IR spectrum of FHTPB, the TBC_{FHTPB} spectrum also showed a clear absorption peak characteristic of the stretching vibrations of the ether bond at 1110 cm^{-1}, which further confirmed the successful ring-opening polymerization reaction. In addition, the SEC trace of TBC_{FHTPB} (Figure 2b) showed a single peak that was consistent with a normal distribution, indicating that the final product contained a single homogeneous polymer. An analysis of the SEC trace gave M_n = 5000 g mol^{-1} and M_w/M_n = 1.63.

Table 1. Assignment of the ^{13}C NMR signals.

Peak	Chemical Shift (ppm)	Carbon
1	142.78	$-\overset{*}{C}H=CH_2$
2	131.37	$-CH_2-CH=\overset{*}{C}H-CH_2-$
3, 4	130.49–130.74, 128.78–130.32	$-CH_2-\overset{*}{C}H=\overset{*}{C}H-CH_2-$
5	128.39–129.78	$-CH_2-\overset{*}{C}H=CH-CH_2-$
6	114.44	$-CH=\overset{*}{C}H_2$
7	43.56	$-CH_2-\overset{*}{C}H-$
8	38.29	$-\overset{*}{C}H_2-CH=CH-CH_2-$
9	34.15	$-\overset{*}{C}H_2-CH(C_2H_4)-\overset{*}{C}H_2-CH=$
10, 12	32.71, 27.44	$-\overset{*}{C}H_2-CH=CH-\overset{*}{C}H_2-$
11, 13	30.15, 24.95	$-CH_2-CH=CH-\overset{*}{C}H_2-$
14, 15	132.59, 130.22	$-CH_2-CH=\overset{*}{C}H-CH_2-OH$
16, 17	63.87, 58.61	$-CH_2-CH=CH-\overset{*}{C}H_2-OH$
18	74.39–74.85	$-CH_2-O-\overset{*}{C}H-CH_3$
19	70.71–71.28	$-O-\overset{*}{C}H_2-CH_2-CH_2-\overset{*}{C}H_2-O-$
20	69.07	$-\overset{*}{C}H_2-O-CH-CH_3$
21	26.52–26.61	$-O-CH_2-\overset{*}{C}H_2-\overset{*}{C}H_2-CH_2-O-$
22	17.48	$-CH_2-O-CH-\overset{*}{C}H_3$

3.2. Synthesis and Characterization of AHTPB and TBC_{AHTPB}

The NMR spectrum of AHTPB is shown in Figure 3c. The peaks at δ = 5.66–5.51 were assigned to the hydrogens on the double-bonded carbon attached to the hydroxyl group of 1,2 units, the peak at δ = 5.40 was from the protons on the double-bonded carbon in the 1,4 units, the peak at δ = 5.09–4.86 was from the protons on the double-bonded terminal carbon of the 1,2 units, the peak at δ = 3.68 was from the protons on the methylene group attached to the terminal hydroxyl group, the peak at δ = 2.06 was from the protons from the methylene groups of the 1,4 units, and the peaks at δ = 1.55–1.17 were from the protons on the methylene group of the 1,2 units. The presence of these peaks in the ^1H NMR spectrum confirmed that AHTPB was successfully synthesized. In addition, the SEC trace of AHTPB showed a single peak with M_n = 3500 g mol^{-1} and M_w/M_n = 1.03 (Figure 3b). Similar to the results for TBC_{FHTPB}, proton peaks from the methylene group attached to the ether bond (δ = 3.74–3.15), peaks from the protons in the methylene group in THF (δ = 1.75–1.51), and peaks from the protons in the methyl group in PO (δ = 1.13) were also present in the ^1H NMR spectrum of TBC_{AHTPB} (Figure 3d). Peaks corresponding to the α-C and β-C in the polyether segments at 70 ppm and 26 ppm, respectively, were also observed in the ^{13}C NMR spectrum of TBC_{AHTPB}, confirming that TBC_{AHTPB} was successfully synthesized (Figure 3f and Table 1). Furthermore, in addition to the vibrational absorption peaks corresponding to trans-1,4 units; cis-1,4 units; and 1,2 units at 966 cm^{-1}, 724 cm^{-1}, and 911 cm^{-1}, respectively, the FT-IR spectrum of TBC_{AHTPB} had a clear vibrational absorption peak corresponding to the ether bond stretching at 1114 cm^{-1}, which further confirmed the completion of the ring-opening polymerization reaction (Figure 3a). The SEC trace of TBC_{AHTPB} showed a single sharp peak with M_n = 4800 g mol^{-1} and M_w/M_n = 1.11 (Figure 3b). Notably, the molecular weight distribution of TBC_{AHTPB} was narrower than that of TBC_{FHTPB}, which was attributed to the fact that the molecular weight distribution of the triblock copolymers was mainly inherited from that of HTPBs, and the molecular weight distribution of AHTPB was narrower than that of FHTPB due to the different polymerization mechanisms.

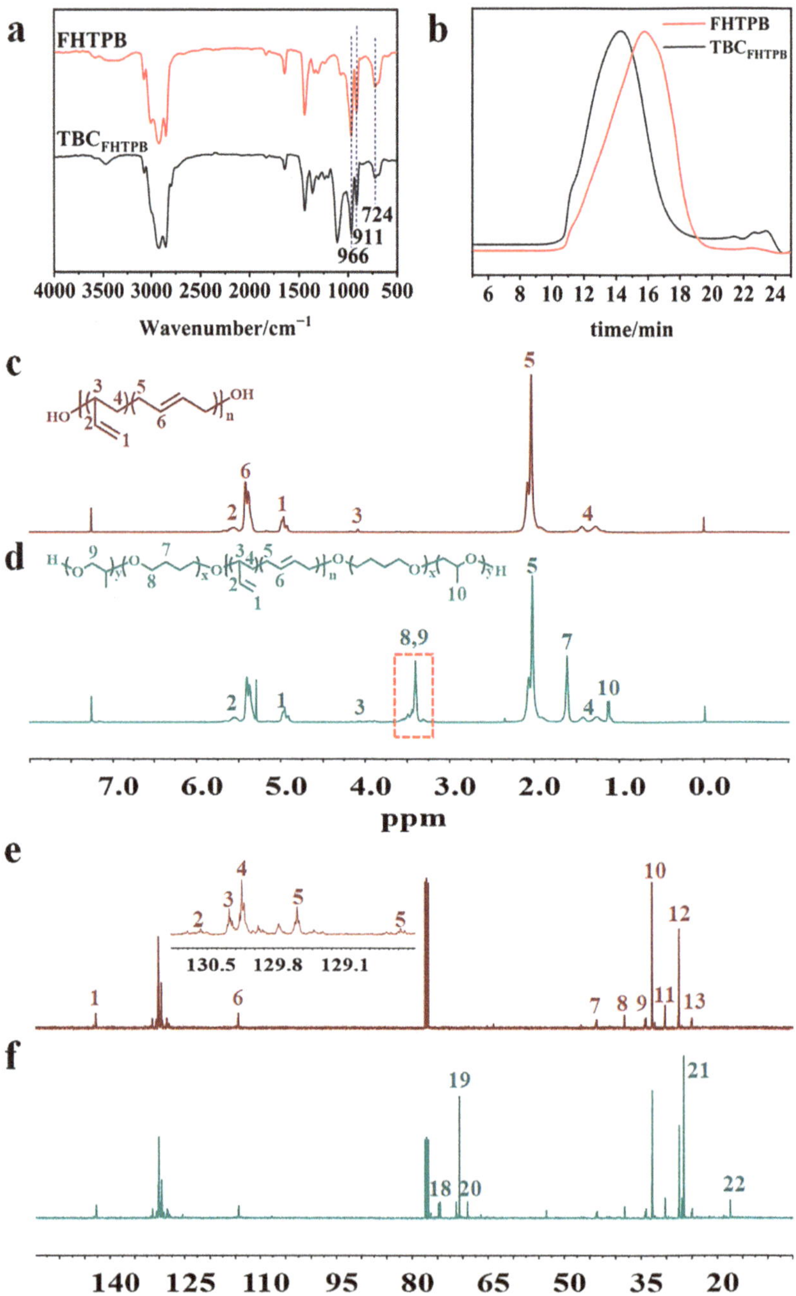

Figure 2. FT-IR spectra of FHTPB and TBC_{FHTPB} (**a**), SEC traces of FHTPB and TBC_{FHTPB} (**b**), ^{1}H NMR of FHTPB (**c**) and TBC_{FHTPB} (**d**), and ^{13}C NMR of FHTPB (**e**) and TBC_{FHTPB} (**f**).

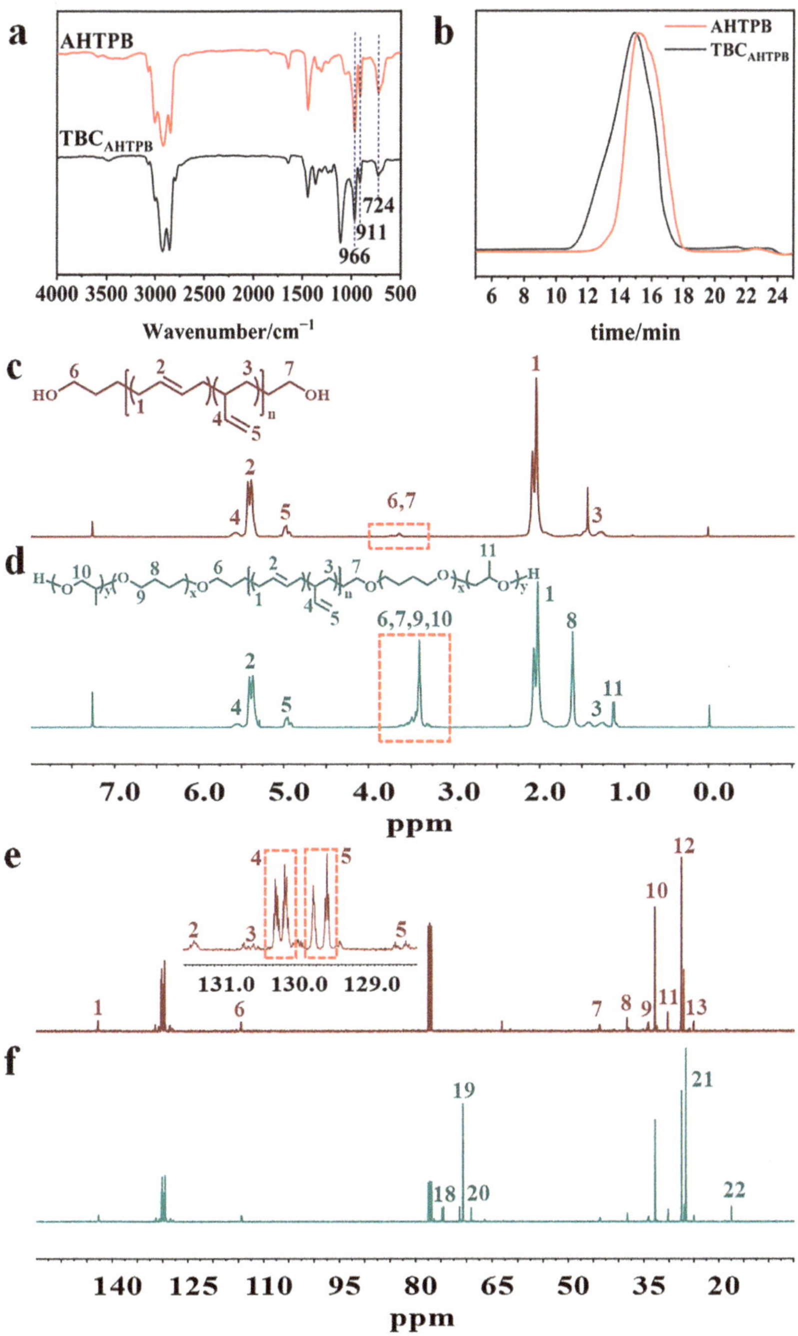

Figure 3. FT-IR spectra of AHTPB and TBC_{AHTPB} (**a**), SEC traces of AHTPB and TBC_{AHTPB} (**b**), ^{1}H NMR of AHTPB (**c**) and TBC_{AHTPB} (**d**), and ^{13}C NMR of AHTPB (**e**) and TBC_{AHTPB} (**f**).

3.3. Synthesis and Characterization of RHTPB and TBC_{RHTPB}

The ^{1}H NMR spectrum of RHTPB is shown in Figure 4c. The peaks at δ = 5.78–5.52 were from the protons on the double-bonded carbon attached to the hydroxymethyl groups, the peak at δ = 5.40 was the proton peak from the double bonds in the 1,4 units, the peak at δ = 4.13 was the proton peak of the methylene groups attached to the terminal hydroxyl groups, and the peaks at δ = 2.31–1.81 were the proton peaks from the methylene groups in the 1,4 units. The ^{1}H NMR spectrum confirmed that RHTPB was successfully synthesized. The SEC trace of RHTPB also contained a single peak with M_n = 3500 g mol^{-1} and M_w/M_n = 1.47 (Figure 4b). Similar to the two previous triblock copolymers, the peaks from the methylene groups attached to the ether bonds (δ = 3.62–3.24), the methylene groups in THF (δ = 1.62), and the methyl groups in PO (δ = 1.11) were also observed in the ^{1}H NMR spectrum of TBC_{RHTPB} (Figure 4d). In addition, in the ^{13}C NMR spectrum of TBC_{RHTPB}, the expected peaks from the α-C and β-C in the polyether segment were also observed at 70 ppm and 26 ppm, respectively (Figure 4f and Table 1). In the FT-IR spectrum of TBC_{RHTPB}, the characteristic absorption peak of the ether bond at 1112 cm^{-1} confirmed the completion of the ring-opening polymerization reaction (Figure 4a). Significantly, the FT-IR spectrum of RHTPB only contained absorption peaks corresponding to trans-1,4 units and cis-1,4 units at 966 cm^{-1} and 724 cm^{-1}, respectively, and no absorption peaks for 1,2 units at 911 cm^{-1} were observed. The difference in microstructure was attributed to the mechanism of RHTPB synthesis, resulting in the absence of 1,2 units. According to the SEC traces, TBC_{RHTPB} contained a single structure with M_n = 4800 g mol^{-1} and M_w/M_n = 1.62 (Figure 4b). All the above results confirmed that TBC_{RHTPB} was successfully synthesized.

3.4. Viscosity of the HTPBs and Their Triblock Copolymers

The overall content of 1,2 units in the HTPB affects the viscosity of the polymers. The viscosities of the HTPBs and the corresponding triblock copolymers at different temperatures are shown in Figure 5. The viscosities of HTPBs and their triblock copolymers decreased gradually with increasing temperature. At a given temperature, the viscosity of FHTPB was the highest, followed by AHTPB, and RHTPB had the lowest viscosity. FHTPB had the highest 1,2 isomeric content, at 30.1%, which was consistent with this polymer having the highest viscosity, while RHTPB contained no 1,2 units and therefore had the lowest viscosity (the isomeric content of each HTPB microstructure is shown in Table 2). The viscosities of the triblock copolymers showed the same trend as the corresponding HTPBs but were significantly higher than the HTPBs. The higher viscosities were attributed to the longer molecular chains and higher molecular weights of the triblock copolymers, which restricted the movements of the chain segments.

Table 2. Microstructures of the FHTPB, AHTPB, and RHTPB.

Polymer	M_n/g mol^{-1}	M_w/M_n	% of Microstructure Determined via FT-IR		
			Cis-1,4	1,2	Trans-1,4
FHTPB	3400	1.67	26.2	30.1	43.7
AHTPB	3500	1.03	58.9	15.7	25.4
RHTPB	3500	1.47	78.0	0	22.0

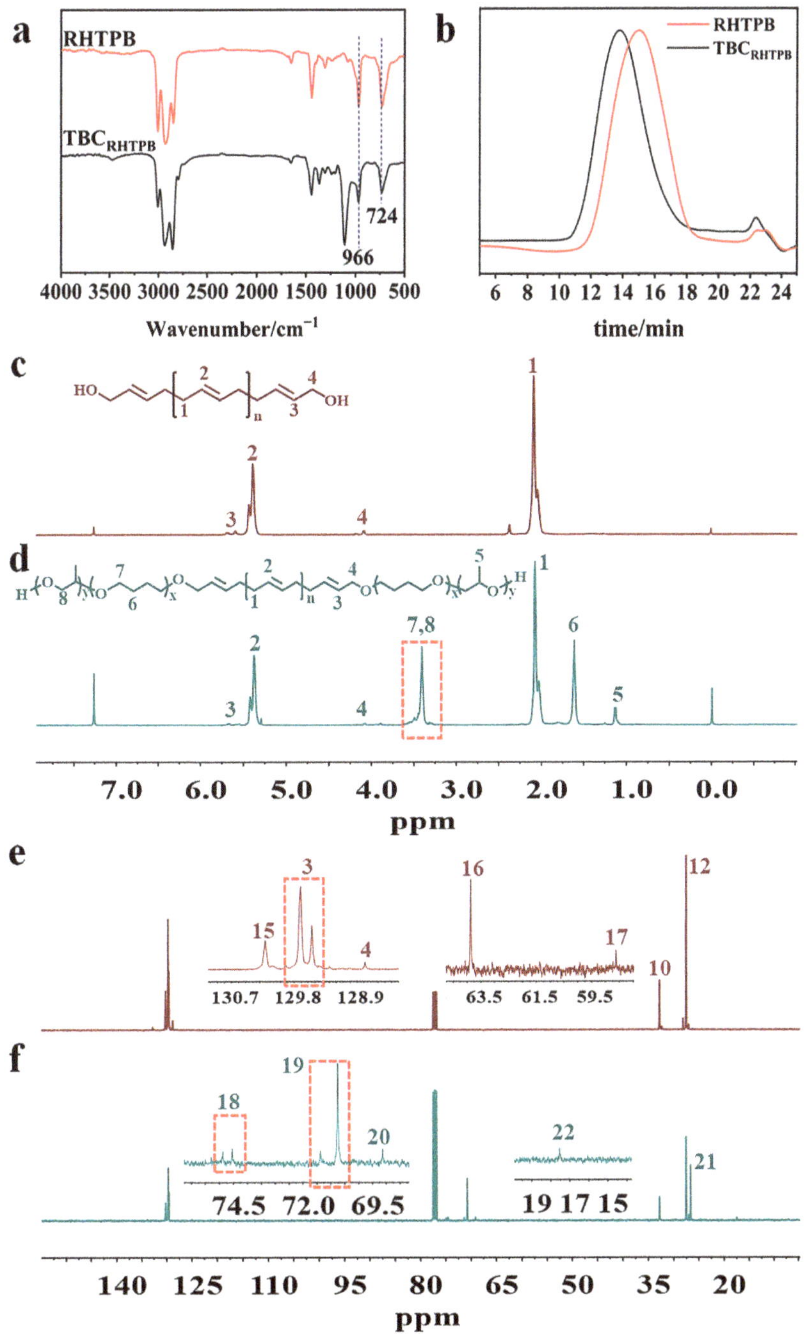

Figure 4. FT-IR spectra of RHTPB and TBC$_{RHTPB}$ (**a**), SEC traces of RHTPB and TBC$_{RHTPB}$ (**b**), ^{1}H NMR of RHTPB (**c**) and TBC$_{RHTPB}$ (**d**), and ^{13}C NMR of RHTPB (**e**) and TBC$_{RHTPB}$ (**f**).

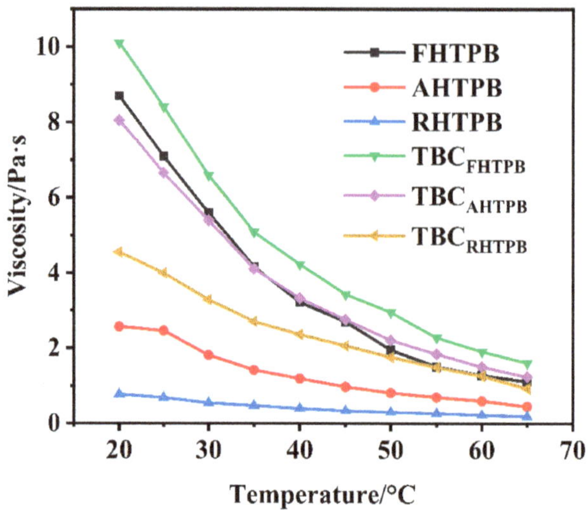

Figure 5. Viscosities of the FHTPB, AHTPB, RHTPB, TBC_{FHTPB}, TBC_{AHTPB}, and TBC_{RHTPB} as a function of temperature.

3.5. Thermal Properties of the HTPBs and Their Triblock Copolymers

The DSC heating curves of PTHF, the HTPBs, and the HTPB triblock copolymers are shown in Figure 6. FHTPB, AHTPB, and RHTPB are amorphous polymers with glass transition temperatures of −79.6 °C, −88.3 °C, and −110.3 °C, respectively. The DSC results showed that T_g gradually decreased as the 1,2 content decreased. The T_g values of TBC_{FHTPB}, TBC_{AHTPB}, and TBC_{RHTPB} were −81.8 °C, −87.6 °C, and −105.6 °C, respectively. The measured T_g values of the HTPB triblock copolymers were similar to the HTPBs, indicating that the triblock copolymers inherited the T_g values of the corresponding HTPBs. In addition, the DSC curves of TBC_{FHTPB}, TBC_{AHTPB}, and TBC_{RHTPB} showed weak melting peaks at −8.3 °C, −18.2 °C, and −36.6 °C, respectively, due to the crystallization of the polyether segments. Remarkably, the melting temperatures (T_m) of the triblock copolymers were substantially lower than that of PTHF (T_m = 28.4 °C). The lower T_m indicated that the crystallization of the polyether segments in the triblock copolymers was constrained by the methyl side groups in the PO units, which disrupted the chain segment ordering and prevented the formation of crystal nuclei. Furthermore, the T_g values of TBC_{FHTPB}, TBC_{AHTPB}, and TBC_{RHTPB} decreased in that order, indicating that the amorphous PB chain segments effectively inhibited the crystallization of the polyether chain segments, and the effects of the HTPB segment on the crystallization increased as the 1,2 content decreased.

The TGA curves of the HTPBs and the triblock copolymers are shown in Figure 7. The TGA curves showed that the thermal decomposition temperatures of FHTPB, AHTPB, and RHTPB were similar and were all higher than those of the triblock copolymers. The decrease in decomposition temperature was because the polyether blocks in the triblock copolymers had a lower thermal decomposition temperature than the HTPB blocks, thus leading to a lower thermal decomposition temperature for the triblock copolymers. The thermal decomposition temperatures of TBC_{FHTPB}, TBC_{AHTPB}, and TBC_{RHTPB} were all approximately 357 °C, indicating that the HTPB microstructure did not affect the thermal stability of the corresponding triblock copolymers.

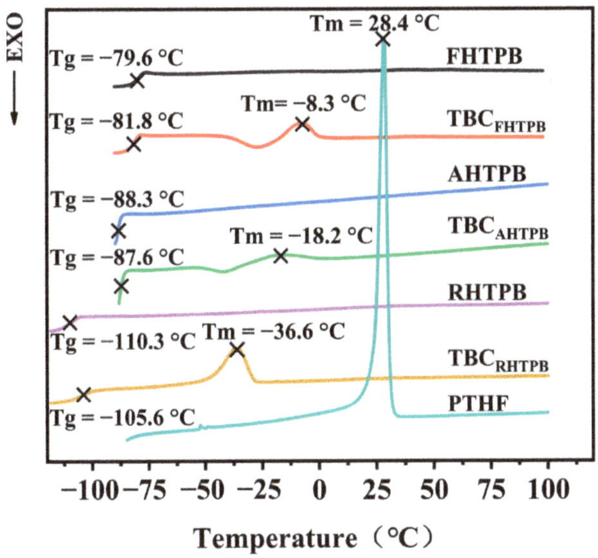

Figure 6. DSC heating curves of the PTHF, FHTPB, AHTPB, RHTPB, TBC_{FHTPB}, TBC_{AHTPB}, and TBC_{RHTPB}.

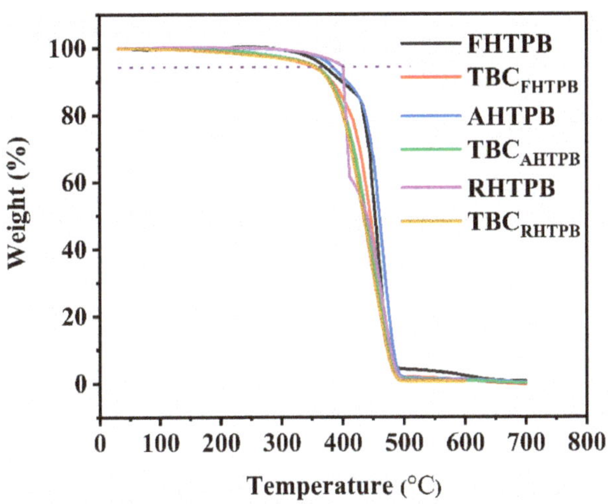

Figure 7. TGA curves of the FHTPB, AHTPB, RHTPB, TBC_{FHTPB}, TBC_{AHTPB}, and TBC_{RHTPB}.

3.6. Property Comparison of Polyurethane Elastomers

3.6.1. Tensile Properties

In order to investigate the effect of the HTPB microstructure on the performance of polyurethane elastomers prepared from the HTPB-based triblock copolymers, the breakage strength and elongation at break of the three polyurethane elastomers were tested, and the test results are shown in Figure 8. The results showed that the tensile strengths of PU-1, PU-2, and PU-3 decreased as the trans-1,4 content deceased, and their tensile strengths were 1.44 MPa, 1.18 MPa, and 0.68 MPa, respectively. Conversely, the elongations at break of PU-1, PU-2, and PU-3 increased as the cis-1,4 content increased and were 405%, 473%, and 516%, respectively. These trends showed that a high cis-1,4 content in the HTPB increased the elongation at break of the material, while a high trans-1,4 content increased the tensile strength of the material.

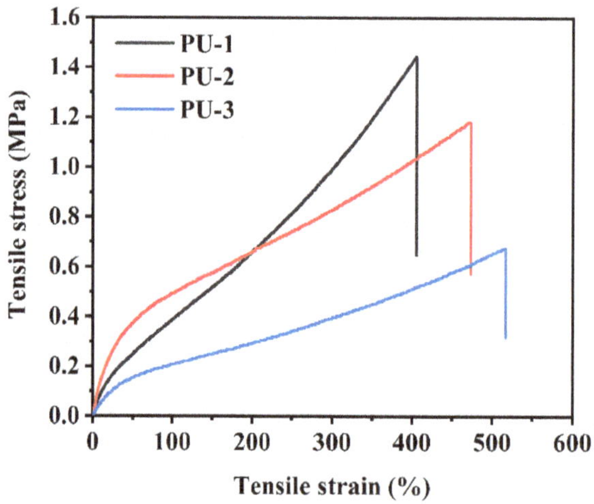

Figure 8. Stress–strain curves of PU-1, PU-2, and PU-3.

3.6.2. XRD Analysis

The XRD curves of the polyurethane elastomers are shown in Figure 9. A broad diffraction peak was seen at $2\theta = 19°\sim21°$ in the curves of PU-1, PU-2, and PU-3, which was from the hard crystalline segments in the PUs. The diffraction peaks were attributed to the structurally ordered regions of the PUs, namely, the hard segments with ordered hydrogen bonding interactions. In addition, a weaker diffraction peak at $2\theta = 39°\sim41°$ was seen in the curves of PU-1, PU -2, and PU-3 and was attributed to the crystallization of the polyether block in the soft segments of the PUs. It was observed that PU-1 was more crystalline than PU-2 and PU-3, which might have been because the HTPB microstructure affected the flexibility of the chains.

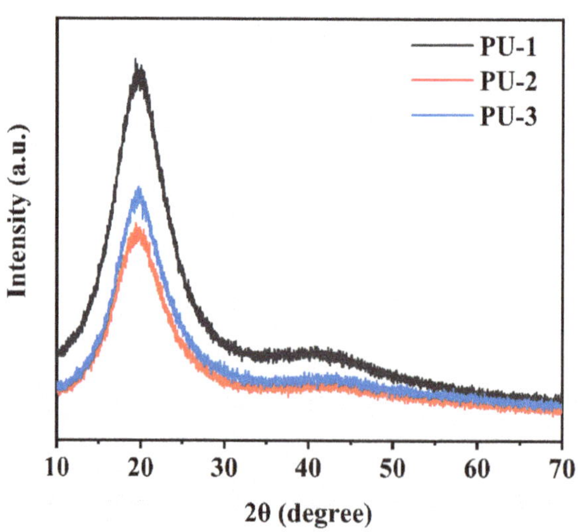

Figure 9. XRD curves of PU-1, PU-2, and PU-3.

3.6.3. Surface Characterization of Polyurethane Elastomers

The water contact angles on the PU-1, PU-2, and PU-3 surfaces are shown in Figure 10 and were found to be 83.6°, 81.9°, and 84.3°, respectively. It can be seen that the contact angles of the three PUs did not differ much, indicating that the microstructural differences of HTPB do not have much effect on the surface hydrophilicity/hydrophobicity of the PUs, and it also shows that the ratio of cis/trans isomers in HTPB does not have much effect on its own polarity. Furthermore, the XPS results clearly showed that the C, N, and O elemental contents on the surfaces of PU-1, PU-2, and PU-3 were similar, also confirming that the HTPB microstructure had little effect on the surface structure of the PUs (Figure 11).

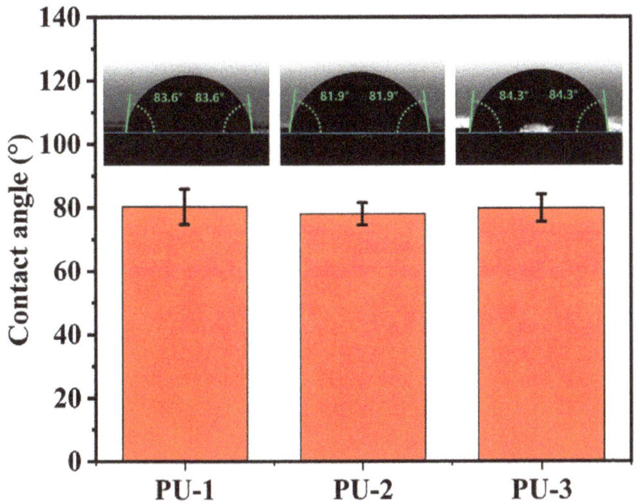

Figure 10. Contact angle test results of PU-1, PU-2, and PU-3.

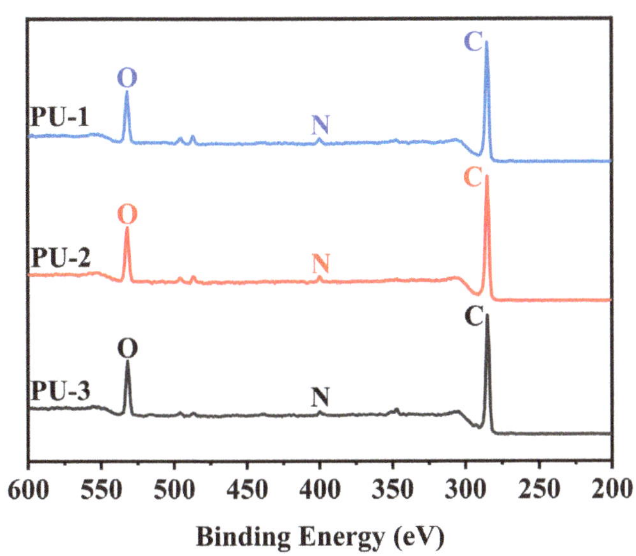

Figure 11. XPS test results of PU-1, PU-2, and PU-3.

4. Conclusions

We successfully synthesized AHTPB and RHTPB via reactive anion polymerization and ring-opening translocation polymerization (ROMP), respectively. To investigate the effects of different HTPB isomers on the performance of polyether-modified triblock copolymers and their elastomers, corresponding triblock copolymers, TBC_{FHTPB}, TBC_{AHTPB}, and TBC_{RHTPB}, were synthesized via a cationic ring-opening copolymerization reaction using FHTPB, AHTPB, and RHTPB as chain transfer agents and THF and PO as copolymer monomers, respectively. The three triblock copolymers were then used to prepare three polyurethane elastomers (PU-1, PU-2, and PU-3), using the polyether-modified triblock copolymers as soft segments and TDI as hard segments. The viscosities of HTPBs and their corresponding triblock copolymers all decreased as the 1,2 unit contents decreased. In parallel, the T_g values of the HTPBs gradually decreased as the contents of 1,2 units decreased and the contents of cis-1,4 units increased. The measured T_g values of the corresponding triblock copolymers followed the same trend. The T_g of RHTPB was −110.3 °C, and that of TBC_{RHTPB} was −105.6 °C, indicating that the introduction of polyether segments did not affect the low-temperature properties of HTPBs. In addition, TGA tests showed that the HTPB microstructure did not affect the thermal stability of the triblock copolymers, and all three triblock copolymers had a decomposition temperature of 357 °C. Studies of the polyurethane elastomer performance showed that PU-1 had the maximum tensile strength of 1.44 MPa, while PU-3 had the maximum elongation at break of 516%. The differences in the polyurethane performance were mainly attributed to the lower content of cis-1,4 units in FHTPB compared to AHTPB and RHTPB. Contact angle measurements showed that the contact angles on PU-1, PU-2, and PU-3 were similar and were around 80°. These results suggested that the microstructural content of the HTPB block did not affect the surface properties of the PUs. In conclusion, the present study clarified the effects of the different HTPB microstructures on the performances of the corresponding polyether-modified triblock copolymers and PUs through a series of comparative performance studies and provides data to support the development and application of polyether-modified HTPB block copolymers. The microstructures of polymers are key factors affecting the properties of materials. We believe the microstructure of a polymer can be a precise characterization with the development of science and technology. Meanwhile, we can obtain polymer materials with better performance by manipulating the microstructures of polymers.

Author Contributions: Conceptualization, X.Z.; Methodology, P.B.; Software, P.B.; Validation, W.Z.; Formal analysis, L.T.; Resources, W.Z.; Data curation, J.H.; Writing–original draft, P.B.; Writing–review & editing, P.B.; Funding acquisition, X.Z. All authors have read and agreed to the published version of the manuscript.

Funding: This work was supported by the Natural Science Foundation of Shandong Province (No. ZR2021QE137); the Project Supported by the Foundation of Key Laboratory of Pulp and Paper Science and Technology of Ministry of Education of China (No. KF201922); the major scientific and technological innovation projects of Shandong Province (2019JZZY020230); the Open Foundation of Key Laboratory of Auxiliary Chemistry and Technology for Chemical Industry, Ministry of Education, Shaanxi University of Science and Technology and Shaanxi Collaborative Innovation Center of Industrial Auxiliary Chemistry and Technology, Shaanxi University of Science and Technology (No. KFKT2022-01); and the Natural Science Foundation of Shaanxi Province (No. 2023-JC-YB-354).

Institutional Review Board Statement: Not applicable.

Data Availability Statement: Not applicable.

Conflicts of Interest: The authors declare no conflict of interest.

References

1. Selvakumar, S.; Rao, G.S.; Reddy, K.A. Diffusion of Labile Chemical Species in HTPB and HTPB-XT Solid Propellants and Its Effect over Solid Rocket Motor Properties on Aging—A Study. *Propell. Explos. Pyrot.* **2021**, *46*, 782–790.
2. Chaturvedi, S.; Dave, P.N. Solid propellants: AP/HTPB composite propellants. *Arab. J. Chem.* **2019**, *12*, 2061–2068.

3. Li, J.; Ning, Z.; Yang, W.; Yang, B.; Zeng, Y. Hydroxyl-Terminated Polybutadiene-Based Polyurethane with Self-Healing and Reprocessing Capabilities. *ACS Omega* **2022**, *7*, 10156–10166. [CrossRef]
4. Ma, L.; Zhu, X.; Zhang, W.; Zhang, H.; Wang, J.; Qu, J. Study on the preparation and performance comparison of side-chain hydroxyl-terminated polybutadiene derivatives with narrowly molecular weight distribution used for polyurethane. *Polym. Test.* **2021**, *104*, 107389.
5. Sikder, B.K.; Jana, T. Effect of Solvent and Functionality on the Physical Properties of Hydroxyl-Terminated Polybutadiene (HTPB)-Based Polyurethane. *ACS Omega* **2018**, *3*, 3004–3013. [CrossRef] [PubMed]
6. Hosseini, S.R.; Nikje, M.M.A. Synthesis and characterization of novel epoxy-urethane coating and its graphene nanocomposites. *Polym. Compos.* **2023**, *44*, 2794–2803. [CrossRef]
7. Zhang, W.; Zhang, T.; Liu, H.; Zheng, Y.; Zhong, Y.; Wang, G.; Zhu, Q.; Liu, X.; Zhang, L.; Li, H. Synthesis and characterization of a novel hydroxy telechelic polyfluoroether to enhance the properties of HTPB solid propellant binders. *Colloids Surf. A* **2022**, *648*, 129199.
8. Zhang, P.; Tan, W.; Zhang, X.; Chen, J.; Yuan, J.; Deng, J. Chemical Modification of Hydroxyl-Terminated Polybutadiene and Its Application in Composite Propellants. *Ind. Eng. Chem. Res.* **2021**, *60*, 3819–3829.
9. Yuan, B.; Wang, G.; Tian, W.; Zhou, L.; Li, C. Fabrication of Hydroxy-Terminated Polybutadiene with Piezoelectric Property by Functionalized Branch Chain Modification. *Molecules* **2023**, *28*, 1810.
10. Zhang, Y.; Zheng, J.; Zhang, X.; Du, Y.; Li, K.; Yu, G.; Jia, Y.; Liu, Y. Effect of chemical copolymerization and mixed chain extenders on mechanical properties of HTPB polyurethane. *IOP Conf. Ser. Mater. Sci. Eng.* **2021**, *1167*, 012012.
11. Liang, J.; Nie, J.; Zhang, H.; Guo, X.; Yan, S.; Han, M. Interaction Mechanism of Composite Propellant Components under Heating Conditions. *Polymers* **2023**, *15*, 2485. [CrossRef]
12. Liu, J.; Yang, D.; Li, S.; Bai, C.; Tu, C.; Zhu, F.; Xin, W.; Li, G.; Luo, Y. Synthesis and characterization of hydroxyl-terminated polybutadiene modified low temperature adaptive self-matting waterborne polyurethane. *RSC Adv.* **2023**, *13*, 7020–7029. [PubMed]
13. Lemos, M.F.; Mendes, L.C.; Bohn, M.A. On the functionalization and characterization of hydroxyl-terminated polybutadiene with octyl-1-azide and the evaluation of polyurethane elastomers based on such modified HTPB. *J. Appl. Polym. Sci.* **2021**, *138*, e49981. [CrossRef]
14. Kumar, D.; Mohammad, S.A.; Kumar, A.; Mane, S.R.; Banerjee, S. An amino acid-derived ABCBA-type antifouling biohybrid with multi-stimuli responsivity and contaminant removal capability. *Polym. Chem.* **2022**, *13*, 1960–1969.
15. Li, L.; Peng, W.; Liu, L.; Zheng, S. Toughening of epoxy by nanostructures with ABA triblock copolymers: An influence of organosilicon modification of block copolymer. *Polym. Eng. Sci.* **2022**, *62*, 392–404. [CrossRef]
16. Wang, Z.; Zhao, Y.; Wei, Y. Syntheses and properties of tri- and multi-block copolymers consisting of polybutadiene and polylactide segments. *RSC Adv.* **2022**, *12*, 29777–29784.
17. Min, X.; Fan, X. A New Strategy for the Synthesis of Hydroxyl Terminated Polystyrene-b-Polybutadiene-b-Polystyrene Triblock Copolymer with High Cis-1, 4 Content. *Polymers* **2019**, *11*, 598. [CrossRef] [PubMed]
18. Dossi, E.; Earnshaw, J.; Ellison, L.; Santos, G.R.D.; Cavaye, H.; Cleaver, D.G. Understanding and controlling the glass transition of HTPB oligomers. *Polym. Chem.* **2021**, *12*, 2606–2617. [CrossRef]
19. Garraza, A.L.R.; Mansilla, M.A.; Depaoli, E.L.; Macchi, C.; Cerveny, S.; Marzocca, A.J.; Somoza, A. Comparative study of thermal, mechanical and structural properties of polybutadiene rubber isomers vulcanized using peroxide. *Polym. Test.* **2016**, *52*, 117–123. [CrossRef]
20. Wrana, C.; Schawe, J.E.K. Isothermal crystallization of cis-1.4-polybutadiene at low temperatures. *Thermochim. Acta* **2020**, *690*, 178669.
21. Eslami, H.; Gharibi, A.; Plathe, F.M. Mechanisms of Nucleation and Solid−Solid-Phase Transitions in Triblock Janus Assemblies. *J. Chem. Theory Comput.* **2021**, *17*, 1742–1754. [PubMed]
22. Sadeghi, G.M.M.; Morshedian, J.; Barikani, M. The effect of solvent on the microstructure, nature of hydroxyl end groups and kinetics of polymerization reaction in synthesize of hydroxyl terminated polybutadiene. *React. Funct. Polym.* **2006**, *66*, 255–266. [CrossRef]
23. Grishchenko, V.K.; Boiko, V.P.; Svistova, E.I.; Yatsimirskaya, T.S.; Valuev, V.I.; Dmitrieva, T.S. Hydrogen-Peroxide-Initiated Polymerization of Isoprene in Alcohol Solutions. *J. Appl. Polym. Sci.* **1992**, *46*, 2081–2087. [CrossRef]
24. Chen, J.; Lu, Z.; Pan, G.; Qi, Y.; Yi, J.; Bai, H. Synthesis of hydroxyl-terminated polybutadiene possessing high content of 1,4-units via anionic polymerization. *Chin. J. Polym. Sci.* **2010**, *28*, 715–720. [CrossRef]
25. Min, X.; Fan, X.; Liu, J. Utilization of steric hindrance of alkyl lithium-based initiator to synthesize high 1,4 unit- containing hydroxyl- terminated polybutadiene. *R. Soc. Open Sci.* **2018**, *5*, 180156. [CrossRef] [PubMed]
26. Zhu, X.; Fan, X.; Zhao, N.; Min, X.; Liu, J.; Wang, Z. Influence of mono-lithium based initiators with different steric volumes on 1,4 unit content of hydroxyl terminated polybutadiene using anionic polymerization. *RSC Adv.* **2017**, *7*, 52712–52718. [CrossRef]
27. Li, L.; Zhang, C.; Zheng, S. Synthesis of POSS-terminated polycyclooctadiene telechelics via ring-opening metathesis polymerization. *J. Polym. Sci. Part A Polym. Chem.* **2017**, *55*, 223–233. [CrossRef]

28. Torres-Rocha, O.L.; Wu, X.; Zhu, C.; Crudden, C.M.; Cunningham, M.F. Polymerization-Induced Self-Assembly (PISA) of 1,5-Cyclooctadiene Using Ring Opening Metathesis Polymerization. *Macromol. Rapid Commun.* **2019**, *40*, 1800326.
29. Zhu, X.; Fan, X.; Zhao, N.; Liu, J.; Min, X.; Wang, Z. Comparative study of structures and properties of HTPBs synthesized via three different polymerization methods. *Polym. Test.* **2018**, *68*, 201–207. [CrossRef]

polymers

MDPI

Article

Three-Dimensional Printing of Shape Memory Liquid Crystalline Thermoplastic Elastomeric Composites Using Fused Filament Fabrication

Peerawat Prathumrat [1,*], Mostafa Nikzad [1,*], Fareed Tamaddoni Jahromi [1], Elnaz Hajizadeh [2] and Igor Sbarski [1]

[1] Department of Mechanical and Product Design Engineering, School of Engineering, Swinburne University of Technology, Hawthorn, VIC 3122, Australia; ftamaddonijahromi@swin.edu.au (F.T.J.); isbarski@swin.edu.au (I.S.)

[2] Department of Mechanical Engineering, Faculty of Engineering and Information Technology, University of Melbourne, Parkville, VIC 3010, Australia; ellie.hajizadeh@unimelb.edu.au

* Correspondence: pprathumrat@swin.edu.au (P.P.); mnikzad@swin.edu.au (M.N.)

Abstract: Liquid crystalline elastomers (LCEs) are stimuli-responsive materials utilised in shape memory applications. The processability of these materials via advanced manufacturing is being paid increasing attention to advance their volume production on an industrial scale. Fused filament fabrication (FFF) is an extrusion-based additive manufacturing (AM) technique that offers the potential to address this. The critical challenge, however, is the rheological characteristics of LCEs that need to be tuned to achieve a facile processability through the extrusion-based method. In this work, new filaments of liquid crystalline thermoplastic elastomer (LCTPE) and its composites with lignin were made by the ternary system of LCE, thermoplastic polyurethane (TPU), and lignin. The results showed that TPU improves the melt flow index of the LCTPE system to approximately 10.01 g/10 min, while adding lignin further enhances the value of this index for the composites up to 21.82 g/10 min. The microstructural analysis indicated that the effective distribution of lignin and reduced domain size of the LCEs in the ternary blend contribute to the enhanced flowability of this filament through 3D printing. Samples of 3D-printed LCTPE and LCTPE/lignin composites maintained their shape memory characteristics via thermomechanical activation. Full shape recovery of the new LCTPE matrix and its composites with lignin was achieved in 39 s and 32 s at 130 °C, followed by 28 s and 24 s at 160 °C, respectively. The successful fabrication of LCTPE and LCTPE/lignin composite samples through 3D printing demonstrates a potential procedure for processing these shape memory materials using the FFF technique, and lignin offers a sustainable and cost-effective material solution that enhances the properties of this composite material.

Keywords: 3D printing; liquid crystalline thermoplastic elastomer (LCTPE); lignin; composites; fused filament fabrication (FFF)

1. Introduction

The additive manufacturing (AM) technique was first developed in 1986 to fabricate a variety of materials with complex geometries from 3D model data [1]. This technology possesses advantages over conventional moulding techniques, such as lead time reduction, low cost, freedom of design, and geometry customisation [2]. Currently, various industries, such as biomedical, prototyping, construction, and polymer composites, exploit this technology to create printed products [3–5]. In polymer science and engineering, 3D printing plays a significant role in fabricating polymer parts, in which these materials can be built by different printing techniques, depending on the nature of the crystalline or amorphous structures. Fused filament fabrication (FFF) is one of the most widespread and popular 3D printing techniques to create polymeric samples, which is built on the melt extrusion method. This technique operates by extrusion of the thermoplastic filament into the desired shape according to the defined CAD models [6].

Liquid crystalline elastomers (LCEs) are a particular class of polymeric materials that possess elastic and anisotropic properties [7]. These elastomers can respond to external stimuli such as heat and UV, which can deform their shapes. This stimuli-responsiveness makes them suitable to utilise in shape memory and shape-changing applications. Furthermore, their molecular structures contain mesogens and flexible tails resembling the hard and soft phases of the typical underlying mechanism of the shape memory polymers (SMPs). In our previous work, LCE based on diacrylate reactive mesogens and their relevant composites with lignin were fabricated and exhibited excellent shape memory properties [8]. These elastomeric composites with lignin exhibited shape fixity and shape recovery of 97.0% and 99.7%, respectively. Furthermore, the presence of lignin content in this composite system could reduce the recovery time to 70 s, while the neat LCE sample took 100 s.

In order to exploit these materials on an industrial scale for mass customisation, the processability of these composite materials through 3D printing technology needs to be investigated. In the last decade, several 3D printing techniques based on vat photopolymerization, such as stereolithography (SLA) and direct ink writing (DIW), have successfully fabricated LCE parts, though often at small scales [9]. However, the 3D printing techniques based on fused filament fabrication (FFF) have been given limited attention due to issues associated with the rheological behaviour of LCEs, especially at the macroscopic scale. To obtain suitable rheological properties of LCEs for extrusion-based 3D printing, fabricating hybrid materials containing hard thermoplastic and soft elastomeric phases is a potential solution that provides an alternative approach to printability [10]. Such thermoplastic elastomers (TPEs) give rise to microstructures with tuneable flow properties like thermoplastics, which allows leveraging a straightforward and faster extrusion process based on the well-developed rheological principles for this class of material.

Thermoplastic polyurethane (TPU) is a widely used candidate as a hard-thermoplastic phase in TPE structures. This material is a linear block copolymer with thermoplasticity, elasticity, shear thinning behaviour, 3D printability, and low cost, making it apposite for extrusion techniques [10]. Furthermore, TPU can exhibit shape memory characteristics because of their hard diisocyanate soft polyol segments. Recently, Wang et al. [11] fabricated shape memory polycaprolactone (PCL)/TPU composites using a FFF technique. These materials showed excellent shape fixity and recovery of 100%. Bi et al. [12] utilised a FFF to build shape memory polyurethane/polycaprolactone blends with dopamine/multi-walled carbon nanotubes (PMWCNTs) filler. These materials responded to near-infrared light and exhibited shape fixity and recovery of 90% and 75% after three cycles. In addition, Huang et al. [13] reported that the addition of TPU enhanced the melt flow index (MFI) of the polymer blends due to its high melt fluidity, resulting in an increased flowability for the processing techniques. Given TPU's suitable molecular mechanism for shape memory applications, these previous works suggest that TPU is an excellent material for improving the rheological properties of polymeric materials for 3D printing through a FFF technique.

Lignin is a bio-based material produced as a by-product of the pulping process. This material has been used as a filler in polymeric composite systems due to its beneficial characteristics—for instance, high thermomechanical properties, high UV stability, renewability, being environmentally benign, and low cost [14]. Recently, Zhou et al. [15] studied the 3D printing of lignin/TPU and carbon fibre (CF)/lignin/TPU composites using a FFF technique. The lignin/TPU composites could be printed and demonstrated a smooth surface up to 50 wt% lignin. The CF was further added at 0.5 wt% to enhance the mechanical properties of the printed composites. Their results showed that the tensile strength and elongation at break of the CF/lignin/TPU composites increased by 67% and 138% compared to lignin/TPU composites.

Considering the superior shape memory characteristics of LCE and lignin/LCE composites, they are great candidates in a variety of industrial and engineering applications, such as in soft robotics, aerospace engineering, smart actuators, and sensors. In order to enable their high-volume production, the processibility of these materials needs to be investigated. Fused filament fabrication (FFF) 3D printing technology has been selected due to being relatively fast, reliable, and inexpensive. However, the rheological properties of LCE are the major challenge that needs to be improved for FFF 3D printing. Thus, this work focuses on investigating the processability of a new liquid crystalline thermoplastic elastomer (LCTPE) filament based on LCE and TPU and their composites containing lignin through the FFF AM technique. The LCE was synthesised using RM257 monomers and PETMP crosslinking agent through a thiol-Michael addition reaction [8]. The TPU was used to improve the printability of the system and contribute to the shape memory behaviour of the printed LCTPE samples. Lignin was further added to create the LCTPE composites. The shape memory behaviour and thermal, rheological, microscopic, and spectroscopic properties of these materials were characterised.

2. Materials and Methods

2.1. Materials

Diacrylate mesogen, named 1,4-bis4-(3-acryloyloxypropypropyloxy) benzoyl-oxy]-2-methylbenzene (RM257), was purchased from Daken Chemical (Zhengzhou, China). Kraft lignin (alkali lignin), pentaerythritol tetra (3-mercaptopropionate) (PETMP) crosslinking agent, and dipropyl amine (DPA) catalyst were purchased from Sigma-Aldrich, Inc. (Castle Hill, NSW, Australia). Thermoplastic polyurethane (WHT-1180EC) was purchased from Wanhua Chemical Group Co., Ltd. (Yantai, China). Toluene (AR grade) was purchased from ChemSupply Australia Pty Ltd. (Gillman, SA, Australia). All materials were used as received without further purification.

2.2. Sample Preparation

Liquid crystalline elastomer (LCE) was synthesised following the steps from our previous work [8]. First, RM257 monomers were dissolved by adding 40 wt% toluene. The mixtures were heated at 80 °C in the oven until becoming a homogeneous solution. The solution was then left to cool down at room temperature. Next, 36 wt% PETMP crosslinking agent and 14 wt% DPA catalyst were added dropwise into the solution. In this work, the DPA was separately diluted in toluene at a weight ratio of 1:50 before adding to the solution. The solution was immediately stirred and vacuumed to remove the bubbles that occurred during the mixing process. The solution was then poured into the mould and left to cure at room temperature for 24 h. Upon completion of the polymerisation process, the LCE samples were heated at 80 °C for 12 h to evaporate the solvent before characterisation.

Liquid crystalline thermoplastic elastomers (LCTPEs) were prepared by mixing LCE with thermoplastic polyurethane (TPU) at a weight ratio of 50:50. TPU was dried at 80 °C for 24 h in a vacuum oven before mixing. These components were mixed using an internal batch mixer (HAAKE Polylab QC, Thermo Scientific, Waltham, MA, USA) at 190 °C with separately different rotor speeds of 60 rpm, 120 rpm, and 180 rpm. Meanwhile, the LCTPE/lignin composites with an LCE:TPU:lignin mass content of 40:40:20 were prepared using a batch mixer in two steps. The details of the mixing process are included in Table 1.

Table 1. The details of the mixing process of LCTPE and LCTPE/lignin composite samples.

Samples	Mixing Process
LCTPE@60	60 rpm for 10 min at 190 °C
LCTPE@120	120 rpm for 10 min at 190 °C
LCTPE@180	180 rpm for 10 min at 190 °C
LCTPE/lignin@180	LCE and TPU were mixed at 180 rpm for 9 min, and lignin was then added to mix for 1 min at 190 °C.

The filaments of the LCTPE and LCTPE/lignin composites were prepared using a single-screw extruder, Filabot EX2 (Filabot HQ, Barre, VT, USA). The LCTPE was extruded at 190 °C with an average diameter of 2.85 mm to be used on an Ultimaker^3 FFF 3D printer. The extruded filament was cooled down using two fans at room temperature and spooled using a Filabot spooler. The processes of filament preparation and 3D printing are shown in Figure 1.

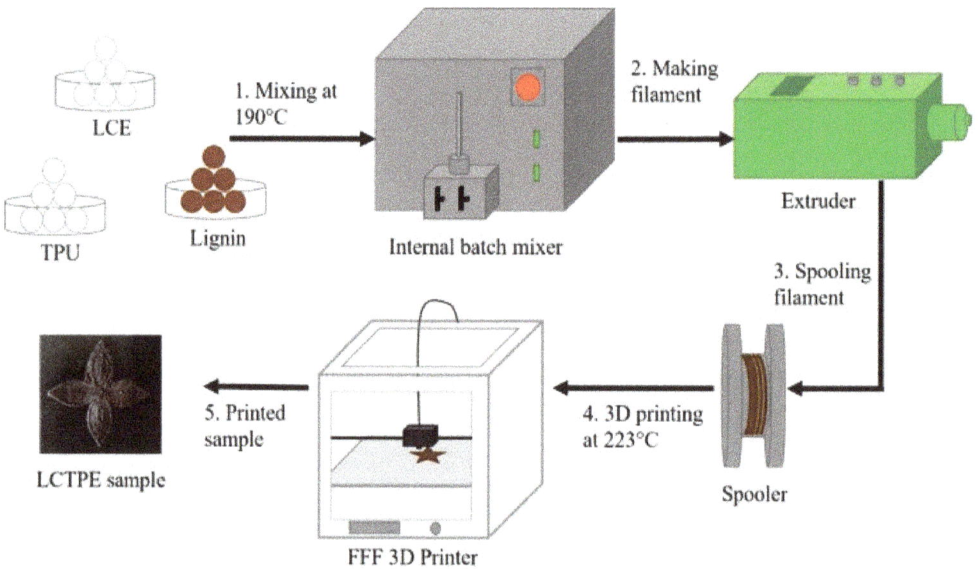

Figure 1. Filament preparation and 3D printing process of LCTPE and LCTPE/lignin composite materials.

2.3. Sample Characterisation

2.3.1. Melt Flow Index

The melt flow index (MFI) of the neat TPU, LCTPE, and LCTPE/lignin composites was measured using the melt flow index tester Modular Line (CEAST, Instron, Pianezza, Italy). A sample mass of 8 mg to 10 mg was used. The test was performed at 200 °C under a specific load of 2.16 kg.

2.3.2. Optical Microscopy

The surface morphology and particle distribution of the LCTPE and LCTPE/lignin composite samples were investigated using optical microscopy (Olympus BX61, Hamburg, Germany). The images of these samples were produced by imaging at 20× magnification. A 50 μm × 50 μm ROI was set for the best focus of the area of each captured image.

2.3.3. Thermal Properties

Differential scanning calorimetry (DSC) measurements were conducted to evaluate phase transition temperatures of the neat TPU, neat LCE, LCTPE samples, and relevant composites using a DSC 2960 (TA Instruments, New Castle, DE, USA). The samples with a mass of 5 mg to 10 mg were sealed in aluminium pans. The heating rate was set at 5 °C/min in a temperature range of −40 °C to 150 °C. A N_2 atmosphere was set with a 110 mL/min flow rate.

2.3.4. Thermal Stability

The thermal stability of all the materials was evaluated using a thermogravimetric analyser (TGA) Q50 (TA Instruments, New Castle, DE, USA). A sample weight of 10 mg to 20 mg was used. All samples were run with a heating rate of 10 °C/min from room temperature to 800 °C. The air was purged with a flow rate of 60 mL/min. Thermal degradation and derivative thermogravimetric (DTG) curves of all the samples were analysed.

2.3.5. Fourier-Transform Infrared (FTIR) Spectroscopy

FTIR analysis was used to measure the structural composition of neat TPU, neat LCE, LCTPE, and their composites with lignin. The FTIR Nicolet iS5 (Thermo Scientific) was utilised with a scan range of 4000 cm^{-1} to 400 cm^{-1}. The spectra were scanned in a transmittance mode with 64 scans per sample.

2.3.6. Shape Memory Behaviours

Shape memory behaviours of the LCTPE and composite samples were observed, including shape fixity and shape recovery. The samples were heated at 130 °C for 10 min and manually deformed to a temporary shape. Then, the samples were cooled down to room temperature to maintain this temporary shape. The samples were reheated at 130 °C to observe their shape recovery. The images of various stages of these samples were taken through a shape memory process. In addition, the effect of temperature on the recovery time of these samples was recorded using a video camera at temperatures of 130 °C and 160 °C.

3. Fabrication of 3D-Printed LCTPEs and Their Composites

The flower-shaped samples of the LCTPE@180 and LCTPE/lignin@180 composites were printed using a FFF machine, Ultimaker^3 (Zaltbommel, The Netherlands). Slicer software Cura 4.13 was used to prepare the G-code file. The process parameters for 3D printing the samples are summarised in Table 2. All samples were printed with 100% infill density using concentric patterns. A nozzle temperature of 223 °C was set to enhance the extrusion ability and avoid clogging [16]. The viscosity of the polymer composite melt was shown to be proportional to the reciprocal of the absolute temperature, which was governed by the activation energy of the system denoted by Arrhenius's Equation [17]. The filaments, printing process, and examples of printed samples are shown in Figure 2.

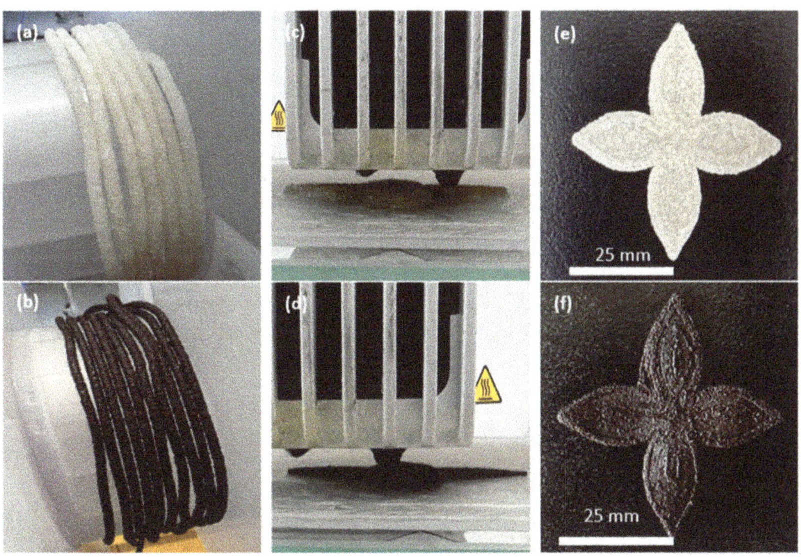

Figure 2. The filaments of the LCTPE@180 and LCTPE/lignin@180 composites (**a**,**b**), FFF 3D printing process (**c**,**d**), and printed specimens with a thickness of 2 mm (**e**,**f**).

Table 2. The parameters utilised for 3D printing of LCTPE and its composite samples.

Parameters	Value
Print nozzle diameter (mm)	0.8
Nozzle temperature (°C)	223
Bed temperature (°C)	50
Layer thickness (mm)	0.15
Printing speed (mm/s)	30

4. Results and Discussion

4.1. The Effect of Rotor Speed on the Melt Flow Index of LCTPE and Their Composites

The melt flow index (MFI) is a measure of the flow characteristics of the molten polymer under a particular shear stress. This property is a crucial parameter in providing insight into the printability of the polymeric samples using the extrusion-based printer. In a previous study, the MFI threshold value of 10 g/10 min was sufficient for printing polylactic acid (PLA) through a FFF technique at a temperature range of 190 °C to 220 °C [18]. In this work, the MFI of neat TPU and LCTPE and composite samples were evaluated at 200 °C using an applied load of 2.16 kg.

Figure 3 demonstrates that the MFI of the LCTPEs increases with the increasing rotor speed in the mixing process. In this study, the LCTPE samples were mixed with different rotor speeds of 60 rpm, 120 rpm, and 180 rpm. The LCTPE samples mixed at 60 rpm (LCTPE@60), 120 rpm (LCTPE@120), and 180 rpm (LCTPE@180) displayed MFI values of 5.84 ± 0.46 g/10 min, 9.52 ± 0.42 g/10 min, and 10.01 ± 0.58 g/10 min, respectively. This increment could be attributed to a decreased particle size of the LCEs at a higher shear rate, which results in an increased MFI value of the LCTPE material at a high temperature. An increase in the MFI value of the LCTPE samples also contributed to their printability through the extruder nozzle of the printer. These results were similar to those reported by Tjong et al. [19]. In addition, the LCTPE@180 sample showed a continuous filament and smooth surface after the filament extrusion process. Meanwhile, the MFI value of neat TPU was 15.23 ± 0.79 g/10 min. This value was higher than the LCTPE samples, in which the strong covalent bonding of the crosslinked LCE structures in the LCTPE matrix reduced their flowability at elevated temperatures. In contrast, the MFI value of the neat LCE material could not be measured, since it could not be melted at the set temperature of the MFI test due to the high stiffness and strength of its crosslinked structures.

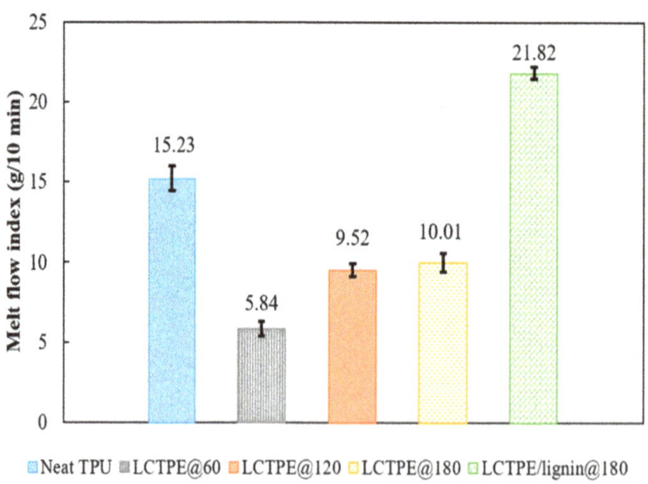

Figure 3. Melt flow index (MFI) test results of the neat TPU, LCTPE@60, LCTPE@120, LCTPE@180, and LCTPE/lignin@180 composite samples at 200 °C.

Furthermore, the LCTPE/lignin composites were prepared with a mass ratio of 40:40:20 in a batch mixer with a rotor speed of 180 rpm. The incorporation of lignin into the LCTPE matrix increased their MFI value. The results showed the MFI of LCTPE/lignin@180 composites of 21.82 ± 0.37 g/10 min. It could be suggested that lignin served as a plasticiser at elevated temperatures, increasing the flowability of the composite materials [20].

4.2. Surface Morphology of LCTPE and Composite Materials

The surface morphology of the LCTPE samples fabricated by different rotor speeds and the LCTPE/lignin composite sample was investigated using an optical microscope. The microscope images of various samples are shown in Figure 4a–d. The image of the LCTPE@60 sample in Figure 4a shows the crystalline morphology of the LCE particles distributed along the TPU region. An increase in rotor speed in the mixing process reduced the particle size of the LCE, as seen in Figure 4b,c. A high centrifugal force of a high rotor speed in the mixing process could improve the shear behaviour of the LCE domain, resulting in reduced particle sizes. It could confirm an increase in the MFI values of these materials, where the reduced LCE particles could enhance the flowability of the molten LCTPE through the nozzle without clogging. The addition of a lignin filler to the LCTPE matrix in Figure 4d showed the agglomeration characteristic due to the strong cohesion force between the lignin particles [21] and moderate dispersion. Nevertheless, lignin tended to distribute along the LCTPE matrix. This filler significantly improved the MFI value of the LCTPE/lignin@180 composite material, further contributing to the flowability of these composites. Thus, the size reduction of the LCE particles and distribution of lignin along the LCTPE matrix could improve the flow behaviour, leading to 3D printability via a FFF technique.

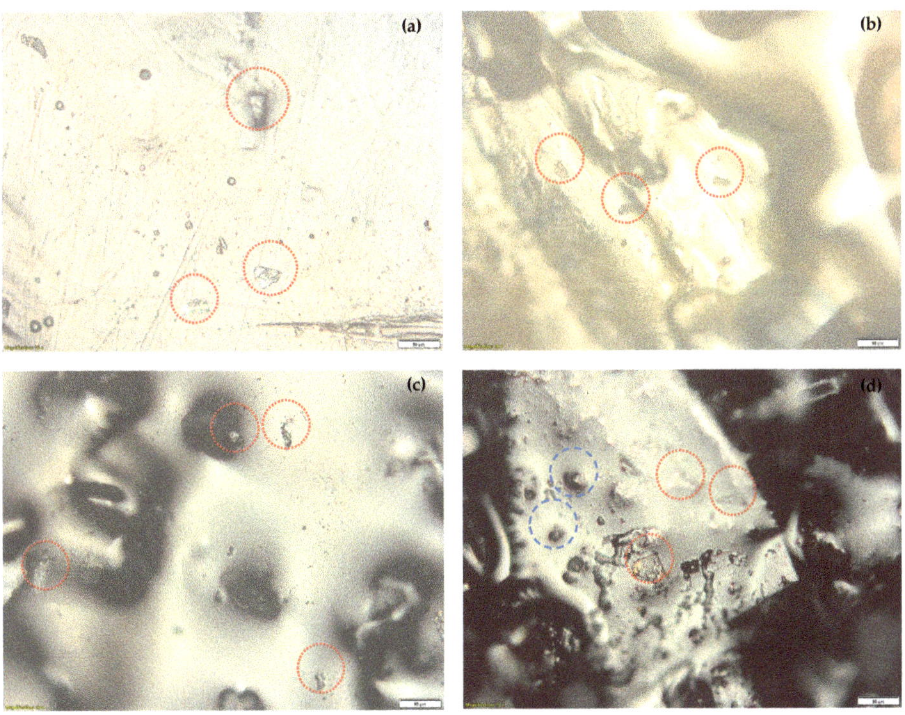

Figure 4. Optical microscopy images of (**a**) the LCTPE@60, (**b**) LCTPE@120, (**c**) LCTPE@180, and (**d**) LCTPE/lignin@180 composites magnified 20×; LCE particles (red dotted circle) and lignin (blue dashed circle).

In addition, the figures indicate the immiscible blending of various compositions. To consider the dynamics of the heterogeneous blending of this composite system, the size and size distribution of the dispersed phase in the polymer matrix play an essential role [22]. The average crystalline morphology size of LCE is likely to decrease with a higher shear rate, which enhances the rheological properties of the composites. The topology and convection flow of the configuration of the interfaces are other influential factors in the dynamics of the blending [23]. The aromatic structure of the lignin filler also enables an increase in the flowability of the blend when replacing the linear structures of the LCE/TPU phase due to the improved imposed flow field. This is similar to the report of Chapman et al. [24]. Meanwhile, the modification of the interface of crystalline blending with lignin as a complex structural filler results in its viscoelastic response, in which the interfacial region is enabled to slip by the applied shear stress, similarly reported by Galeski [25] and Saha et al. [26]. In addition, the interfacial dynamics of the polymers can be varied by a degree of stretching and molecular weight [27].

4.3. Thermal Properties of LCTPEs and Their Composites

The thermal properties of the neat LCE, neat TPU, LCTPE@180, and LCTPE/lignin@180 composites were evaluated using DSC analysis. Figure 5 illustrates the thermogram of these samples over well-defined temperatures. The neat LCE exhibited two phase transitions at −6 °C and 96 °C, corresponding to a glass transition (T_g) and a nematic–isotropic transition (T_{NI}) temperature. The T_{NI} is a particular transition temperature of the LCE class of polymers that allows the molecular alignment of liquid crystalline structures from nematic to isotropic states. This temperature can be used as a switching temperature in a shape memory process without interference by ambient conditions. The DSC curve of the neat TPU exhibited a small endothermic peak at approximately 70 °C. This peak can be defined as the melting temperature of hard-segment domains of diisocyanate structures [28].

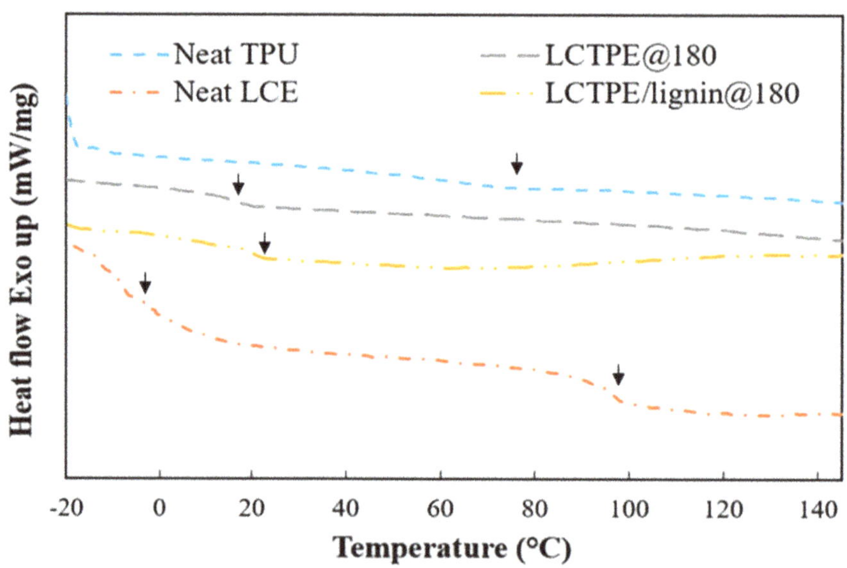

Figure 5. DSC thermograms of neat LCE, neat TPU, and LCTPE materials in oxidative conditions.

In addition, a DSC curve of the LCTPE@180 sample was evaluated. This material indicated a T_g at 16 °C, which was higher than the neat LCE sample. This enhancement could be attributed to the effect created by the rigid diisocyanate segments of urethane structures. However, the absence of an endothermic peak of the neat TPU in the LCTPE@180 curve was possibly due to the lack of intermolecular interactions between hard segments of TPU, since this segment could form a hydrogen bond with lignin [29]. Furthermore, the LCTPE/lignin@180 composites exhibited two inflexion points of stepwise transition at 25 °C and 95 °C, corresponding to a T_g and T_{NI}, respectively. The T_g of the LCTPE/lignin@180 composites was slightly shifted higher than the LCTPE@180 sample. This could be explained by the fact that the three-dimensional network structures of the lignin content could obstruct the movement of the polymeric chains of the LCTPE matrix, resulting in an increased T_g.

4.4. Thermal Stability of LCTPEs and Their Composites

The thermal stability of the neat LCE, neat TPU, LCTPE@180, and LCTPE/lignin@180 composite samples were examined under oxidative conditions. The TGA and DTG thermograms of all the samples are depicted in Figures 6 and 7. All the samples exhibited two degradation stages. The neat LCE sample showed the degradation temperature at 10 wt% loss ($T_{d,10}$) at 379 °C and the first maximum DTG peak at 403 °C. This degradation stage was ascribed to the breakages of the thioether linkages of the LCE structures and the decomposition of the aliphatic chains of RM257. The second DTG peak of LCE was indicated at 590 °C, corresponding to the fragmentation of the nematic mesogens of the LCE structures. Meanwhile, the TGA curve of the neat TPU showed the $T_{d,10}$ at 333 °C and the first DTG peak at 400 °C, which corresponded to the scission of the hard segment with the formation of diisocyanate and diol groups. The second DTG peak was in the range of 530 °C to 598 °C, corresponding to the degradation of the soft diol segment of TPU [30,31].

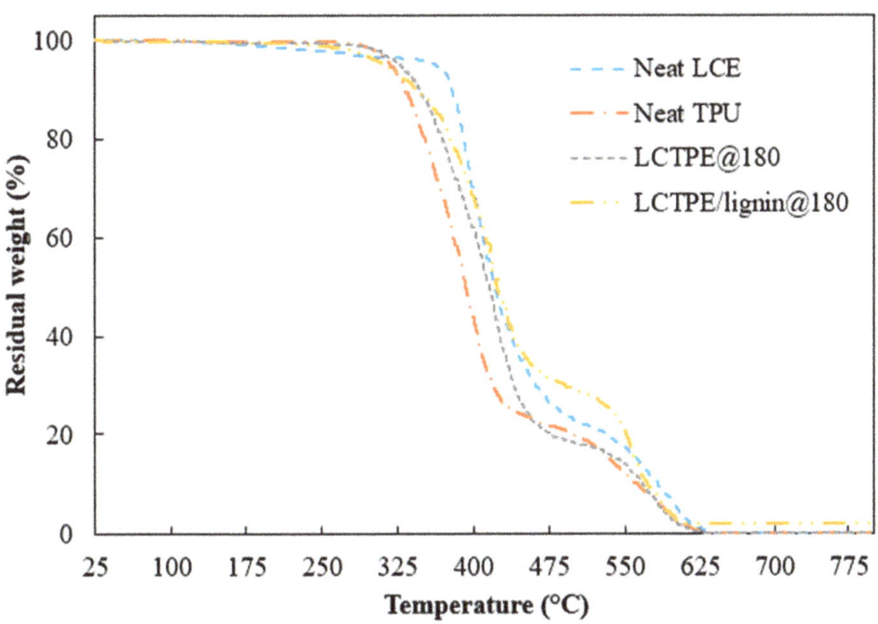

Figure 6. Thermal degradation of neat LCE, neat TPU, and LCTPE materials in oxidative conditions.

The LCTPE@180 sample demonstrated a similar pattern of degradation of LCE and TPU. However, the first prominent peak of this material exhibited a synergistic characteristic compared to their parental materials, which could be due to the topological entanglement of the LCE network structures and aliphatic backbones of TPU. The LCTPE/lignin@180 composites exhibited a similar trend as the LCTPE@180 sample. However, the second DTG peak of the composites was higher than the neat LCE, the neat TPU, and LCTPE@180 samples, which could be due to the fragmentation of the phenylpropanoid structures of lignin.

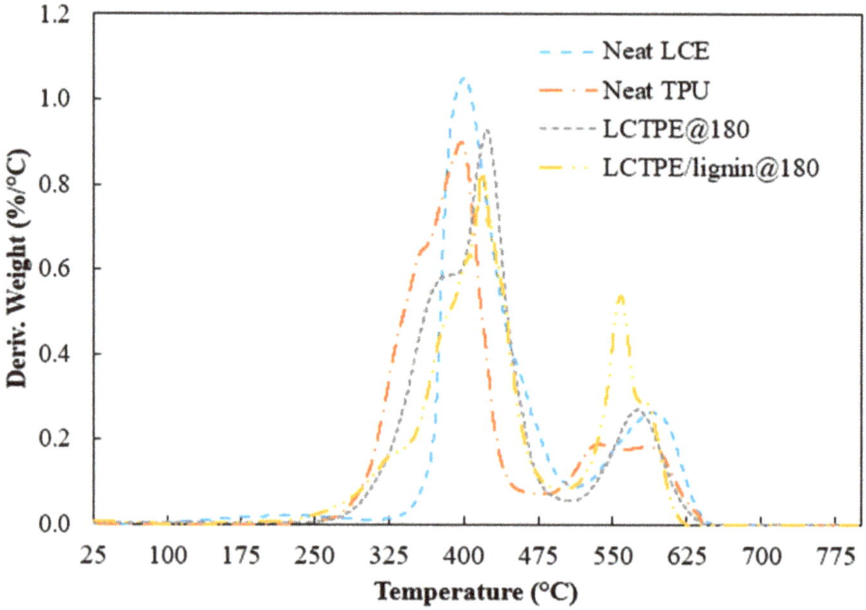

Figure 7. DTG thermograms of neat LCE, neat TPU, LCTPE@180, and LCTPE/lignin@180 composite materials in oxidative conditions.

4.5. FTIR Analysis of the Neat LCE, Neat TPU, and LCTPE Materials

The FTIR spectra of the neat LCE, neat TPU, LCTPE@180, and LCTPE/lignin@180 composite samples are shown in Figure 8. The spectrum of neat TPU showed peaks in the range of 2997 cm^{-1} to 2810 cm^{-1}, indicating asymmetric and symmetric C-H bond stretching vibrations [32]. The peaks at 1456 cm^{-1} and 1375 cm^{-1} belonged to the C-C stretching of aromatic rings and the C-N bond stretching, respectively. The peak observed at 1166 cm^{-1} corresponded to the NH–(C=O)–O bond stretching vibration of urethane groups [33].

The FTIR curve of neat LCE exhibited a broad peak in a range of 2992 cm^{-1} to 2842 cm^{-1}, corresponding to intramolecular C-H bond stretching. The peak at 1723 cm^{-1} indicated a bond stretching of C=O in RM257 structures. The peaks between 1300 cm^{-1} and 1000 cm^{-1} specified the C-O-C stretching vibration of the ester groups in the LCE structures. Meanwhile, the LCTPE@180 sample exhibited combined FTIR peaks of LCE and TPU. That combination has no further significant peaks that have been observed. It suggests that the mixture of LCE and TPU was only a result of the physical entanglement between their respective macromolecules. Such an inter-dispersion, however, contributed to a higher shift of the DTG peak of LCTPE@180, which was higher than the neat LCE and TPU, as seen in Section 4.4. A small broad peak at 3330 cm^{-1} of the LCTPE@180 spectrum was further observed, demonstrating a stretching vibration of the N-H groups of the urethane linkages. The spectrum of LCTPE/lignin@180 composites showed a similar curve to the LCTPE@180 sample. However, a broad peak observed between 3674 cm^{-1} and 3177 cm^{-1} corresponded to the O−H stretching of the hydrogen-bonded aromatic and aliphatic groups of lignin structures.

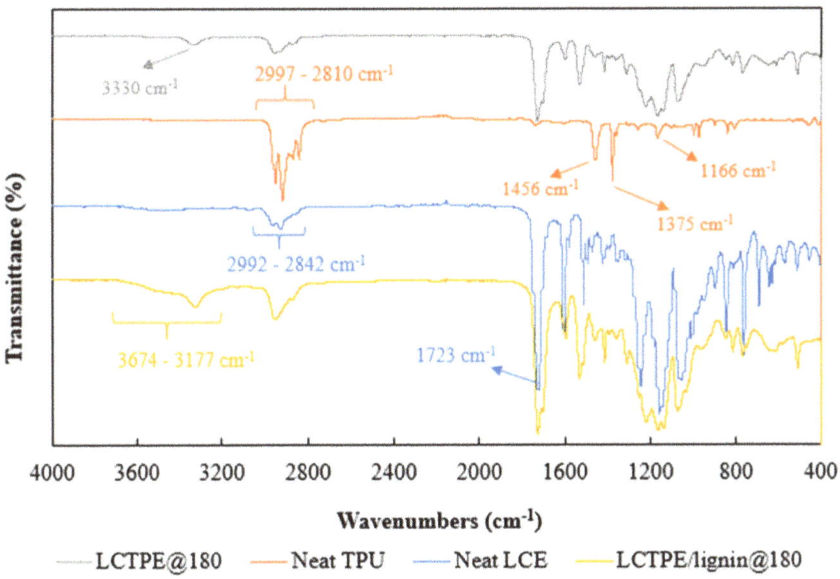

Figure 8. FTIR spectra of the neat LCE, neat TPU, and LCTPE materials.

4.6. Shape Memory Behaviours of LCTPE Materials

The shape memory behaviours of the 3D-printed LCTPE sample and LCTPE/lignin@180 composites with a thickness of 2 mm were observed through a thermomechanical process, including shape fixity and shape recovery. The molecular structure of LCTPE could underpin the typical molecular mechanism of shape memory polymers (SMPs), in which they consist of netpoints (hard phase) and switch units (soft phase). In the LCTPE system, the mesogen of the LCE structure, together with the diisocyanate segment of TPU, work as netpoints, which are responsible for determining the original shape of the shape memory elastomers. These netpoints can store the elastic energy during the shape memory process. In contrast, the flexible aliphatic chains of the LCE structure and polyol segments of TPU work as switch units in the mechanism, which are responsible for deforming and recovering shapes. Therefore, the combined structures of LCE and TPU contribute to the underlying shape memory mechanism of the LCTPE system. A similar observation was reported by Ghosh et al. [34] that the interpenetrated network of polystyrene/thermoplastic polyurethane-based nanocomposites could decrease the recovery time. This composite system exhibited a full recovery in 22 s and 98 s when activated by a temperature of 50 °C and sunlight at 42 °C, respectively. In addition, the molecular entanglement between LCE and TPU structures could improve the shape recovery of this polymeric system, in which the entanglement could serve as the physical crosslinks and increase the orientational entropy of this system, resulting in the fast and facile recovery over the transition temperature. Similar observations were also made by Hao et al. [35] and Hoy et al. [36].

Figures 9 and 10 illustrate the snapshots of various stages of shape memory behaviours of the 3D-printed LCTPE@180 and LCTPE/lignin@180 composite samples. First, the original blooming shape was heated in the oven at 130 °C for 10 min, where this temperature was above the T_{NI} of this LCTPE system. Second, the sample was then manually folded and cooled to room temperature to maintain the temporary shape. Third, the temporary folded shape was reheated at the same temperature without the applied force. Fourth, the folded sample gradually recovered to a bloomed shape again.

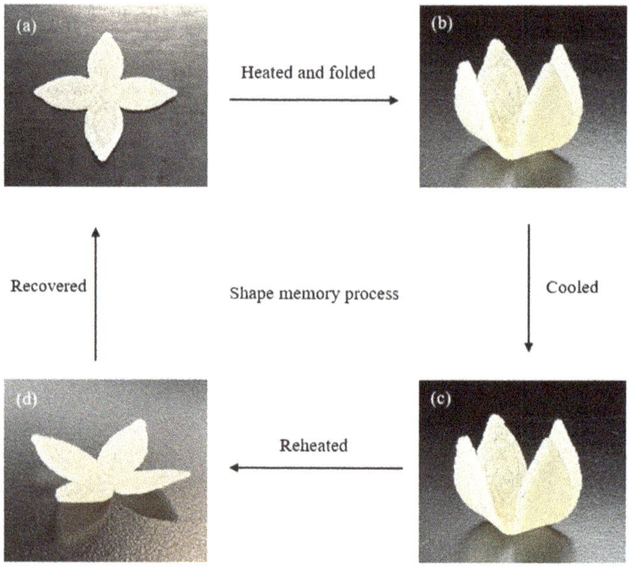

Figure 9. Snapshots of the shape memory process of the LCTPE@180 sample: (**a**) the original blooming shape, (**b**) the temporary shape folded at 130 °C, (**c**) the temporary shape cooled at room temperature, and (**d**) the recovered shape at 130 °C.

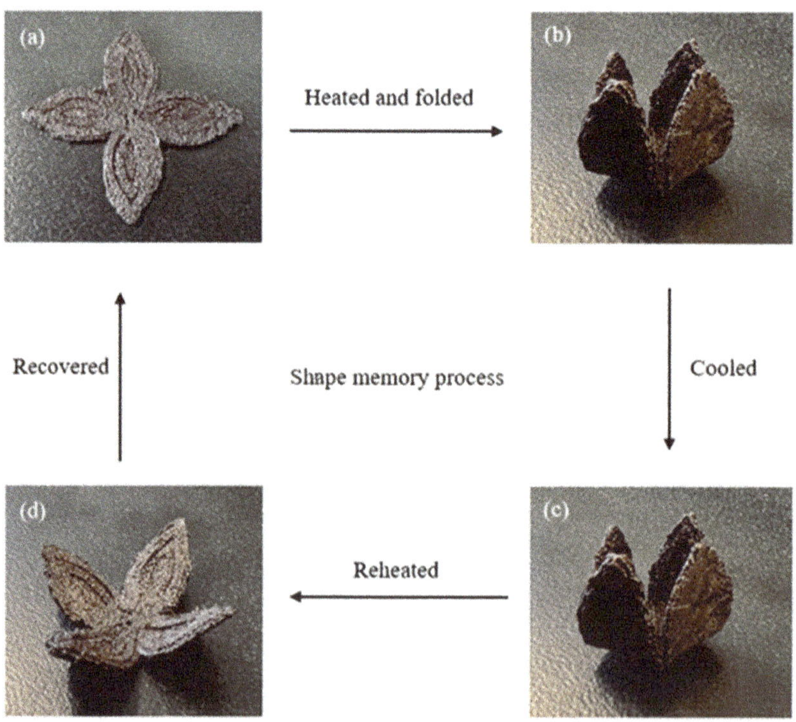

Figure 10. Snapshots of the shape memory process of LCTPE/lignin@180 composite sample: (**a**) the original blooming shape, (**b**) the temporary shape folded at 130 °C, (**c**) the temporary shape cooled at room temperature, and (**d**) the recovered shape at 130 °C.

The recovery time of these samples was also investigated by the influence of well-defined temperatures using a video camera. Temporarily folded samples were placed in the oven at different temperatures of 130 °C and 160 °C. The snapshots of the shape recovery behaviour at a temperature of 130 °C are shown in Figure 11. Moreover, the results of the recovery time as a function of temperature are plotted in Figure 12. The LCTPE@180 sample exhibited a full recovery in 39 s and 28 s when activated at 130 °C and 160 °C, respectively. However, the LCTPE/lignin@180 composites could fully recover at 130 °C and 160 °C in 32 s and 24 s, respectively. These results indicate that lignin served as the extra netpoint density, which could store the higher elastic energy in the deformation process and release this energy during the recovery process, resulting in the faster recovery of the composites.

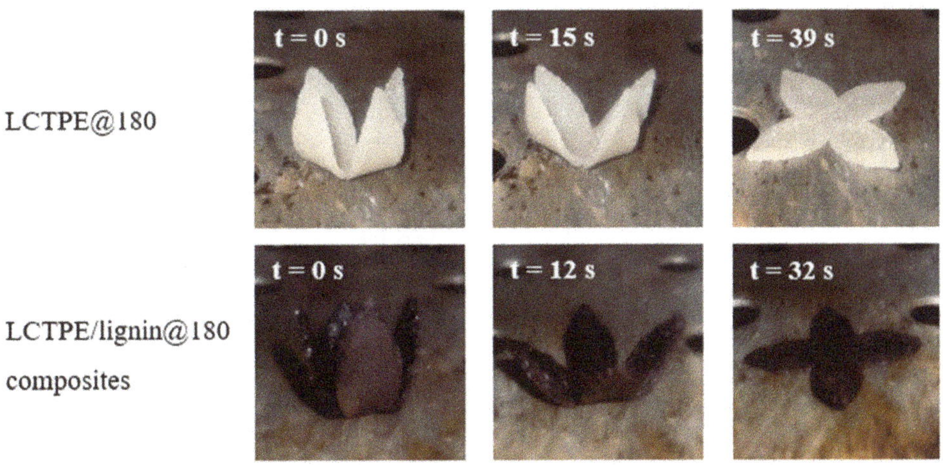

Figure 11. The snapshots of the shape recovery behaviour of LCTPE@180 and LCTPE/lignin@180 composite samples at a temperature of 130 °C.

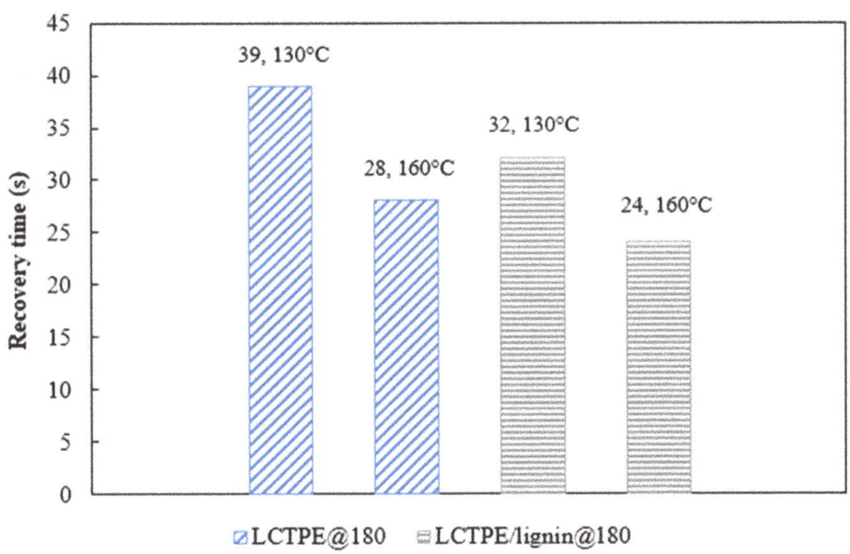

Figure 12. The recovery time of LCTPE@180 and LCTPE/lignin@180 composite samples at different testing temperatures of 130 °C and 160 °C.

5. Conclusions

This study succeeded in the 3D printing of a new liquid crystalline thermoplastic elastomer (LCTPE) and their composites with lignin using a FFF technique. The MFI results of LCTPE filaments showed an increasing trend with the increasing rotor speeds of the mixer. The LCTPE@180 filament exhibited a MFI value above 10 g/10 min, corresponding to the threshold value for printing with a FFF printer. Moreover, the addition of a lignin filler into this material further increased the MFI value, confirming the printability of this composite material by using this printing technique. The optical microscopic results demonstrated that an increase in the rotor speed reduced the particle size of LCE crystalline morphology and improved the distribution of lignin along the LCTPE matrix, ensuring the flowability of molten LCTPE/lignin composite materials through the nozzle without clogging. The presence of a lignin filler also increased the glass transition temperature of the composites compared to the neat LCTPE counterpart.

The samples of 3D-printed LCTPE@180 and LCTPE/lignin@180 composites demonstrated excellent shape memory behaviour through thermomechanical activation. A variation in temperature also influenced the recovery time of these materials, in which they could take a shorter time to recover when the temperature increased. Hence, incorporating lignin in the fabrication of LCTPE and its composite samples can be an effective way to improve the printability and processability of this class of shape memory materials for industrial and engineering applications, such as soft robotics, smart sensors and actuators, and high-end space engineering. The utilisation of abundant lignin enables the prospect of developing interesting applications of this exciting class of material in a sustainable and cost-effective fashion. To further advance the functionalities of these materials in future works, the variations of the compositions of this ternary system should be considered.

Author Contributions: P.P.: Conceptualisation, Methodology, Validation, Investigation, Formal analysis, Writing—Original Draft, and Writing—Review and Editing. M.N.: Conceptualisation, Methodology, Writing—Review and Editing, Supervision, Resources, and Project Administration. F.T.J.: Methodology and Writing—Review and Editing. E.H.: Writing—Review and Editing and Supervision. I.S.: Supervision. All authors have read and agreed to the published version of the manuscript.

Funding: This research received no external funding.

Institutional Review Board Statement: Not applicable.

Data Availability Statement: The data presented in this study are available on request from the corresponding authors.

Acknowledgments: P. Prathumrat would like to acknowledge Shammi Sultana Nisha of the Swinburne University of Technology for the assistance in optical microscopy of the specimens.

Conflicts of Interest: The authors declare no conflict of interest.

References

1. Hull, C.W. Apparatus for Production of Three-Dimensional Objects by Stereolithography. U.S. Patent 638905, 11 March 1986.
2. Singh, S.; Ramakrishna, S.; Berto, F. 3D Printing of polymer composites: A short review. *Mater. Des. Process. Commun.* **2020**, *2*, e97. [CrossRef]
3. Lai, J.; Wang, C.; Wang, M. 3D printing in biomedical engineering: Processes, materials, and applications. *Appl. Phys. Rev.* **2021**, *8*, 021322. [CrossRef]
4. Macdonald, E.; Salas, R.; Espalin, D.; Perez, M.; Aguilera, E.; Muse, D.; Wicker, R.B. 3D printing for the rapid prototyping of structural electronics. *IEEE Access* **2014**, *2*, 234–242. [CrossRef]
5. El-Sayegh, S.; Romdhane, L.; Manjikian, S. A critical review of 3D printing in construction: Benefits, challenges, and risks. *Arch. Civ. Mech. Eng.* **2020**, *20*, 34. [CrossRef]
6. Melnikova, R.; Ehrmann, A.; Finsterbusch, K. 3D printing of textile-based structures by Fused Deposition Modelling (FDM) with different polymer materials. In Proceedings of the IOP Conference Series: Materials Science and Engineering, Ningbo, China, 27–29 May 2014; p. 012018.
7. Saed, M.O.; Torbati, A.H.; Nair, D.P.; Yakacki, C.M. Synthesis of programmable main-chain liquid-crystalline elastomers using a two-stage thiol-acrylate reaction. *JoVE (J. Vis. Exp.)* **2016**, *107*, e53546.

8. Prathumrat, P.; Nikzad, M.; Hajizadeh, E.; Okhawilai, M.; Sbarski, I. Lignin-liquid crystalline elastomeric composites for shape memory applications and their thermomechanical properties. *Express Polym. Lett.* **2023**, *17*, 104–117. [CrossRef]
9. Kotikian, A.; Truby, R.L.; Boley, J.W.; White, T.J.; Lewis, J.A. 3D printing of liquid crystal elastomeric actuators with spatially programed nematic order. *Adv. Mater.* **2018**, *30*, 1706164. [CrossRef]
10. Awasthi, P.; Banerjee, S.S. Fused deposition modeling of thermoplastic elastomeric materials: Challenges and opportunities. *Addit. Manuf.* **2021**, *46*, 102177. [CrossRef]
11. Wang, Y.; Zhang, J.; Li, M.; Lei, M.; Wang, Y.; Wei, Q. 3D printing thermo-responsive shape memory polymer composite based on PCL/TPU blends. *J. Polym. Res.* **2022**, *29*, 243. [CrossRef]
12. Bi, H.; Xu, M.; Ye, G.; Guo, R.; Cai, L.; Ren, Z. Mechanical, Thermal, and Shape Memory Properties of Three-Dimensional Printing Biomass Composites. *Polymers* **2018**, *10*, 1234. [CrossRef]
13. Huang, J.; Lu, X.; Zhang, G.; Qu, J. Study on the rheological, thermal and mechanical properties of thermoplastic polyurethane/poly (butylene terephthalate) blends. *Polym. Test.* **2014**, *36*, 69–74. [CrossRef]
14. Li, H.; Sivasankarapillai, G.; McDonald, A.G. Lignin valorization by forming toughened thermally stimulated shape memory copolymeric elastomers: Evaluation of different fractionated industrial lignins. *J. Appl. Polym. Sci.* **2015**, *132*, 41389. [CrossRef]
15. Zhou, X.; Ren, Z.; Sun, H.; Bi, H.; Gu, T.; Xu, M. 3D printing with high content of lignin enabled by introducing polyurethane. *Int. J. Biol. Macromol.* **2022**, *221*, 1209–1217. [CrossRef]
16. León-Calero, M.; Reyburn Valés, S.C.; Marcos-Fernández, Á.; Rodríguez-Hernandez, J. 3D printing of thermoplastic elastomers: Role of the chemical composition and printing parameters in the production of parts with controlled energy absorption and damping capacity. *Polymers* **2021**, *13*, 3551. [CrossRef]
17. Peleg, M. Temperature–viscosity models reassessed. *Crit. Rev. Food Sci. Nutr.* **2018**, *58*, 2663–2672. [CrossRef]
18. Wang, S.; Capoen, L.; D'hooge, D.R.; Cardon, L. Can the melt flow index be used to predict the success of fused deposition modelling of commercial poly(lactic acid) filaments into 3D printed materials? *Plast. Rubber Compos.* **2018**, *47*, 9–16. [CrossRef]
19. Tjong, S.C.; Xu, S.A. Mechanical properties of glass fiber and liquid crystalline polymer reinforced polypropylene hybrid composites toughened with elastomers. *J. Appl. Polym. Sci.* **2004**, *94*, 1539–1546. [CrossRef]
20. Domínguez-Robles, J.; Larrañeta, E.; Fong, M.L.; Martin, N.K.; Irwin, N.J.; Mutjé, P.; Tarrés, Q.; Delgado-Aguilar, M. Lignin/poly(butylene succinate) composites with antioxidant and antibacterial properties for potential biomedical applications. *Int. J. Biol. Macromol.* **2020**, *145*, 92–99. [CrossRef]
21. Klapiszewski, L.; Jamrozik, A.; Strzemiecka, B.; Matykiewicz, D.; Voelkel, A.; Jesionowski, T. Activation of magnesium lignosulfonate and kraft lignin: Influence on the properties of phenolic resin-based composites for potential applications in abrasive materials. *Int. J. Mol. Sci.* **2017**, *18*, 1224. [CrossRef]
22. Sadiku-Agboola, O.; Sadiku, E.R.; Adegbola, A.T.; Biotidara, O.F. Rheological properties of polymers: Structure and morphology of molten polymer blends. *Mater. Sci. Appl.* **2011**, *2*, 30. [CrossRef]
23. Roths, T.; Friedrich, C.; Marth, M.; Honerkamp, J. Dynamics and rheology of the morphology of immiscible polymer blends–on modeling and simulation. *Rheol. Acta* **2002**, *41*, 211–222. [CrossRef]
24. Chapman, C.D.; Shanbhag, S.; Smith, D.E.; Robertson-Anderson, R.M. Complex effects of molecular topology on diffusion in entangled biopolymer blends. *Soft Matter* **2012**, *8*, 9177–9182. [CrossRef]
25. Galeski, A. Dynamic mechanical properties of crystalline polymer blends. The influence of interface and orientation. *e-Polymers* **2002**, *2*, 346–374. [CrossRef]
26. Saha, S.; Xu, D.; Gersappe, D. Effect of Compatibilizers on the Structure and Dynamics at Polymer Blend Interfaces. *Tribol. Lett.* **2021**, *69*, 61. [CrossRef]
27. Holt, A.P.; Bocharova, V.; Cheng, S.; Kisliuk, A.M.; White, B.T.; Saito, T.; Uhrig, D.; Mahalik, J.P.; Kumar, R.; Imel, A.E.; et al. Controlling Interfacial Dynamics: Covalent Bonding versus Physical Adsorption in Polymer Nanocomposites. *ACS Nano* **2016**, *10*, 6843–6852. [CrossRef]
28. Somdee, P.; Lassú-Kuknyó, T.; Kónya, C.; Szabó, T.; Marossy, K. Thermal analysis of polyurethane elastomers matrix with different chain extender contents for thermal conductive application. *J. Therm. Anal. Calorim.* **2019**, *138*, 1003–1010. [CrossRef]
29. Ortiz-Serna, P.; Carsí, M.; Culebras, M.; Collins, M.N.; Sanchis, M.J. Exploring the role of lignin structure in molecular dynamics of lignin/bio-derived thermoplastic elastomer polyurethane blends. *Int. J. Biol. Macromol.* **2020**, *158*, 1369–1379. [CrossRef]
30. Dong, M.; Li, Q.; Liu, H.; Liu, C.; Wujcik, E.K.; Shao, Q.; Ding, T.; Mai, X.; Shen, C.; Guo, Z. Thermoplastic polyurethane-carbon black nanocomposite coating: Fabrication and solid particle erosion resistance. *Polymer* **2018**, *158*, 381–390. [CrossRef]
31. Liu, H.; Dong, M.; Huang, W.; Gao, J.; Dai, K.; Guo, J.; Zheng, G.; Liu, C.; Shen, C.; Guo, Z. Lightweight conductive graphene/thermoplastic polyurethane foams with ultrahigh compressibility for piezoresistive sensing. *J. Mater. Chem. C* **2017**, *5*, 73–83. [CrossRef]
32. Pan, R.; Yang, L.; Zheng, L.; Hao, L.; Li, Y. Microscopic morphology, thermodynamic and mechanical properties of thermoplastic polyurethane fabricated by selective laser sintering. *Mater. Res. Express* **2020**, *7*, 055301. [CrossRef]
33. Haryńska, A.; Gubanska, I.; Kucinska-Lipka, J.; Janik, H. Fabrication and Characterization of Flexible Medical-Grade TPU Filament for Fused Deposition Modeling 3DP Technology. *Polymers* **2018**, *10*, 1304. [CrossRef] [PubMed]
34. Ghosh, T.; Voit, B.; Karak, N. Polystyrene/thermoplastic polyurethane interpenetrating network-based nanocomposite with high-speed, thermo-responsive shape memory behavior. *Polymer* **2020**, *200*, 122575. [CrossRef]

35. Hao, X.; Kaschta, J.; Liu, X.; Pan, Y.; Schubert, D.W. Entanglement network formed in miscible PLA/PMMA blends and its role in rheological and thermo-mechanical properties of the blends. *Polymer* **2015**, *80*, 38–45. [CrossRef]
36. Hoy, R.S.; Robbins, M.O. Strain hardening of polymer glasses: Entanglements, energetics, and plasticity. *Phys. Rev. E* **2008**, *77*, 031801. [CrossRef]

polymers

MDPI

Communication

Angle-Independent Color Change in Thermoresponsive Gel-Immobilized Colloidal Amorphous Film Attached to PET Substrate

Sato Nakagawa, Yuna Hirano, Mikako Tanaka and Toshimitsu Kanai *

Graduate School of Engineering Science, Yokohama National University, 79-5 Tokiwadai, Hodogaya-ku, Yokohama 240-8501, Japan; satoshiba310@gmail.com (S.N.); hirano-yuna-gc@ynu.jp (Y.H.); tanaka-mikako-hw@ynu.ac.jp (M.T.)
* Correspondence: tkanai@ynu.ac.jp

Abstract: Gel-immobilized colloidal amorphous structures comprise short-range-ordered monodisperse submicrometer particles embedded into a soft polymer gel. They exhibit an angle-independent structural color that is tunable in response to external stimuli via a volume change in the gel, which has significant potential for the development of sensors that respond to stimuli via angle-independent color changes. In this study, the amorphous structure of a charged colloidal suspension in water was immobilized in a thermoresponsive poly(*N*-isopropylacrylamide) (PNIPAM) gel film and simultaneously attached to a polyethylene terephthalate (PET) substrate. The gel film exhibited a uniform angle-independent color that changed in response to changes in temperature (i.e., thermosensitivity). Attachment to the PET substrate suppressed changes in the gel film area and film distortion, despite significant volume changes in the gel. Consequently, the degree of thermosensitivity was enhanced. The PET-attached gel-immobilized colloidal amorphous film was easy to handle and had excellent flexibility, allowing it to wrap around the surfaces of curved objects. These features are advantageous for sensor applications.

Keywords: stimuli-responsive gel; structural color; colloidal amorphous structure; sensor; film

1. Introduction

Nanostructures at the scale of the wavelength of visible light can generate color via the reflection or scattering of a specific wavelength of light. This color is called structural color [1–4]. In contrast to general dyes and pigments, structural colors have unique properties such as nonfading against ultraviolet (UV) light and color tuning according to structural changes; hence, they are expected to be employed as novel colorants and sensors [5–7]. Extensive research has been conducted on the preparation and application of colloidal crystals with artificial structural colors [8–14]. Colloidal crystals utilize periodically arranged monodisperse submicrometer particles to exhibit vivid reflective colors. As the color or reflected light wavelength changes according to Bragg's law, various colors can be generated by adjusting the diameter of the colloidal particles, the interparticle distance, and the refractive index of the colloidal crystal. Furthermore, colloidal crystals embedded into stimuli-responsive polymers allow for the on-demand tuning of the interparticle distance and Bragg wavelength via the external deformation of the polymer [15–20]. However, the color of the colloidal crystals changes with the viewing angle, which may limit their application.

Recently, colloidal amorphous arrays with angle-independent structural colors have been developed [21–23]. These colloidal amorphous arrays, which comprise monodisperse submicrometer particles with a short range order, have received increasing attention. Several excellent preparation methods have been reported, including rapid solvent evaporation from a monodisperse colloidal suspension [24–26] and drying of a suspension containing two different sizes of monodisperse particles [27–29]. However, in these previously reported colloidal amorphous arrays, the colloidal particles are in contact with each other in the dry state. Consequently, monodisperse particles of different sizes must be prepared to create different colors, which is laborious and time-consuming.

By contrast, we previously achieved the formation of a colloidal amorphous structure in a charged colloidal suspension, in which monodisperse submicrometer-size particles were separated from each other using a liquid medium (water) [30]. When the electrorepulsion between the charged particles was weakened, such as by the addition of salt, the crystalline structure transformed into an amorphous one with an angle-independent color. This amorphous structure with separated particles has a great practical advantage in that the color can be easily altered by simply changing the particle volume fraction or amount of solvent. Furthermore, we successfully immobilized the colloidal amorphous structure in a stimulus-responsive hydrogel film while maintaining a low particle volume fraction and good optical quality. The angle-independent color could be altered on demand by changing the ethanol concentration of the swelling solvent because the resultant volume change in the gel changed the interparticle distance. Thus, the film has the potential to be used as a sensor for detecting external stimuli using angle-independent color changes. However, there are still several issues for the practical application of this material. For example, the thin wet gel film requires careful handling, and the volume changes in the gel (required to alter the color) were accompanied by a significant reduction in the film area and film distortion.

In this study, we immobilized a charged colloidal amorphous structure with a low particle volume fraction in a thermoresponsive poly(*N*-isopropylacrylamide) (PNIPAM) gel film and simultaneously attached the film to a polyethylene terephthalate (PET) substrate. The PET-attached PNIPAM-immobilized colloidal amorphous film exhibited a uniform angle-independent color that changed in response to changes in temperature. Moreover, owing to the attachment of the film to the PET substrate, the reduction in gel film area and film distortion were suppressed, even during significant gel volume changes. Furthermore, the degree of color change in response to temperature changes, that is, the thermosensitivity, was enhanced. Finally, the attachment of the film to the PET substrate improved the ease of handling and flexibility of the material. These features are advantageous for sensor applications.

2. Materials and Methods

An aqueous suspension of monodisperse polystyrene particles with diameters of 160 nm (5016 B, Thermo Fisher Scientific, Waltham, MA, USA) was used to prepare a colloidal amorphous film. The suspension was deionized by adding an ion exchange resin (AG 501-X8(D), Bio-Rad, Hercules, CA, USA) and stirring gently for at least two weeks. The obtained charge-stabilized colloidal crystals were centrifuged to prepare concentrated colloidal crystals. A thermoresponsive *N*-isopropylacrylamide monomer (NIPAM, FUJIFILM Wako Pure Chemical Corp., Tokyo, Japan), *N,N′*-methylenebisacrylamide crosslinker (BIS, FUJIFILM Wako Pure Chemical Corp., Tokyo, Japan), and 2,2′-azobis [2-methyl-*N*-(2-hydroxyethyl)propionamide] photoinitiator (VA, FUJIFILM Wako Pure Chemical Corp., Tokyo, Japan) were dissolved in ultrapure water (Milli-Q system, Merck KGaA, Darmstadt, Germany) to prepare the gelation reagent. The concentrated charge-stabilized colloidal crystals, gelation reagent, and ultrapure water were added to a square cuvette such that the concentrations of particles, NIPAM, BIS, and VA were 10.7 vol%, 800 mM, 40 mM, and 0.4 mM, respectively, and the total volume was 0.5 mL. To achieve the amorphous structure, a 10 mM aqueous NaCl solution (FUJIFILM Wako Pure Chemical Corp., Tokyo, Japan) was added to the cuvette in 0.5 μL increments, and the reflection spectrum was measured using a fiber spectrometer (Fastevert S-2630, Soma Optics, Ltd., Tokyo, Japan) after each addition. As the salt concentration increased, the intensity of the Bragg reflection peak from the charge-stabilized colloidal crystals was reduced until an amorphous structure with an angle-independent color was achieved.

Figure 1 shows a schematic of the preparation process of the PET-attached PNIPAM-immobilized colloidal amorphous film. The surface of the PET substrate (Lumirror T60, thickness: 100 μm, Toray Industries, Inc., Tokyo, Japan) was treated using plasma ion bombardment (PIB-20, Vacuum Device, Mito, Japan) to attach the gel film to the PET substrate using graft polymerization [31]. The PET substrate was then placed on a quartz substrate along with a silicone separator sheet with a channel (channel height: 0.1 mm; width: 9 mm; length: 50 mm) and quartz substrate with holes and sandwiched within a metal frame. The frame was then fixed in place with screws.

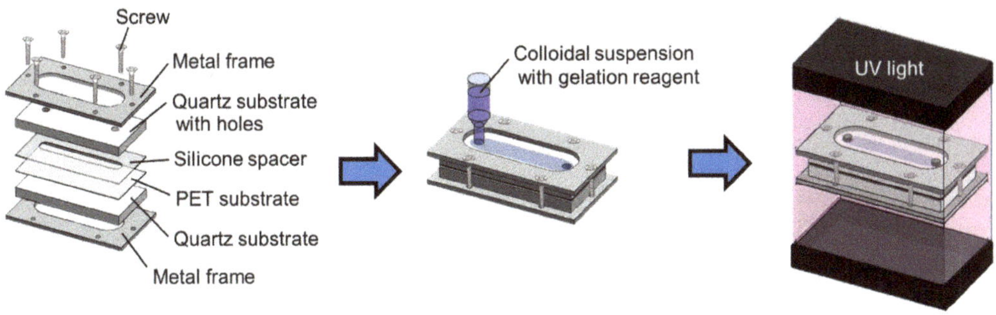

Figure 1. Schematic diagram of the preparation process of the polyethylene terephthalate (PET)-attached poly(*N*-isopropylacrylamide) (PNIPAM)-immobilized colloidal amorphous film.

The colloidal amorphous suspension containing the gelation reagent was bubbled with Ar gas for 5 min and infused into the cell. The suspension was uniformly irradiated with UV light (MBRL-CUV7530, MORITEX Corporation, Saitama, Japan) from the top and bottom surfaces of the cell for 1 h to photopolymerize the gelation reagent. The transmission spectra before and after UV irradiation were measured using a spectrophotometer (V-670, Jasco Corp., Tokyo, Japan) and photographs were taken using a digital camera (IXY Digital 800IS, Canon Inc., Tokyo, Japan). Finally, the cells were disassembled to obtain the PET-attached gel-immobilized colloidal amorphous film. Ultrapure water was placed on the PET-attached gel film and covered with another PET substrate using a 0.3 mm thick silicone spacer.

Further samples were prepared without plasma treatment of the PET substrate. The prepared PNIPAM-immobilized colloidal amorphous film was readily separated from the PET substrate in these samples; therefore, the unattached gel film was sandwiched between two PET substrates along with ultrapure water using a 0.3 mm thick silicone spacer.

To determine the temperature-dependent changes in the PET-attached and unattached samples, they were placed on a thermal stage, and the transmission spectrum and color were recorded using a spectrophotometer (V-670, Jasco Corp., Tokyo, Japan) and digital camera (IXY Digital 800IS, Canon Inc., Tokyo, Japan), respectively. Changes in the film morphology were also monitored from the digital photographs.

3. Results and Discussion

Figure 2a demonstrates the change in the reflection spectrum at normal incidence for the colloidal suspension when the salt concentration C_s was increased. A strong Bragg reflection peak of the charge-stabilized colloidal crystals was observed at approximately 690 nm for the suspensions with C_s = 0, 40, and 100 μM. Above C_s = 120 μM, the peak was significantly reduced and only a very small peak owing to the amorphous phase was observed, as shown in the 20× magnified spectra displayed in Figure 2b. The peak was further reduced when the salt concentration was further increased and was not observable above C_s = 180 μM to transform the random phase.

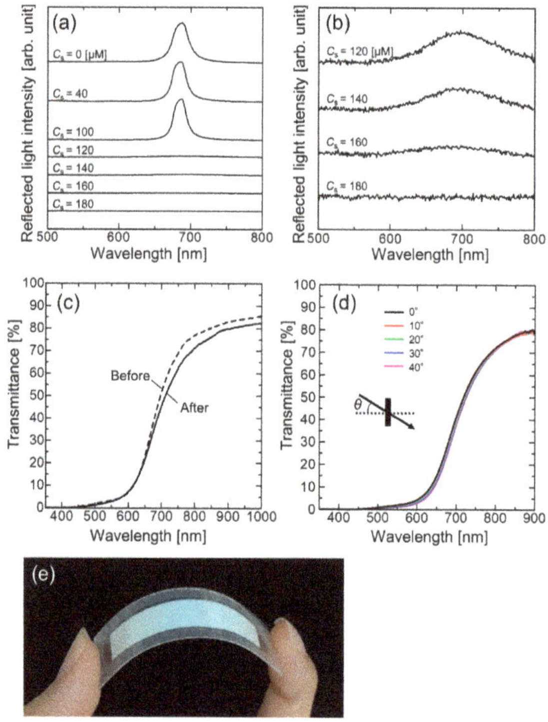

Figure 2. (**a**) Reflection spectra at normal incidence for the colloidal suspensions at different salt concentrations. (**b**) 20-times-magnified reflection spectra at salt concentrations between 120 μM and 180 μM. (**c**) Transmission spectra of colloidal amorphous suspension in the cell before and after ultraviolet (UV) light irradiation. (**d**) Transmission spectra of PET-attached PNIPAM-immobilized colloidal amorphous film at various incident light angles. (**e**) Photograph of PET-attached PNIPAM-immobilized colloidal amorphous film at 31 °C.

Figure 2c shows the transmission spectra of the colloidal amorphous suspension in the cell at a salt concentration of 120 μM before and after UV light irradiation. Before UV irradiation, the spectrum was characteristic of the amorphous structure, that is, the transmittance was low in the short-wavelength range and sharply increased at around the Bragg reflection wavelength of the charge-stabilized colloidal crystal with the same particle diameter and volume fraction [30]. A high transmittance of more than 70% was achieved in the longer wavelength range. After UV irradiation, the spectral profile was largely maintained, with a less than 5% decrease in transmittance, which is probably due to the slight deterioration of the amorphous structure. Furthermore, the spectrum was stable upon changing the angle of incident light (Figure 2d). These results indicate that the amorphous structure was successfully immobilized in the PNIPAM gel film. After disassembling the fabricated cell to obtain the PET-attached PNIPAM-immobilized colloidal amorphous film, the sample was easy to handle and had excellent flexibility, as shown in Figure 2e. In contrast to the colloidal crystals, the film showed a uniform color, even on a curved surface.

Because the volume of the PNIPAM gel changes significantly at a transition temperature of approximately 32 °C [32], the angle-independent color of the PNIPAM-immobilized colloidal amorphous film can be altered by changing the temperature. The photographs and transmission spectra in Figure 3a,b, respectively, show the temperature-dependent changes in the PNIPAM-immobilized colloidal amorphous film without PET attachment.

As the temperature increased, the film shrank, resulting in a color change from light purple at 24 °C to light green at 31 °C and light blue at 34 °C. The colors originate from the coherent scattering of light, and the color change was mainly derived from the reduction in the interparticle distance of the colloidal amorphous structure as the PNIPAM gel in which it was embedded shrank. However, the film distorted and tore during shrinkage. As the temperature increased, the transmission spectrum shifted to shorter wavelengths and the transmittance intensity in the long-wavelength range decreased. The decrease in transmittance in the long-wavelength range was probably caused by the turbulence of the amorphous structure and increased area density of the particles upon gel shrinkage.

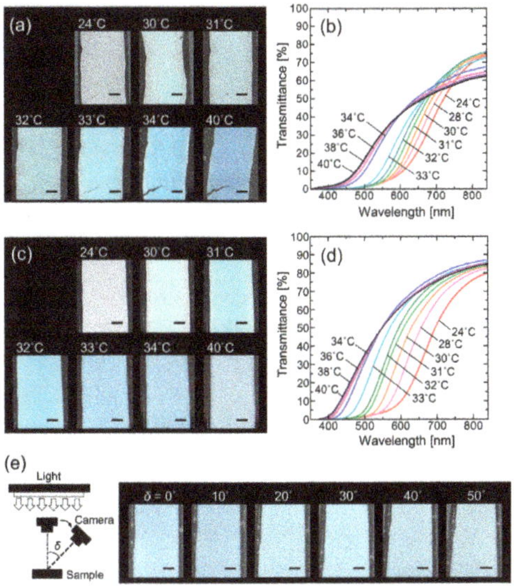

Figure 3. (**a**) Photographs and (**b**) transmission spectra at normal incidence of unattached PNIPAM-immobilized colloidal amorphous film at various temperatures. (**c**) Photographs and (**d**) transmission spectra at normal incidence of PET-attached PNIPAM-immobilized colloidal amorphous film at various temperatures. (**e**) Photographs of PET-attached PNIPAM-immobilized colloidal amorphous film at 33 °C from different camera angles. Scale bars in (**a**,**c**,**e**): 2 mm.

By contrast, the PET-attached PNIPAM-immobilized colloidal amorphous film did not exhibit a reduction in the film area without waving or warping as the temperature increased, and the film shrank in the thickness direction. Therefore, it exhibited a more uniform and strong color change (Figure 3c). This was clearly observed in the transmission spectra; the degree to which the spectrum of the PET-attached gel-immobilized colloidal amorphous film blue-shifted was greater than that of the unattached film (Figure 3d). In addition, attachment to the PET substrate preserved the excellent transmission properties, that is, high transmittance in the long-wavelength range, during the temperature change. The enhancement in thermosensitivity is probably caused by an increase in shrinkage in the thickness direction of the PNIPAM gel film, which is complementarily caused by the suppression of in-plane shrinkage [33,34]. Such preservation of the transmittance in the long-wavelength range suggests the amorphous ordering and particle area density are not deteriorated in the in-plane-shrunken PET-attached gel film.

Notably, the PET-attached PNIPAM-immobilized colloidal amorphous film exhibited a uniform angle-independent color at each temperature. For instance, the film appeared a uniform light blue color at 33 °C, regardless of the viewing angle, as shown in Figure 3e.

4. Conclusions

A charged colloidal amorphous structure suspended in water was successfully immobilized in a thermoresponsive PNIPAM gel film and simultaneously attached to a PET substrate via photopolymerization. The PNIPAM-immobilized colloidal amorphous film with a low particle volume fraction exhibited a uniform angle-independent color that changed in response to temperature changes. Attachment to the PET substrate suppressed the reduction in film area and film distortion, even when the gel underwent high volume changes. Furthermore, the thermosensitivity was enhanced by an increase in shrinkage in the thickness direction of the PET-attached PNIPAM gel film. The thermosensitivity was controlled with concentrations of monomer and colloids, similar to that of gel-immobilized colloidal crystals. The PET-attached gel-immobilized colloidal amorphous film was easy to handle and had excellent flexibility, allowing its attachment to curved surfaces for sensing applications. These features are advantageous for sensor applications.

Author Contributions: Conceptualization, T.K.; validation, S.N. and Y.H.; investigation, S.N. and Y.H.; data curation, S.N., Y.H., and M.T.; writing—original draft preparation, S.N. and T.K.; writing—review and editing, T.K. and M.T.; visualization, S.N., Y.H., and M.T.; supervision, T.K.; funding acquisition, T.K. All authors have read and agreed to the published version of the manuscript.

Funding: This research was funded by a KAKENHI Grant-in-Aid for Scientific Research (grant number 18H01721) from the Japan Society for the Promotion of Science.

Institutional Review Board Statement: Not applicable.

Informed Consent Statement: Not applicable.

Data Availability Statement: The data presented in this study are available upon request from the corresponding author.

Conflicts of Interest: The authors declare no conflict of interest. The funders had no role in the design of the study; in the collection, analyses, or interpretation of data; in the writing of the manuscript; or in the decision to publish the results.

References

1. Rezaei, S.D.; Dong, Z.; Chan, J.Y.E.; Trisno, J.; Ng, R.J.H.; Ruan, Q.; Qiu, C.-W.; Mortensen, N.A.; Yang, J.K.W. Nanophotonic Structural Colors. *ACS Photonics* **2021**, *8*, 18–33. [CrossRef]
2. Xuan, Z.; Li, J.; Liu, Q.; Yi, F.; Wang, S.; Lu, W. Artificial Structural Colors and Applications. *Innovation* **2021**, *2*, 100081. [CrossRef] [PubMed]
3. Chen, G.; Hong, W. Mechanochromism of Structural-Colored Materials. *Adv. Opt. Mater.* **2020**, *8*, 2000984. [CrossRef]
4. Quan, Y.-J.; Kim, Y.-G.; Kim, M.-S.; Min, S.-H.; Ahn, S.-H. Stretchable Biaxial and Shear Strain Sensors Using Diffractive Structural Colors. *ACS Nano* **2020**, *14*, 5392–5399. [CrossRef] [PubMed]
5. Fenzl, C.; Hirsch, T.; Wolfbeis, O.S. Photonic Crystals for Chemical Sensing and Biosensing. *Angew. Chem. Int. Ed.* **2014**, *53*, 3318–3335. [CrossRef] [PubMed]
6. Qin, M.; Sun, M.; Hua, M.; He, X. Bioinspired structural color sensors based on responsive soft materials. *Curr. Opin. Solid State Mat. Sci.* **2019**, *23*, 13–27. [CrossRef]
7. Li, S.; Kou, D.; Zhang, S.; Ma, W. Large-Area Fabrication of Structurally Colored and Humidity Sensitive Composite Nanofilm via Ultrasonic Spray-Coating. *Polymers* **2021**, *13*, 3768. [CrossRef] [PubMed]
8. Pariente, J.Á.; Blanco, Á.; López, C. Colloidal photonic crystals formation studied by real-time light diffraction. *Nanophotonics* **2022**, *11*, 3257–3267. [CrossRef]
9. Wang, L.; Xu, Y.; Chu, Z.; Tang, W.; Qiu, Y.; Zhao, X.; Jiang, W.; Ye, J.; Chen, C. Rapid Coating of Ultraviolet Shielding Colloidal Crystals. *Crystals* **2020**, *10*, 502. [CrossRef]
10. Lai, C.-F.; Li, J.-S. Self-assembly of colloidal Poly(St-MMA-AA) core/shell photonic crystals with tunable structural colors of the full visible spectrum. *Opt. Mater.* **2019**, *88*, 128–133. [CrossRef]
11. Gao, W.; Rigout, M.; Owens, H. Self-assembly of silica colloidal crystal thin films with tuneable structural colours over a wide visible spectrum. *Appl. Surf. Sci.* **2016**, *380*, 12–15. [CrossRef]
12. Takeoka, Y. Structural Colored Gel. *J. Photopolym. Sci. Technol.* **2009**, *22*, 123–132. [CrossRef]
13. Liu, J.; Nero, M.; Jansson, K.; Willhammar, T.; Sipponen, M.H. Photonic crystals with rainbow colors by centrifugation-assisted assembly of colloidal lignin nanoparticles. *Nat. Commun.* **2023**, *14*, 3099. [CrossRef] [PubMed]
14. Kobori, M.; Hirano, Y.; Tanaka, M.; Kanai, T. Practical Preparation of Elastomer-Immobilized Nonclose-Packed Colloidal Photonic Crystal Films with Various Uniform Colors. *Polymers* **2023**, *15*, 2294. [CrossRef] [PubMed]

15. Wang, Y.; Cui, H.; Zhao, O.; Du, X. Chameleon-Inspired Structural-Color Actuators. *Matter* **2019**, *1*, 626–638. [CrossRef]
16. Liao, J.; Zhu, C.; Gao, B.; Zhao, Z.; Liu, X.; Tian, L.; Zeng, Y.; Zhou, X.; Xie, Z.; Gu, Z. Multiresponsive Elastic Colloidal Crystals for Reversible Structural Color Patterns. *Adv. Funct. Mater.* **2019**, *29*, 1902954. [CrossRef]
17. Qi, F.; Meng, Z.; Xue, M.; Qiu, L. Recent advances in self-assemblies and sensing applications of colloidal photonic crystals. *Anal. Chim. Acta* **2020**, *1123*, 91–112. [CrossRef]
18. Yu, S.; Dong, S.; Jiao, X.; Li, C.; Chen, D. Ultrathin Photonic Polymer Gel Films Templated by Non-Close-Packed Monolayer Colloidal Crystals to Enhance Colorimetric Sensing. *Polymers* **2019**, *11*, 534. [CrossRef]
19. Kredel, J.; Gallei, M. Compression-Responsive Photonic Crystals Based on Fluorine-Containing Polymers. *Polymers* **2019**, *11*, 2114. [CrossRef]
20. Ruan, J.-L.; Chen, C.; Shen, J.-H.; Zhao, X.-L.; Qian, S.-H.; Zhu, Z.-G. A Gelated Colloidal Crystal Attached Lens for Noninvasive Continuous Monitoring of Tear Glucose. *Polymers* **2017**, *9*, 125. [CrossRef]
21. Yin, H.; Dong, B.; Liu, X.; Zhan, T.; Shi, L.; Zi, J.; Yablonovitch, E. Amorphous diamond-structured photonic crystal in the feather barbs of the scarlet macaw. *Proc. Natl. Acad. Sci. USA* **2012**, *109*, 10798–10801. [CrossRef]
22. Shi, L.; Zhang, Y.; Dong, B.; Zhan, T.; Liu, X.; Zi, J. Amorphous Photonic Crystals with Only Short-Range Order. *Adv. Mater.* **2013**, *25*, 5314–5320. [CrossRef] [PubMed]
23. Hu, Y.; Zhang, Y.; Yang, D.; Ma, D.; Huang, S. Self-assembly of colloidal particles into amorphous photonic crystals. *Mater. Adv.* **2021**, *2*, 6499–6518.
24. Takeoka, Y.; Yoshioka, S.; Takano, A.; Arai, S.; Khanin, N.; Nishihara, H.; Teshima, M.; Ohtsuka, Y.; Seki, T. Production of Colored Pigments with Amorphous Arrays of Black and White Colloidal Particles. *Angew. Chem. Int. Ed.* **2013**, *52*, 7261–7265. [CrossRef] [PubMed]
25. Li, Q.; Zhang, Y.; Shi, L.; Qiu, H.; Zhang, S.; Qi, N.; Hu, J.; Yuan, W.; Zhang, X.; Zhang, K.-Q. Additive Mixing and Conformal Coating of Noniridescent Structural Colors with Robust Mechanical Properties Fabricated by Atomization Deposition. *ACS Nano* **2018**, *12*, 3095–3102. [CrossRef] [PubMed]
26. Wen, X.; Lu, X.; Wei, C.; Li, J.; Li, Y.; Yang, S. Bright, Angle-Independent, Solvent-Responsive, and Structurally Colored Coatings and Rewritable Photonic Paper Based on High-Refractive-Index Colloidal Quasi-Amorphous Arrays. *ACS Appl. Nano Mater.* **2021**, *4*, 9855–9865. [CrossRef]
27. Harun-Ur-Rashid, M.; Imran, A.B.; Seki, T.; Ishii, M.; Nakamura, H.; Takeoka, Y. Angle-Independent Structural Color in Colloidal Amorphous Arrays. *ChemPhysChem* **2010**, *11*, 579–583. [CrossRef] [PubMed]
28. Forster, J.D.; Noh, H.; Liew, S.F.; Saranathan, V.; Schreck, C.F.; Yang, L.; Park, J.-G.; Prum, R.O.; Mochrie, S.G.J.; O'Hern, C.S.; et al. Biomimetic Isotropic Nanostructures for Structural Coloration. *Adv. Mater.* **2010**, *22*, 2939–2944. [CrossRef]
29. Chen, G.; Yi, B.; Huang, Y.; Liang, Q.; Shen, H. Development of bright and low angle dependence structural colors from order-disorder hierarchical photonic structure. *Dyes Pigment.* **2019**, *161*, 464–469. [CrossRef]
30. Kanai, T.; Nakagawa, S.; Takeshima, H.; Hirano, Y. Amorphous photonic structure in a charged colloidal system showing angle-independent uniform color. *J. Colloid Interface Sci.* **2023**, *629*, 225–232. [CrossRef]
31. Sugiyama, K.; Kato, K.; Kido, M.; Shiraishi, K.; Ohga, K.; Okada, K.; Matsuo, O. Grafting of vinyl monomers on the surface of a poly(ethylene terephthalate) film using Ar plasma-post polymerization technique to increase biocompatibility. *Macromol. Chem. Phys.* **1998**, *199*, 1201–1208. [CrossRef]
32. Saunders, B.R.; Vincent, B. Microgel particles as model colloids: Theory, properties and applications. *Adv. Colloid Interface Sci.* **1999**, *80*, 1–25. [CrossRef]
33. Kanai, T.; Yano, H.; Kobayashi, N.; Sawada, T. Enhancement of Thermosensitivity of Gel-Immobilized Tunable Colloidal Photonic Crystals with Anisotropic Contraction. *ACS Macro Lett.* **2017**, *6*, 1196–1200. [CrossRef] [PubMed]
34. Kanai, T.; Kobayashi, N.; Tajima, H. Enhanced linear thermosensitivity of gel-immobilized colloidal photonic crystal film bound on glass substrate. *Mater. Adv.* **2021**, *2*, 2600–2603. [CrossRef]

polymers

MDPI

Communication

Transparent and Flexible Actuator Based on a Hybrid Dielectric Layer of Wavy Polymer and Dielectric Fluid Mixture

Mallappa Mahanthappa, Hyun-U Ko and Sang-Youn Kim *

Advanced Research Technology Center, Korea University of Technology and Education, 1600, Chungjeol-ro, Byeongcheon-myeon, Dongnam-gu, Cheonan-si 31253, Chungchungnam-do, Republic of Korea; mallappa.m@koreatech.ac.kr (M.M.); lostmago@koreatech.ac.kr (H.-U.K.)

* Correspondence: sykim@koreatech.ac.kr; Tel.: +82-(0)41-560-1484

Abstract: Transparent and flexible vibrotactile actuators play an essential role in human–machine interaction applications by providing mechanical stimulations that can effectively convey haptic sensations. In the present study, we fabricated an electroactive, flexible, and transparent vibrotactile actuator with a dielectric layer including a dielectric elastomer and dielectric fluid mixture. The dielectric fluid mixture of propylene carbonate (PC) and acetyl tributyl citrate (ATBC) was injected to obtain a transparent dielectric layer. To further improve the haptic performance, different weight ratios of dielectric fluid (PC: ATBC) were injected. The fabricated vibrotactile actuators based on a transparent dielectric layer were investigated for their electrical and electromechanical behavior. The proposed actuators generate a large vibrational intensity (~2.5 g) in the range of 200–250 Hz. Hence, the proposed actuators open up a new class of vibrotactile actuators for possible use in various domains, including robotics, smart textiles, teleoperation, and the metaverse.

Keywords: vibrotactile; transparent and flexible actuator; dielectric fluid; propylene carbonate; acetyl tributyl citrate

1. Introduction

Beyond the foldable smartphone, lightweight mobile devices such as laptops with flexible displays are competitively announced to attract the interest of potential users. Although the devices appeal to their future-oriented characters and design, they are lacking compared to traditional devices in terms of tactile user experience, for example, typing the keyboard due to the absence of a rigid interface. To this point, vibrotactile haptic actuators have been considered a candidate to apply tactile user experience [1].

Vibrotactile haptic actuators are a kind of haptic actuators that can evoke vibration stimuli in the human skin for inducing haptic feedback in human–machine interactions. For generating vibrational stimuli, a wide variety of actuators, such as eccentric motor-based tactile actuators, linear resonance actuators, and piezoelectric transducers, are commercially used [2–4]. However, the actuators are not applicable for flexible devices due to their rigid body. So, the flexible actuators based on electroactive polymers (EAPs) are highlighted for flexible devices [5–8].

The EAPs are polymers that are deformed by responding to electrical stimuli [9]. Since they convert the electrical signal to a mechanical reaction without external rigid structures, they are investigated for the material of flexible actuators [10–12]. Stalbaum et al. reported a soft robot based on an ionic polymer metal composite [10]. Kim et al. proposed an eco-friendly electroactive paper (EAPap)-based actuator [11]. Choi et al. suggested a soft lens based on PVC gel [12].

Among the EAPs, dielectric elastomers (DEs) are widely investigated for vibrotactile actuators due to their fast response and large deformation [13,14]. Ozsecen et al. reported the vibrotactile display based on DEs in 2010 [13]. Heo et al. designed a haptic actuator based on a wavy shape to improve actuation [14]. However, the low transparency due to the diffuse reflection by complex structures limits the application of the actuator for flexible display. To increase the transparency by removing the diffuse reflection effect, adding a dielectric fluid with a similar refractive index might be one option. A hydraulically amplified self-healing electrostatic (HESEL) actuator is a kind of actuator that achieves high transparency and high performance by adding dielectric fluid [15]. However, the actuator is not applicable for a haptic actuator flexible display due to its low actuation frequency (<100 Hz) and curved shape.

In this paper, we proposed a flexible and transparent actuator with a dielectric layer including a wavy elastomer and dielectric fluid mixture of propylene carbonate (PC) and acetyl tributyl citrate (ATBC). In this mixture, PC plays the role of the high dielectric liquid. ATBC was used as an additive for increasing the dielectric short strength. The dielectric behavior of the dielectric layer was investigated using analysis of capacitance. The transparency of the dielectric layer was around 80%. The optimal concentration ratio of the dielectric mixture was defined from the vibrotactile behavior of actuators with them.

2. Experimental Section

2.1. Materials and Reagents

For the fabrication of the dielectric layer, polydimethylsiloxane (PDMS, Sylgard184, Dow Corning Corp., Midland, TX, USA), Ecoflex (Ecoflex-0010™, Smooth-On Inc., Easton, PA, USA), propylene carbonate anhydrous (PC, 99.7%, Sigmaldrich, St. Louis, MO, USA), and acetyl tributyl citrate (ATBC, 28–30%, Merck, Rahway, NJ, USA) were used.

2.2. Instrumentation

The UV-visible spectrometer (Cary 60, Agilent Technology, Santa Clara, CA, USA) was used to evaluate transparency. The capacitance of the dielectric layer was investigated with an impedance and phase analyzer (SI-1296, AMETEK Scientific Instruments, Oak Ridge, TN, USA). For analyzing vibration performance, the input signal was generated with a function generator (9310, Protek, Gwangmyeong-si, Gyeonggi-do, South Korea) and amplified with a high-voltage amplifier (Trek 10/40A, Denver, CO, USA). The vibration signal was corrected by an accelerometer (Type 4393, Brüel & Kjær, Nærum, Denmark) with a signal amplifier (NEXUS, Brüel & Kjær, Nærum, Denmark). The acceleration signal was displayed and recorded on an oscilloscope (TDS 1012B, Tektronix, Beaverton, OR, USA).

2.3. Fabrication of Dielectric Wavy Polymer

PDMS-Ecoflex wavy dielectric elastomer was fabricated by a simple molding method. In this procedure, the polydimethylsiloxane and Ecoflex solution were placed into a beaker and then stirred for 10 min. The resulting solution was poured carefully onto a wavy-patterned Teflon dish, and this was placed in a vacuum chamber to eliminate any bubbles in the mixture. Finally, the mixture was kept for drying in an electric oven at 100 °C for 3 h, and then, it was allowed to cool down to ambient temperature to obtain a PDMS-Ecoflex wavy-shaped layer. The resulting layer detaches from the Teflon dish and is used for further study.

2.4. Fabrication of Dielectric Fluid Mixture

PC-ATBC dielectric fluid was prepared by mixing a definite amount of polycarbonate and ATBC. The various weight ratios of PC to ATBC (P:A) were 10:0, 9.5:0.5, 9:1, 8:2, 7:3, 6:4, 5:5, and 0:10 and are denoted as PC, $PA_{9.5:0.5}$, $PA_{9:1}$, $PA_{8:2}$, $PA_{7:3}$, $PA_{6:4}$, $PA_{5:5}$, and ATBC, respectively (as shown in Table 1).

Table 1. Code of actuator.

Code	Polymer	Dielectric Liquid Ratio	
		PC	ATBC
B_Air	-	-	-
F_PC	Flat	10	0
W_Air	Wavy	-	-
W_PC	Wavy	10	0
$W\_PA_{9.5:0.5}$	Wavy	9.5	0.5
$W\_PA_{9:1}$	Wavy	9	1
$W\_PA_{8:2}$	Wavy	8	2
$W\_PA_{7:3}$	Wavy	7	3
$W\_PA_{6:4}$	Wavy	6	4
$W\_PA_{5:5}$	Wavy	5	5
W_ATBC	Wavy	0	10

2.5. *Design of Vibrotactile Actuator*

Figure 1a represents the fabrication of a soft and transparent vibrotactile actuator based on a dielectric elastomer. The proposed actuator was composed of a top layer, a dielectric layer, an adhesive pillar, and a bottom layer. The top and bottom layers were made up of indium tin oxide/poly(ethylene terephthalate) (ITO/PET), which is an actuation component, and a flexible supporter. The top layer acts as a cathode, whereas the bottom layer acts as an anode. Then, a PDMS-Ecoflex-based wavy dielectric polymer layer is placed between two electrodes (top and bottom layer) and is referred to as a dielectric layer. Using VHB tape, the edges of the topmost layer are connected to the edges of the bottom layer, which is called the adhesive layer. The dielectric fluid (PC: ATBC) was injected through VHB tape using a syringe, and the wavy-shaped dielectric layer disappeared and obtained a transparent layer (as shown in Figure 1). The sizes of the top and bottom layers are 50 mm (W) × 36 mm (L) × 100 μm (T), whereas the size of the dielectric layer is 23 mm (L) × 23 mm (W) × 1.2 mm (T). Figure 1c shows the flexibility of the actuator. The proposed actuator is highly flexible and transparent; hence, it is used in haptic performance. To improve the transparency of the haptic performance, the dielectric fluid weight ratio is optimized. The haptic performance of the designed actuator greatly depends on the mechanical and dielectric properties of the dielectric layer including elastomer and dielectric liquid. To achieve the best haptic performance, it is necessary to optimize the mechanical properties of the dielectric layer. According to our previous report, the highest vibration performance of the dielectric layer was observed at the weight ratio of PDMS to Ecoflex is 1:5 [14]; hence, it was selected for further studies. To further improve the transparency and vibration performance of the dielectric layer, dielectric fluid with different weight ratios of PC to ATBC was injected. The codes of the actuator are denoted in Table 1.

2.6. *The Vibrotactile Actuator Performance Evaluation*

The actuation performance was investigated by using the accelerometer. The input signal was generated by the function generator and amplified by the high-voltage amplifier. The signal and ground line from the high-voltage amplifier were connected to the bottom and top layers as electrodes, respectively. The accelerometer was located on the top of the vibrotactile actuator. The acceleration signal was amplified by the signal amplifier and collected, displayed, and saved by an oscilloscope. To evaluate the durability of the actuator, 10,000 cycle actuation test was also conducted at normal state and then after 100 times of bending according to ASTM D790 [16].

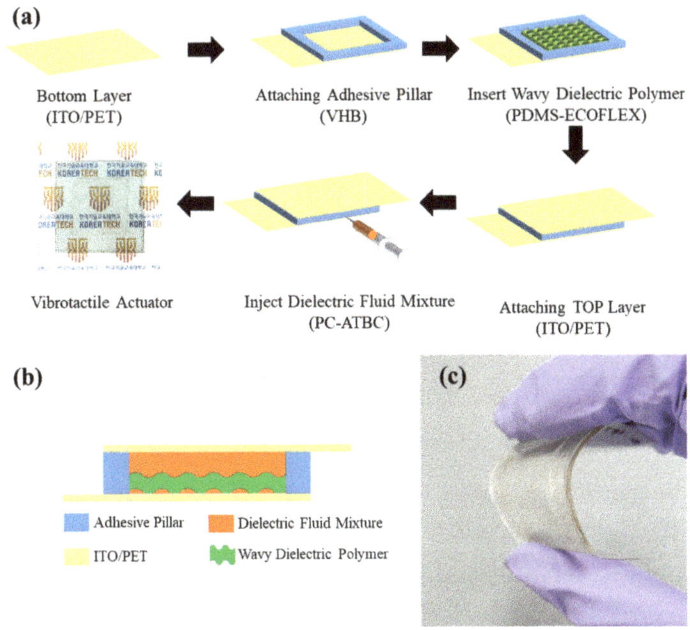

Figure 1. (**a**) Fabrication process, (**b**) cross-section view, and (**c**) bent image of the vibrotactile actuator.

3. Results and Discussion

3.1. Transparency of Dielectric Layers

Prior to the haptic performance, the optical transmittance of the proposed actuators was measured using UV-visible spectroscopy. Figure 2 represents the optical transmittance spectra of the dielectric samples in the absence and presence of dielectric fluid. In the absence of dielectric fluid, the dielectric layer exhibited an optical transmittance of about 40% due to the diffuse reflection by the wavy structure. In the presence of PC and ATBC, the dielectric layer is clearly visible (transparent) and exhibited 84.18% and 87.42% of optical transmittance, respectively. The dramatic improvement is due to liquid with a similar refractive index (PC: 1.42, ATBC:1.44, PDMS: 1.41) covering the wavy structure. In addition, in the presence of dielectric fluid with different weight ratios of PC to ATBC, the dielectric layer exhibited 83.95–86.24% of the optical transmittance observed in the visible region. Although the transparency is comparably lower than the dielectric layer, including the flat polymer and PC, it is sufficiently high for a transparent actuator. The obtained results are summarized in Table 2. Hence, these results suggest that the prepared dielectric samples have high optical transmittance and are suitable for applications in flexible and deformable displays.

Table 2. Averaged transmittance of the actuator.

Sample	T% $_{avg.}$
W_ATBC	87.43
W_Air	40.18
W_PC	84.18
W_PA$_{9.5:0.5}$	83.95
W_PA$_{9:1}$	84.39
W_PA$_{8:2}$	84.46
W_PA$_{7:3}$	85.01
W_PA$_{6:4}$	85.77
W_PA$_{5:5}$	86.24
F_PC	92.59

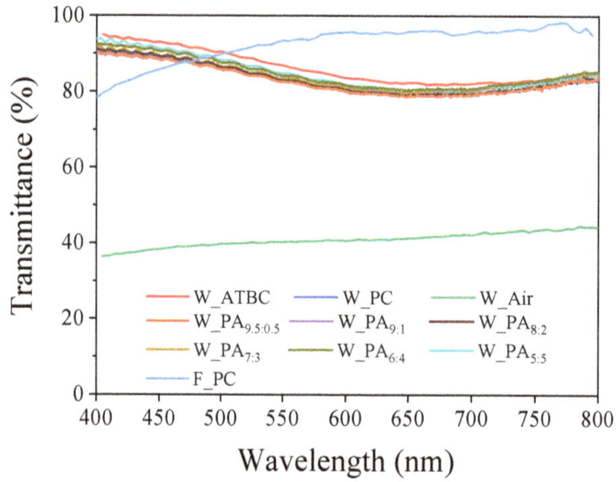

Figure 2. Transmittance of actuators.

3.2. *Capacitance of the Dielectric Layer*

To understand the behavior of the dielectric layer, capacitance-frequency measurements were performed over a frequency range of 10–10^6 Hz, with an AC amplitude of 1 V. Figure 3 depicts the capacitance values for the vibrotactile actuators samples in the absence and presence of dielectric fluid. The capacitance value decreases with increasing frequency for all samples. This feature is commonly observed in propylene carbonate with high purity due to the mobility of ions and interfacial change [17]. At low frequencies, the movement of free anions is directed towards the anode, while free cations move towards the cathode. As a result, oppositely charged ions accumulate at the top and bottom layers. This accumulation leads to an electrostatic attraction between the two surfaces. As the frequency of the AC signal increases, the ion mobility imposes limitations on the polarization response time. At the time, fewer ions adsorbed on the top and bottom layer led to a noticeable decrease in the capacitance value. The gradual decrease in the capacitance at high frequencies is considered normal for capacitors due to the dipolar relaxation of the ions in the dielectric layer. In addition, donor–acceptor interactions on the interface of a solid surface and an organic liquid lead to the generation of surface charge and counterions in the liquid [18]. The surface charge increases the capacitance of the material.

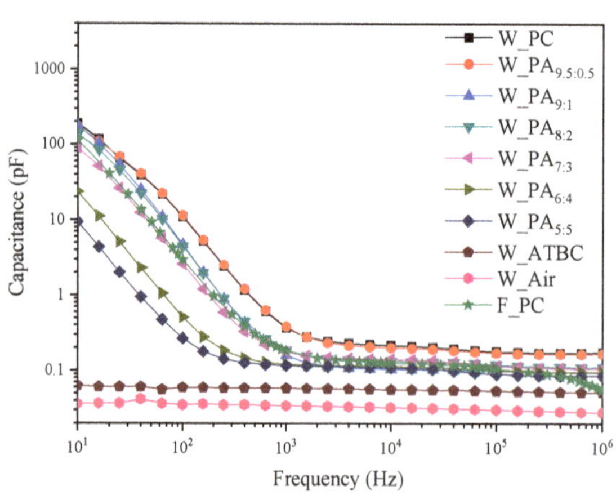

Figure 3. Capacitance of actuators.

When comparing the capacitance values of the wavy-shaped dielectric layer in the absence and presence of ATBC fluid, the capacitance is increased by around two times. However, the capacitance of the W_PC dielectric layer at 10 Hz frequency is four orders higher than the capacitance of W_ATBC due to the high polarity of PC, which leads to improved ion mobility. The capacitance of W_PC is also four times higher than F_PC, caused by the increase in the surface charge by an increase in the surface area. The data show that the actuator with W_PC will perform drastically higher actuation than actuators with W_ATBC and F_PC according to the following equation.

$$a = \frac{CV^2}{md} \quad (1)$$

Here, a, C, V, m, and d are acceleration, capacitance, input voltage, mass, and thickness of the dielectric layer, respectively. However, the high capacitance probably leads to an easy electrical short [19]. So, we propose a dielectric liquid mixture of PC and ATBC in this paper to increase electric stability. The capacitance of dielectric layers including low content of ATBC as $W\_PA_{9.5:0.5}$ and $W\_PA_{9:1}$ is similar to W_PC. As the content of ATBC increases over the ratio, however, the capacitance decreases. From the result, it is estimated that the highest actuation performance might be observed at the $W\_PA_{9.5:0.5}$ and $W\_PA_{9:1}$.

3.3. Haptic Performance of Proposed Vibrotactile Actuators

Further, the acceleration or amplitude or vibration of a dielectric layer of the as-designed actuator was analyzed by acceleration—frequency and acceleration—voltage measurements. In the first experiment, the frequency response of the proposed actuator was performed by sweeping the frequency of the input signal from 0 to 300 Hz using a fixed peak-to-peak input voltage of 2 kV except for W_PC. Since the W_PC was short at the 2 kV, it was measured under a 1 kV input voltage. The $W\_PA_{9.5:0.5}$ and F_PC cases also measured under 1 kV input voltage. The vibrational acceleration of the designed actuator in the absence and presence of dielectric fluid is shown in Figure 4. In the presence of ATBC and PC, the dielectric layer shows negligible and maximum changes in the vibrational amplitude of the proposed actuator. Similarly, increasing the weight ratio of ATBC decreases the vibrational amplitude of the dielectric layer over the frequency range from 200 to 250 Hz as a function of applied frequency due to capacitance variation and the damping effect from the viscosity variation (PC: 2.0 cP, ATBC: 20 cP). The damping effect is also the reason for the broadened resonance peak, whereas the peak frequency shows an increasing trend as the increasing content of ATBC might be due to a lower density of ATBC (1.04 g/L) than PC (1.20 g/L). However, the actuator with a high content of PC shows an easy short at a lower voltage. Moreover, the actuator with a wavy shape of the polymer in the dielectric layer shows a higher vibrational amplitude than the actuator with a flat polymer due to the capacitance variation.

Additionally, the haptic performance of an actuator is influenced not only by its frequency response but also by the amplitude of the input voltage. While the frequency response determines how well the actuator can reproduce different vibration frequencies, the amplitude of the input voltage plays a crucial role in determining the strength or intensity of the haptic feedback. In a second experiment, Figure 5a shows that the haptic measurements were performed by varying the amplitude of the input voltage at the resonant frequencies of actuators. The vibration strength is directly related to the amplitude of the input voltage. At the low input voltage under 1.5 kV, W_PC shows the highest vibration due to high capacitance according to Equation (1). But it was short under 2 kV input voltage with 200 Hz frequency. On the other hand, the W_ATBC bears the highest input voltage over 3 kV at a higher frequency (250 Hz). The bearable input voltage increases as the content of ATBC increases. Even though the resonance frequency used for actuation increases depending on the increasing content of ATBC, the same trend is shown. Considering the proportional relation between frequency and AC conductivity, the result suggests the enhancement of stability by adding ATBC. By the combination of capacitance and electric stability, the actuator with a dielectric fluid of weight ratio $PA_{9:1}$ exhibits the highest performance among all the prepared vibrotactile actuators under 2.5 kV input voltage.

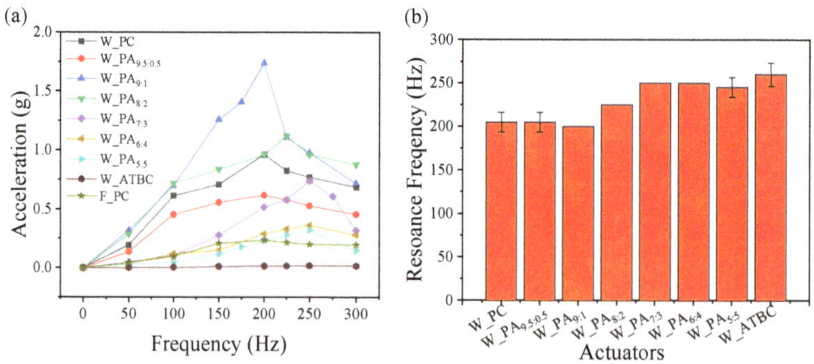

Figure 4. (**a**) Vibration acceleration of vibrotactile actuator under various input frequencies and (**b**) the averaged resonance frequencies for actuators.

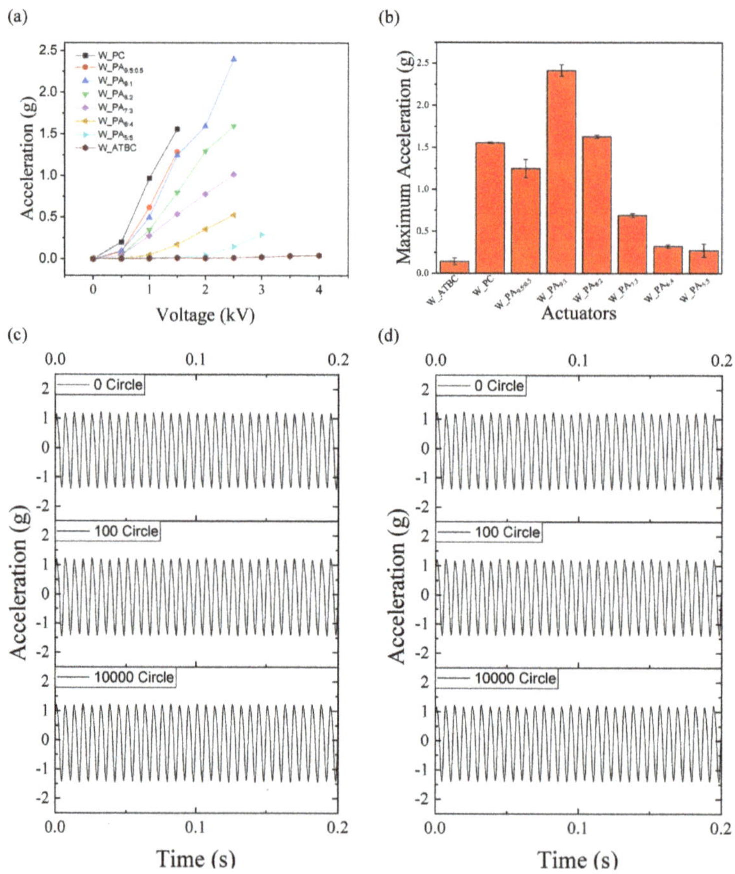

Figure 5. (**a**) Vibration acceleration of vibrotactile actuator under various input voltage, (**b**) maximum acceleration of various actuators, and 10,000 cycle actuation test result (**c**) at normal state and (**d**) after 100 times bending.

4. Conclusions

In summary, the fabrication of a thin, flexible, transparent, and lightweight vibrotactile actuator based on a transparent dielectric layer including dielectric elastomer and dielectric fluid mixture. The proposed vibrotactile actuator is composed of a top layer, an adhesive layer, a wavy-shaped dielectric layer, and a bottom layer. To obtain a transparency of the dielectric layer, the dielectric fluid mixture of PC and ATBC was injected. By optimizing the various dielectric fluid weight ratios of PC:ATBC, a better vibrotactile performance was determined. The fabricated vibrotactile actuator-based transparent dielectric layers were subjected to capacitance, transparency, resonance frequency, and vibration strength measurements. The results obtained from the electrical measurement, the high capacitance value at low frequencies, and the gradual decrease in the capacitance value at high frequencies indicate the proposed dielectric layer is a suitable choice for actuators with a large deformation. Additionally, the proposed actuator produces large vibrational intensity in the range of 200–250 Hz and a high vibrational strength at lower applied voltages. Among the different weight ratios of PC:ATBC, $PA_{9:1}$ exhibited the maximum vibrational intensity at 200 Hz and vibrational strength at 2.5 g. Therefore, the proposed haptic actuator is set to become a fundamental component in the future of smart devices. With its advanced tactile feedback capabilities, this technology will revolutionize the user experience across a range of devices.

Author Contributions: Conceptualization, H.-U.K. and S.-Y.K.; Investigation, H.-U.K. and M.M.; Supervision, S.-Y.K.; Writing—original draft, H.-U.K. and M.M.; Writing—review and editing, H.-U.K. and M.M. All authors contributed to the article and approved the submitted version. H.-U.K. and M.M. equally contributed. All authors have read and agreed to the published version of the manuscript.

Funding: This work was supported by Priority Research Centers Program through the National Research Foundation of Korea (NRF) funded by the Ministry of Education (NRF-2018R1A6A1A03025526) and supported by Materials and Components Technology Development Program (20011013) funded by the Ministry of Trade, Industry and Energy (MOTIE, Korea).

Institutional Review Board Statement: Not applicable.

Data Availability Statement: Data are contained within this manuscript.

Acknowledgments: Thanks to Hyun-Jeong Kim for experimental support. Thanks to the Cooperative Equipment Center at Koreatech for analytic assistance.

Conflicts of Interest: The authors declare no conflict of interest.

References

1. Park, G.; Cha, H.; Choi, S. Haptic enchanters: Attachable and detachable vibrotactile modules and their advantages. *IEEE Trans. Haptics* **2018**, *12*, 43–55. [CrossRef] [PubMed]
2. Kim, S.Y.; Kim, K.Y.; Soh, B.S.; Yang, G.; Kim, S.R. Vibrotactile rendering for simulating virtual environment in a mobile game. *IEEE Trans. Consum. Electron.* **2006**, *52*, 1340–1347. [CrossRef]
3. Huang, H.; Li, T.; Antfolk, C.; Enz, C.; Justiz, J.; Koch, V.M. Experiment and investigation of two types of vibrotactile devices. In Proceedings of the 6th IEEE International Conference on Biomedical Robotics and Biomechatronics, Singapore, 26–29 June 2016.
4. Logothetis, I.; Matsouka, D.; Vassiliadis, S.; Vossou, C.; Siores, E. Optimum operating conditions for PZT actuators for vibrotactile wearables. *J. Electron. Mater.* **2018**, *47*, 3709–3716. [CrossRef]
5. Poncet, P.; Casset, F.; Latour, A.; Domingues Dos Santos, F.; Pawlak, S.; Gwoziecki, R.; Devos, A.; Emery, P.; Fanget, S. Static and dynamic studies of electro-active polymer actuators and integration in a demonstrator. *Actuators* **2017**, *618*, 18. [CrossRef]
6. Choi, D.-S.; Kim, S.-Y. Conceptual design of soft and transparent vibrotactile actuator. In Proceedings of the International Asia Haptics Conference, Singapore, 11–14 November 2018.
7. Lotz, P.; Matysek, M.; Schlaak, H.F. Fabrication and application of miniaturized dielectric elastomer stack actuators. *IEEE ASME Trans. Mechatron.* **2010**, *16*, 58–66. [CrossRef]
8. Park, W.H.; Shin, E.J.; Yun, S.; Kim, S.Y. An enhanced soft vibrotactile actuator based on ePVC gel with silicon dioxide nanoparticles. *IEEE Trans. Haptics* **2018**, *11*, 22–29. [CrossRef] [PubMed]
9. Bar-Cohen, Y.; Anderson, I.A. Electroactive polymer (EAP) actuators—Background review. *Mech. Soft Mater.* **2019**, *1*, 5. [CrossRef]

10. Stalbaum, T.; Hwang, T.; Trabia, S.; Shen, Q.; Hunt, R.; Olsen, Z.; Kim, K.J. Bioinspired travelling wave generation in soft-robotics using ionic polymer-metal composites. *Int. J. Mech. Eng. Robot. Appl.* **2017**, *1*, 167–179. [CrossRef]
11. Kim, J.; Seo, Y.B. Electro-active paper actuators. *Smart Mater. Struct.* **2022**, *11*, 355. [CrossRef]
12. Choi, D.S.; Jeong, J.; Shin, E.J.; Kim, S.Y. Focus-tunable double convex lens based on non-ionic electroactive gel. *Optics Express* **2017**, *25*, 20133–20141. [CrossRef] [PubMed]
13. Ozsecen, M.Y.; Sivak, M.; Mavroidis, C. Haptic Interfaces Using Dielectric Electroactive Polymer. In Proceedings of the SPIE Smart Structures and Materials + Nondestructive Evaluation and Health Monitoring, San Diego, CA, USA, 21 April 2010; Volume 7647.
14. Heo, Y.H.; Choi, D.S.; Kim, D.E.; Kim, S.Y. Flexible vibrotactile actuator based on dielectric elastomer for smart handheld devices. *Appl. Sci.* **2021**, *11*, 12020. [CrossRef]
15. Won, P.; Kim, K.K.; Kim, H.; Park, J.J.; Ha, I.; Shin, J.; Jung, J.; Cho, H.; Kwon, J.; Lee, H.; et al. Transparent soft actuators/sensors and camouflage skins for imperceptible soft robotics. *Adv. Mater.* **2021**, *33*, 2002397. [CrossRef] [PubMed]
16. *ASTM D790*; Standard Test Methods for Flexural Properties of Unreinforced and Reinforced Plastics and Electrical Insulating Materials. ASTM International: West Conshohocken, PA, USA,, 2017.
17. Choudhary, S.; Dhatarwal, P.; Sengwa, R.J. Characterization of conductivity relaxation processes induced by charge dynamics and hydrogen-bond molecular interactions in binary mixtures of propylene carbonate with acetonitrile. *J. Mol. Liq.* **2017**, *231*, 491–498. [CrossRef]
18. Labib, M.E. The origin of the surface charge on particles suspended in organic liquids. *Colloids Surf.* **1998**, *29*, 293–304. [CrossRef]
19. Hossick-Schott, J. Prospects for the ultimate energy density of oxide-based capacitor anodes. In Proceedings of the 21st Capacitor and Resistor Symposium, Barcelona, Spain, 10 January 2007.

polymers

MDPI

Article

Agarose-Based Hydrogel Film with Embedded Oriented Photonic Nanochains for Sensing pH

Dunyi Xiao [1], Huiru Ma [2,*], Wei Luo [1,*] and Jianguo Guan [3,4]

1 School of Materials Science and Engineering, Wuhan University of Technology, Wuhan 430070, China; 317288@whut.edu.cn
2 School of Chemistry, Chemical Engineering and Life Science, Wuhan University of Technology, Wuhan 430070, China
3 State Key Laboratory of Advanced Technology for Materials Synthesis and Processing, International School of Materials Science and Engineering, Wuhan University of Technology, Wuhan 430070, China; guanjg@whut.edu.cn
4 Wuhan Institute of Photochemistry and Technology, 7 North Bingang Road, Wuhan 430083, China
* Correspondence: mahr@whut.edu.cn (H.M.); rowell@whut.edu.cn (W.L.)

Abstract: Responsive photonic crystal hydrogel sensors are renowned for their colorimetric sensing ability and can be utilized in many fields such as medical diagnosis, environmental detection, food safety, and industrial production. Previously, our group invented responsive photonic nanochains (RPNCs), which improve the response speed of photonic crystal hydrogel sensors by at least 2 to 3 orders of magnitude. However, RPNCs are dispersed in a liquid medium, which needs a magnetic field to orient them for the generation of structural colors. In addition, during repeated use, the process of cleaning and redispersing can cause entanglement, breakage, and a loss of RPNCs, resulting in poor stability. Moreover, when mixing with the samples in liquid, the RPNCs may lead to the contamination of the samples being tested. In this paper, we incorporate one-dimensional oriented RPNCs with agarose gel film to prepare heterogeneous hydrogel films. Thanks to the non-responsive and porous nature of the agarose gel, the protons diffuse freely in the gel, which facilitates the fast response of the RPNCs. Furthermore, the "frozen" RPNCs in agarose gel not only enable the display of structural colors without the need for a magnet but also improve the cycling stability and long-term durability of the sensor, and will not contaminate the samples. This work paves the way for the application of photonic crystal sensors.

Keywords: responsive photonic nanochains; pH-responsive photonic crystals; fast response; photonic crystal heterogeneous gels

1. Introduction

Integrating stimulus-responsive hydrogels or polymers with photonic crystals allows the conversion of stimulus signals into visible color patterns [1]. Previous researchers have made substantial contributions in related studies, achieving ion responsiveness [2–6], molecular responsiveness [7–13], temperature responsiveness [14–17], and electrical responsiveness [18–21], playing a significant role in sensing, medical, environmental, and display fields. However, the use of traditional homogeneous hydrogels or polymers poses two challenges: first, hindered diffusion of analyte molecules through the hydrogel [22,23]; second, the relaxation time of the gel is directly proportional to the square of the gel size [24]. Both issues contribute to prolonged response times for responsive photonic crystals, thus affecting their practical applicability. Our team has previously reported the preparation of various responsive photonic nanochains (PNCs) by using hydrogen-bond-guided template polymerization to orient steric-repulsion stabilized magnetic nanoparticles and lock them into one-dimensional structures [25–27]. The surface-responsive gel shell layer, only tens of nanometers thick, overcomes both issues, ensuring rapid response times for responsive photonic crystals in practical applications.

However, currently developed responsive PNCs, when utilized, can only be suspended in liquid, requiring a continuous application of the magnetic field to maintain the orientation for the revealing of structural colors. Additionally, during repeated use, the cleaning and redispersion process poses a risk of chain loss, entanglement, or breakage, leading to a deterioration in optical performance. Lastly, the suspended nature of the chain structure in the liquid makes it easy to cause sample contamination as it drifts with the solution.

To address the above issues, we propose the utilization of responsive PNCs as chromogenic units and their immobilization within a non-responsive porous network gel, resulting in the fabrication of heterogeneous gel films. The non-responsive porous gel chosen for immobilization of the PNCs is agarose gel, known for its tunable pore structure, with the diffusion rate of most analytes within agarose gel nearly equivalent to that in pure water [28,29]. This characteristic ensures the rapid response capability of responsive nanochains in agarose gel. Additionally, agarose is primarily derived from red algae, where it serves as the main polysaccharide component. Agarose gel exhibits excellent biocompatibility, making it commonly used in biomedical fields such as tissue engineering and drug delivery systems. The microstructure and composition of the film were characterized using scanning electron microscopy (SEM), optical microscopy, thermogravimetric analysis (Tg), and Fourier-transform infrared spectroscopy (FTIR). The optical and responsive properties of the heterogeneous film were investigated using a fiber optic spectrometer. Furthermore, the mechanical properties of the heterogeneous gel film were examined using a rotational rheometer.

2. Materials and Methods

2.1. Materials

Acrylic acid (AA, 99%), 2-hydroxethyl methacrylate (HEMA, 99%), 2-hydroxy-2-methylpropiophenone (HMPP, 97%), ethylene glycol dimethacrylate (EGDMA, 98%), 3-Acrylamide phenylboronic acid (AAPBA, 99%), N-(2-Hydroxyethyl) acrylamide (HEAAm, 98%), N,N′-methylenebis(2-propenamide) (BIS, 99%), Acrylic amide (AM, 99%), and agarose were purchased from Aladdin Reagent Co., Ltd. (Shanghai, China). Ethanol, dimethyl sulfoxide (DMSO), and ethylene glycol (EG) were purchased from Sinopharm Chemical Reagent Co., Ltd. (Shanghai, China). All chemicals were used directly as received without further purification. Deionized water (18.20 MΩ·cm) was produced in a Milli-Q system (Millipore, Burlington, MA, USA). Superparamagnetic Fe_3O_4@PVP nanoparticles were synthesized according to the previously reported method [30].

2.2. Preparation of pH RPNCs and Responsive Photonic Nanoparticles

pH RPNCs were synthesized according to the previously reported method [25]. First, we took 4 mL of ethanol dispersion containing Fe_3O_4@PVP nanoparticles, removed the supernatant after centrifugation, added 2 mL of EG, and sonicated until it was uniformly dispersed. The particle concentration of this solution was 15 mg/mL. A total of 720 μL of the above-mentioned dispersion was taken, into which 6.3 g of EG, 19.4 mg of the photoinitiator HMPP, 14.4 mg of the crosslinker EGDMA, 0.26 g of the monomer AA, 0.31 g of the monomer HEMA, and 2.04 g of deionized water were added. After ultrasonic mixing, a brown precursor solution was obtained, which was stored in the dark under refrigeration. After that, the above precursor solution was transferred into a 500 mL glass beaker, which was then placed in a uniform magnetic field of 180 Gs. The polymerization reaction was initiated by exposing the solution to UV light (365 nm wavelength, manufactured by Shenzhen Sankun Technology Co., Ltd. (Shenzhen, China)) for 5 min. Subsequently, RPNCs were washed several times with an ethanol–water mixed solution (50 vol%). The supernatant was decanted, and then 600 μL of the NaOH solution (ionic concentration of 50 mM) was added, mixed well, and left for magnetic settling. Finally, after decanting the supernatant, the RPNCs were dispersed in deionized water for magnetic settling and stored for later use.

2.3. Preparation of pH-Responsive Photonic Nanochains/Agarose Heterogeneous Gel Films

Firstly, 1 g of agarose was added to 10 mL of DMSO and heated to 80 °C with stirring for 1 h to completely dissolve the agarose, preparing a stock solution with a concentration of 0.1 g/mL. Then, 40 μL of the agarose containing the DMSO stock solution, 280 μL of deionized water dispersion containing pH RPNCs, and 90 μL of DMSO were mixed uniformly at 80 °C. Subsequently, adhesive pads with different thicknesses (having a circular cavity with a diameter of 20 mm in the middle) were attached to a glass substrate, and the glass substrates were heated to 60 °C. The above-mentioned mixture solution at 80 °C was transferred to the circular cavity, and a cover glass was placed on top to evenly fill the cavity with the mixture solution. A uniform magnetic field of 600 Gs was applied to orient the RPNCs along the magnetic field direction. Once the glass substrate cooled to room temperature, the pH-responsive heterogeneous hydrogel film was obtained. Finally, the pH-responsive heterogeneous hydrogel film was washed off the glass substrate with deionized water and rinsed three times with deionized water.

2.4. Characterizations

All digital photos in this paper were taken using an iPhone. Optical microscopy images were recorded using an optical microscope (Zeiss Axio Observer 5 M, Gottingen, Germany). The field emission SEM (FE-SEM) images were captured on a Hitachi S-4800 (Tokyo, Japan) scanning electron microscope. A NETZSCH-STA2500 (Yokohama, Japan) Regulus instrument was used to conduct thermal analysis in the air at room temperature to 1000 °C and a heating rate of 5 °C/min. A Nicolet 6700 (Massachusetts, America) spectrometer was used to collect Fourier transform infrared (FTIR) spectra in the range of 400–4000 cm^{-1} with a resolution of 4 cm^{-1}. The reflectance spectra were all obtained by using a fiber optic spectrometer USB 2000+ (Shanghai, China). The shear stress–strain curve data were obtained using the Anton paar MCR 302 (Graz, Austria) rotational rheometer at room temperature with an angular frequency of 5 rad/s. Data processing and plotting were performed using Origin 2021, and the spectral analysis software used was BiaoQi SpecSuite 2018.

3. Results and Discussion

Figure 1 depicts the preparation process of the pH-responsive heterogeneous hydrogel film. Initially, a mixed solution containing pH RPNCs, agarose gel, and solvents was heated to 70 °C and then dropped into the cavity of the double-sided adhesive. Subsequently, it was covered with a glass cover (to ensure that the solution would not solidify too quickly, the glass substrate remained on the heating plate throughout the process). A uniform magnetic field was then applied to orient the RPNCs along the magnetic field lines, after which the heating was turned off and the temperature was allowed to decrease to room temperature. During cooling, the agarose molecules bound together via hydrogen bonding, intertwining to form an agarose gel network with a double helix structure, thereby immobilizing the oriented RPNCs. Due to the fixed particle spacing of the RPNCs, the structural color from PNCs remains stable during the solidifying process of agarose gel.

In the prepared pH-responsive heterogeneous hydrogel film, the agarose gel matrix itself is not pH-responsive. Changes in pH do not cause the agarose gel matrix to expand or contract, and its volume remains constant. On the other hand, after being immobilized within the agarose gel matrix, the RPNCs still exhibit pH responsiveness. Their chain length undergoes expansion or contraction with changes in pH, thereby altering the interparticle spacing and ultimately resulting in color changes. The agarose gel matrix serves merely as a scaffold in the system, maintaining the one-dimensional orientation of the pH RPNCs. The RPNCs, on the other hand, play a role in forming structural colors and responding to external stimuli. Specifically, when the pH-responsive heterogeneous hydrogel film transitions from a high-pH environment to a low-pH environment, or vice versa, the agarose gel matrix, unaffected by pH changes, remains unchanged in volume. However, the pH-responsive gel shell layers on the immobilized RPNCs are influenced by changes in the proton concentration, undergoing protonation or deprotonation processes. The transition between carboxylic acid and carboxylate groups leads to the contraction or expansion of the gel shell layer, ultimately resulting in changes in the interparticle spacing of the RPNCs, as described by the Bragg diffraction formula:

$$m\lambda = 2nd\sin\theta \quad (1)$$

where m represents the diffraction order, λ is the wavelength of the diffracted light, n stands for the effective refractive index, d denotes the lattice spacing, and θ is the angle between the incident light and the lattice plane. Under the same testing conditions, for the same sample, the diffraction order, effective refractive index, and angle between the incident light and the lattice plane remain consistent. The wavelength of the diffracted light is only influenced by the lattice spacing, while changes in particle spacing result in changes in the diffracted wavelength. Macroscopically, this leads to observable changes in the overall color of the pH-responsive heterogeneous hydrogel film.

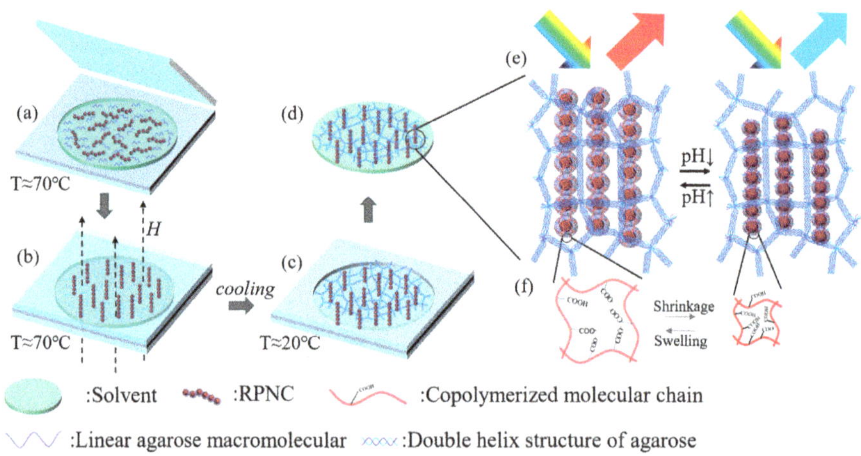

Figure 1. (**a–d**) are schematic of the preparation of pH-responsive heterogeneous gel film, the solution in the figure is a DMSO–water mixed solution (with DMSO 30 vol%), and the color shown in the figure is not the actual color of the solution; it is chosen merely for illustrative purposes. (**e**,**f**) are schematic diagrams illustrating the pH response mechanism.

As shown in Figure 2a, the optical microscope and SEM images indicate that the product is composed of responsive chains and a gel network, with the RPNCs being oriented and fixed within the gel under a magnetic field. Optical microscopy reveals that the typical sample's membrane thickness is around 100 μm, and the RPNCs in the solution are not freely floating and dispersed like those in a liquid. Instead, they are orderly arranged together, indicating the presence of a gel network structure that immobilizes them. The cross-sectional SEM images in Figure 2b also show that the RPNCs are attached to the walls of the macroporous structure, which may occur when the gel loses water and the chains adhere to the walls of the matrix gel pores. This observation indicates the presence of a gel structure, in addition to the chains, that forms aerogels after dehydration.

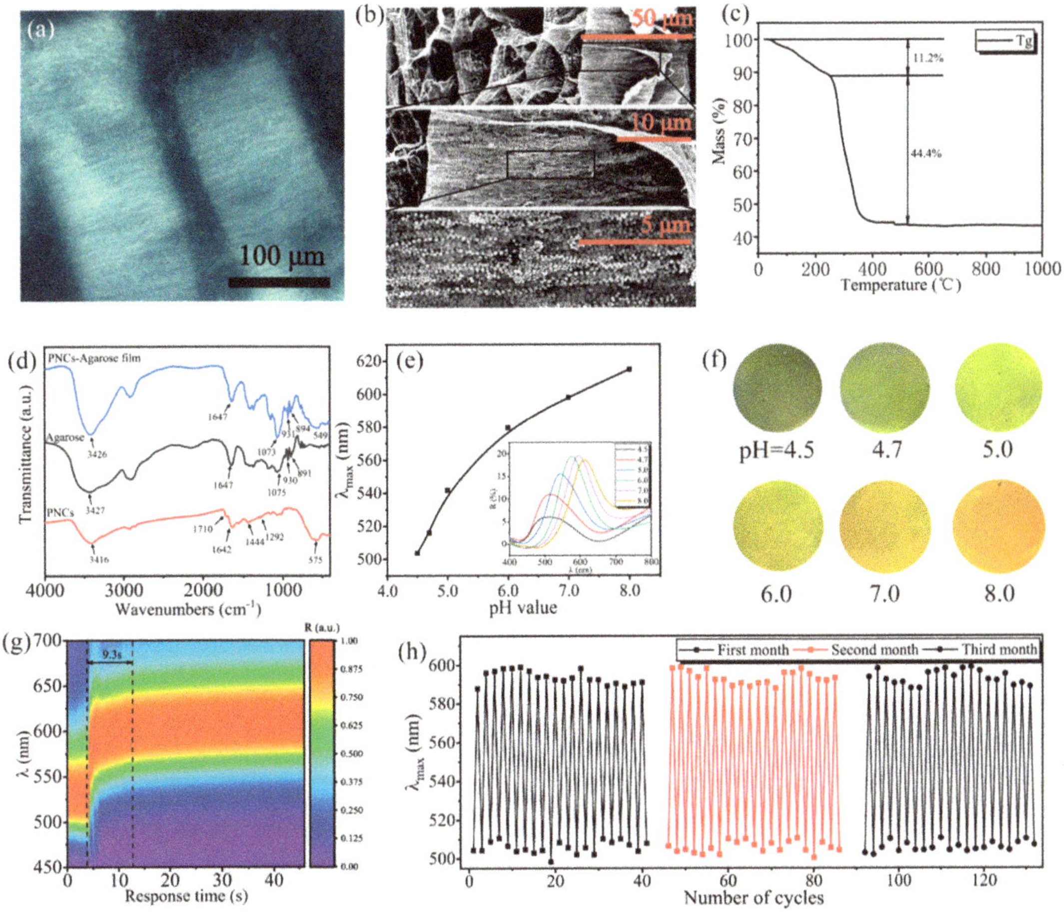

Figure 2. (**a**) Optical microscope images for cross–sectional view of oriented pH–responsive heterogeneous gel film; (**b**) SEM images and localized magnifications of the freeze–dried pH–responsive heterogeneous gel films; (**c**) TG curve of the dried pH–responsive heterogeneous gel film; (**d**) Infrared spectra of dried pH-responsive photonic nanochains, agarose gel film, and pH–responsive heterogeneous gel film; (**e**) diffraction peak position curves and diffraction spectra change with pH values in different buffer solutions for the pH–responsive heterogeneous gel film; (**f**) photographs for pH–responsive heterogeneous gel film in buffer solutions with different pH values; (**g**) the diffraction spectra of the 100 μm thick pH–responsive heterogeneous gel film change over time as it transitions from pH 4.5 to 8.0 buffer solution; (**h**) stability tests for the pH–responsive heterogeneous gel film in pH 4.5 and 8.0 buffer solutions; stability tests were conducted after one month and two months of placement.

From the thermogravimetric curve shown in Figure 2c, it can be observed that during the process from 250 °C to around 400 °C, the sample undergoes the decomposition of organic components, resulting in a final residue content of 44.4% inorganic material. This result is compared with the thermogravimetric curve of the RPNCs (in Figure S1), where the final residue content of inorganic material in the PRNCs is 67.64%. With a chain concentration of 9 mg/mL in the pH-responsive heterogeneous gel film, the calculated inorganic content in the RPNCs is 6.084 mg/mL. With an agarose gel concentration of 7.5 mg/mL, the calculated final residue inorganic content matches the measured value at 44.16%. From the infrared spectrum of RPNCs in Figure 2d, the absorption band at 3416 cm^{-1} corresponds to the O-H stretching vibration peaks of HEMA and AA, while the peak at 1710 cm^{-1} represents the C=O stretching vibration peaks of carboxyl and ester groups. The absorption peaks at 1642 cm^{-1}, 1444 cm^{-1}, and 1292 cm^{-1} originate from the PVP shell layer of the particles, and the peak at 575 cm^{-1} arises from the stretching vibration of Fe-O. The smaller area and weaker organic absorption peaks indicate a lower content of the gel shell layer in the responsive photonic nanochains. In the infrared spectrum of agarose, the absorption band at 1647 cm^{-1} corresponds to the bending vibration peak of O-H, while the bands at 1075 cm^{-1} represent the stretching vibration peaks of C-O. The characteristic peaks at 930 cm^{-1} and 891 cm^{-1} are attributed to the characteristic absorption peaks of 3,6-anhydrogalactose and the C-H bending vibration absorption peaks of pyranose. The consistency of the infrared spectrum peaks among the pH-responsive heterogeneous hydrogel film, RPNCs, and pure agarose indicates that there are no new chemical bonds formed between PNCs and agarose gel, but rather a physical blend.

Figure 2e,f depicts the optical performance testing of the pH-responsive heterogeneous hydrogel film. The film exhibits excellent responsiveness in different pH environments. Typical samples were prepared and placed in different pH buffer solutions without applying a magnetic field, with the solution ion strength at 150 mM. The peak position of pH-responsive heterogeneous gel film shifted by nearly 110 nm, indicating a color change across the green–yellow–orange–red spectrum.

Achieving a rapid response to external analytes has always been a crucial process in advancing the practical application of responsive photonic crystal hydrogel materials. Figure 2g demonstrates that the pH-responsive heterogeneous gel film exhibits rapid responsiveness based on its unique structural advantages. During the process of changing the solution environment, the peak shift occurs very quickly. When typical samples are exchanged from a pH 4.5 to 8.0 buffer solution environment, the time taken to record the wavelength at which the peak reaches 90% equilibrium (referred to as τ_{90}) is measured. The results indicate that the typical samples reach a stable τ_{90} time of approximately 9.3 s, which is several orders of magnitudes shorter compared to the time taken for other pH-responsive homogeneous films of similar thickness [5]. Although reducing the thickness of pH-responsive films to several micrometers may also accelerate the response time, they could not form a free-standing film due to the insufficient mechanical properties [31,32]. We also tested the response time curves of the pH-responsive heterogeneous gel film in different pH environments, as shown in Figure S2.

In addition, using a non-responsive gel to immobilize RPNCs brings another advantage, enhancing stability and convenience. Figure 2h demonstrates that over 20 cycles, the pH-responsive heterogeneous gel film can still maintain good responsiveness, with the peak position remaining fixed. Moreover, after being tested under the same conditions one month and two months later, it still exhibits good responsiveness. In comparison to the current development where the process of washing and redispersing RPNCs during repeated use can easily lead to losses, entanglement, or breakage, thereby deteriorating their optical performance, the introduction of agarose gel enables the pH-responsive heterogeneous gel film to overcome these issues. Photonic crystal heterogeneous gel films should be stored in a sealed container containing pure water to prevent evaporation. It is best to store them at room temperature (recommended temperature: 10–30 °C). Temperatures exceeding 50 °C may cause the non-responsive agarose matrix gel to undergo a solid–liquid transition. As shown in Figure S3, the storage modulus (G′) and loss modulus (G″) curves start to intersect when the temperature rises from room temperature to 68 °C, indicating that the gel film transitions from a solid state to a flowing sol state at this temperature. This behavior is related to the properties of the agarose used. We also tested the effect of different temperatures on the optical performance of the pH-responsive heterogeneous hydrogel film. In addition to examining the effect of temperature on the heterogenous gel film through rheological properties, we also tested its optical properties. During the temperature increase from 7 °C to 52 °C, no changes in optical performance were observed, particularly in the position of the diffraction peaks, as shown in Figure S4. Before use, the photonic crystal heterogeneous gel film can be rinsed 2–3 times with deionized water and then directly placed into the solution environment for testing.

Figure 3a–f depicts experimental investigations by immobilizing PNCs in agarose gels of varying concentrations and subjecting them to buffer solutions at different pH values. From Figure 3a–e, it can be inferred that the diffraction peak positions remain consistent in pH 8.0 buffer solutions, with only the peak width continuously increasing. This indicates that the portion of RPNCs embedded remains the same under different gel concentrations. However, during the process of immobilization, the ordered arrangement of RPNCs is slightly disturbed by the agarose gel matrix, leading to an increase in peak width.

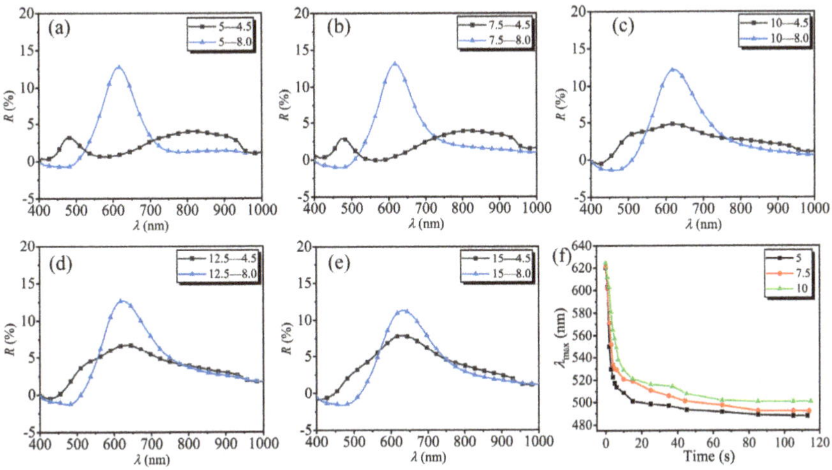

Figure 3. (**a–e**) Digital diffraction spectra images for pH−responsive HGF in buffer solutions with pH 4.5 and 8.0, at different agarose gel concentrations. In the legend, the first digit (5–15) represents the agarose gel concentration in pH−responsive HGF (in mg/mL) and the second digit (4.5 and 8.0) represents the pH of the solution environment where pH−responsive HGF is present; (**f**) curve showing the variation (5, 7.5, and 10 mg/mL) of diffraction peak positions over time as pH−responsive HGF transitions from a buffer solution with pH 8.0 to pH 4.5.

In addition, the size of non-responsive gel pores also affects the response time of pH-responsive heterogeneous hydrogel films. The curve showing the peak position changes over time in Figure 3f provides a more intuitive demonstration of the effect of gel concentration on response time. At the same time point, hydrogel films with lower gel concentrations achieve equilibrium in a shorter time, with a shorter τ_{90}. When transitioning the pH-responsive heterogeneous hydrogel film from a pH 8.0 buffer solution to a pH 4.5 buffer solution, hydrogen ions rapidly diffuse into the interior of the agarose gel network. However, in hydrogel films with higher gel concentrations, the pores are smaller, resulting in stronger confinement of the embedded RPNCs. Consequently, the peak shift in the diffraction of RPNCs is smaller upon pH changes. This phenomenon is more pronounced in pH-responsive heterogeneous hydrogel films with gel concentrations exceeding 10 mg/mL. There exists a significantly low reflectance peak around 620 nm, indicating the presence of particle spacing like that at pH 8.0.

Table 1 presents the comparison of diameters for pure gel films of different concentrations in buffer solutions with varying pH values. From the table, it can be observed that there is no significant volume change for the same agarose gel film in buffer solutions with different pH values (the same gel film was used for each gel concentration). Actual images can be found in Figure S5. Unlike responsive gel films, agarose gel, as a non-responsive gel material, does not undergo volume changes in different pH environments due to the absence of pH-responsive groups in its network structure. Combining the observed phenomenon, when RPNCs are embedded in agarose gel, a shift in diffraction peak occurs. It can be inferred that the only components undergoing deformation in the system are the RPNCs and most of the volume remains unchanged, thus avoiding the phenomenon reported in the literature where gel expansion leads to slower responsiveness.

Table 1. Table comparing the diameters measured for pure gel films of different agarose gel concentrations placed in pH 4.5 and pH 8.0 buffer solutions.

pH Value \ Gel Concentration (mg/mL)	5	7.5	10
4.5	19.27	19.60	20.45
8.0	19.23	19.65	20.50

The unit for diameter in the table is mm.

Figure 4a,b demonstrates the diffraction spectra of pH-responsive heterogeneous hydrogel films formed by low-concentration gelatin gel-fixing responsive particles, which vary with the magnetic field. We synthesized responsive particles under conditions without applying a magnetic field (the preparation process is consistent with RPNCs, except that no magnetic field was applied during polymerization) and used the same method with gelatin gel to prepare heterogeneous gel particle films under the same applied magnetic field. We found that under conditions without applying a magnetic field, structural colors could not be diffracted, indicating that RPNCs are indispensable in the formation of pH-responsive heterogeneous hydrogel films. Due to the lack of fixed-chain structures, responsive particles can move freely between pores, resulting in structural colors appearing with the application of a magnetic field. As the concentration of gelatin gel increases, the pore structure decreases, and the regular arrangement of particles under the magnetic field is restricted by spatial hindrance, leading to a decrease in reflectance. At a high gel concentration, as the pore structure begins to shrink, some of the particles' orientation states can be fixed, allowing peaks to appear even without applying a magnetic field, as shown in Figure 4b. These observations are consistent with those reported by previous researchers [33].

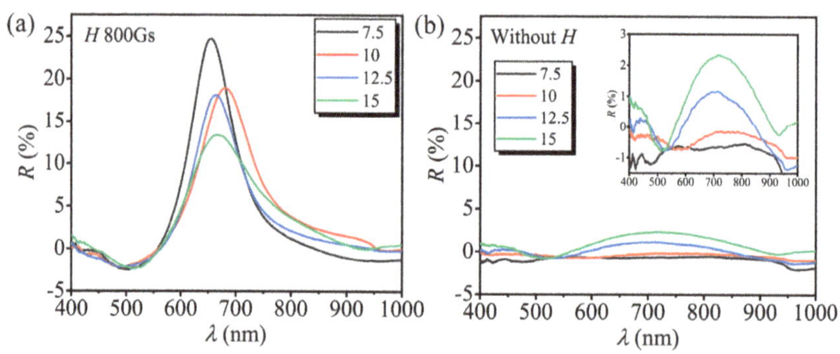

Figure 4. Diffraction spectra of heterogeneous gel films prepared with pH−responsive particles fixed in agarose gel at different concentrations (in mg/mL) under an applied magnetic field (**a**) and without (**b**). The illustration shows a modified version of (**b**) with an adjusted Y−axis to make the diffraction peaks more clearly visible.

We also investigated the effects of the chain concentration and film thickness on the optical, responsive, and mechanical properties of pH-responsive heterogeneous hydrogel films. Firstly, as shown in Figure 5a, it can be observed that with increasing chain concentration from 3 mg/mL to 9 mg/mL, the spectral reflectance increases by approximately 13%. This is because an increase in chain concentration leads to an increase in diffraction units within the pH-responsive heterogeneous hydrogel film. However, further increasing the chain concentration only results in slight fluctuations in reflectance, indicating that when RPNCs reach 9 mg/mL, the maximum limit of RPNCs that can diffract structural colors within the gel has been reached. During the incorporation of RPNCs, it was found that as the chain concentration increases, the pH-responsive heterogeneous hydrogel film becomes increasingly prone to damage. In Figure 5b, the shear stress–strain curves obtained under different chain concentrations by orienting RPNCs parallel to the direction of shear stress under a magnetic field show that pH-responsive heterogeneous hydrogel films with lower chain concentrations can withstand greater stress at the same strain, indicating that they are less likely to rupture during use. After considering the impact of the chain concentration on both optical and mechanical performance, we ultimately chose 9 mg/mL as the most suitable concentration for our needs.

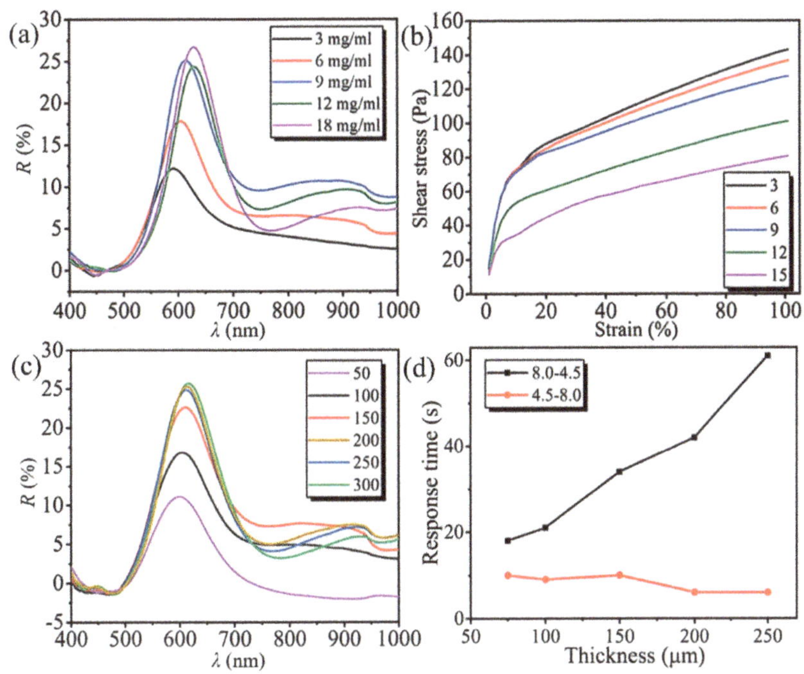

Figure 5. (**a**) Diffraction spectra for pH response of HGF with different chain concentrations, with a membrane thickness of 150 μm; (**b**) shear stress−strain curves for pH response of HGF with different chain concentrations; (**c**) diffraction spectra for pH response of HGF with different membrane thicknesses, with a chain concentration of 9 mg/mL; (**d**) response times for pH response of HGF with different membrane thicknesses in pH buffer solutions ranging from 8.0 to 4.5 (4.5 to 8.0).

Subsequently, an exploration of the influence of film thickness was conducted, with the chain concentration fixed at 9 mg/mL and the gel concentration at 7.5 mg/mL. As shown in Figure 5c, with increasing film thickness, the reflectance of the pH-responsive heterogeneous hydrogel film gradually increases until it reaches 200 μm, after which the rate of increase decreases. The increase in film thickness also represents an increase in the diffraction of structural color units. When the film thickness is small, the pH-responsive heterogeneous hydrogel film appears relatively transparent, allowing most of the light to pass through. As the film thickness increases, more RPNCs are present per unit area, leading to an increase in reflectance. However, the increase in film thickness leads to an increase in the distance for ion diffusion, requiring a longer time for ions to completely penetrate the pH-responsive heterogeneous hydrogel film and respond to the internal RPNCs. By recording the time required for the response to reach equilibrium when transferring pH-responsive heterogeneous hydrogel films of different thicknesses from a pH 8.0 to a pH 4.5 buffer solution, it was found that thicker films require a longer equilibrium time. In Figure 5d, the 50 μm thick film with a chain concentration of 9 mg/mL could not withstand the replacement of the buffer solution due to its low mechanical strength, resulting in fracture, and only the response time data for the 75 μm film were obtained. From the perspective of response time, a film thickness of 100 μm is required to achieve equilibrium within 30 s.

We further tested the mechanical properties of pH-responsive heterogeneous gel films with different thicknesses. The shear stress–strain curves are shown in Figure S6. It is evident that the mechanical performance significantly improves with increasing film thickness, which is consistent with the observations from our experiments.

The difference in trends between 8.0 to 4.5 and 4.5 to 8.0 shown in Figure 5d may be due to the fact that the responsive chains, when undergoing responsive deformation, induce an elastic deformation in the surrounding agarose gel. When the heterogeneous gel film is in the 8.0 buffer solution, the interparticle spacing within the chains is closer to the fixed state. Thus, during the response process from 8.0 to 4.5, the responsive chains need to overcome a certain amount of elastic deformation, increasing the response time. Conversely, in the transition from 4.5 to 8.0, this elastic deformation aids the responsive chains in quickly returning to their initial state.

4. Conclusions

In this study, RPNCs were utilized as both responsive and chromogenic units, while non-responsive gels with adjustable pore sizes served as matrices for immobilizing the responsive materials. Through chemical or physical fixation, two types of PNCs with distinct responsive properties were successfully anchored post-magnetic orientation, leading to the fabrication of fast-response heterogeneous gel films embedding one-dimensional oriented photonic PNCs. The mechanism underlying the fixation of RPNCs in low-concentration gels was elucidated, and the responsive performance, mechanical properties, and stability of photonic crystal heterogeneous gel films were thoroughly investigated. Compared to materials prepared using traditional homogeneous gel films, those fabricated with heterogeneous gel films exhibit the advantage of faster response speed. Furthermore, unlike the RPNCs previously developed by our group, the heterogeneous gel films do not necessitate the application of a magnetic field to display structural color during use. They also demonstrate significant improvements in stability when placed and cycled repeatedly. Typically, a 100 μm pH-responsive heterogeneous gel film can achieve a color change in just 9.3 s as the pH of the environment changes from 4.5 to 8.0. When compared to their homogeneous counterparts with the same thickness, the response rate has been greatly improved. The peak shift of pH-responsive heterogeneous gel films exceeds 100 nm across the green–yellow–orange–red color spectrum when placed in different pH environments. After 20 cycles, the responsiveness of the heterogeneous gel films remained unchanged, showing consistent performance even after several months of intermittent cycling tests, demonstrating good cycle stability and long-term durability. The fast-response heterogeneous gel films embedding one-dimensional-oriented PNCs offer a novel structural model for developing rapid-response, high-performance photonic crystal films, providing a method for the preparation of subsequent photonic crystal heterogeneous gel detection materials.

Supplementary Materials: The following supporting information can be downloaded at: https://www.mdpi.com/article/10.3390/polym16111530/s1, Figure S1: TG curve of the pH-responsive photonic nanochains; Figure S2: Response time of varying pH values; Figure S3: The variation of G' and G'' for the pH-responsive heterogeneous gel film, with a shear strain set at 1%. G' and G'' represent the storage modulus and the loss modulus, respectively; Figure S4: Dependence of λ on temperature of pH-responsive heterogeneous gel film; Figure S5: The digital photographs of pure gel films with agarose concentrations of 5, 7.5, and 10 mg/mL in buffer solutions with pH values of 4.5

and 8.0; Figure S6: Shear stress–strain curves for pH-responsive heterogeneous hydrogel film with different thicknesses.

Author Contributions: Methodology, D.X., H.M., W.L., and J.G.; resources, J.G.; writing—original draft preparation, D.X.; writing—review and editing, H.M., W.L., and J.G.; supervision, H.M., W.L., and J.G. All authors have read and agreed to the published version of the manuscript.

Funding: This research was funded by the National Natural Science Foundation of China (51573144, 21474078 and 51521001), the Natural Science Foundation of Hubei Province (2019CFB572) and the Innovation Team in Key Areas of the Innovation Talent Promotion Plan (2021) of MOST, China.

Institutional Review Board Statement: Not applicable.

Data Availability Statement: The original contributions presented in the study are included in the article/Supplementary Materials, further inquiries can be directed to the corresponding authors.

Conflicts of Interest: The authors declare no conflicts of interest.

References

1. Holtz, J.H.; Asher, S.A. Polymerized colloidal crystal hydrogel films as intelligent chemical sensing materials. *Nature* **1997**, *389*, 829–832. [CrossRef] [PubMed]
2. Lee, K.; Asher, S.A. Photonic Crystal Chemical Sensors: pH and Ionic Strength. *J. Am. Chem. Soc.* **2000**, *122*, 9534–9537. [CrossRef]
3. Debord, J.D.; Lyon, L.A. Thermoresponsive Photonic Crystals. *J. Phys. Chem. B* **2000**, *104*, 6327–6331. [CrossRef]
4. Goponenko, A.V.; Asher, S.A. Modeling of Stimulated Hydrogel Volume Changes in Photonic Crystal Pb2+ Sensing Materials. *J. Am. Chem. Soc.* **2005**, *127*, 10753–10759. [CrossRef] [PubMed]
5. Xu, X.; Goponenko, A.V.; Asher, S.A. Polymerized PolyHEMA Photonic Crystals: pH and Ethanol Sensor Materials. *J. Am. Chem. Soc.* **2008**, *130*, 3113–3119. [CrossRef] [PubMed]
6. Lee, Y.-J.; Braun, P.V. Tunable Inverse Opal Hydrogel pH Sensors. *Adv. Mater.* **2003**, *15*, 563–566. [CrossRef]
7. Zhao, Y.-J.; Zhao, X.-W.; Hu, J.; Li, J.; Xu, W.-Y.; Gu, Z.-Z. Multiplex Label-Free Detection of Biomolecules with an Imprinted Suspension Array. *Angew. Chem. Int. Ed.* **2009**, *48*, 7350–7352. [CrossRef] [PubMed]
8. Asher, S.A.; Alexeev, V.L.; Goponenko, A.V.; Sharma, A.C.; Lednev, I.K.; Wilcox, C.S.; Finegold, D.N. Photonic Crystal Carbohydrate Sensors: Low Ionic Strength Sugar Sensing. *J. Am. Chem. Soc.* **2003**, *125*, 3322–3329. [CrossRef] [PubMed]
9. Zhang, C.; Cano, G.G.; Braun, P.V. Linear and Fast Hydrogel Glucose Sensor Materials Enabled by Volume Resetting Agents. *Adv. Mater.* **2014**, *26*, 5678–5683. [CrossRef]
10. MacConaghy, K.I.; Geary, C.I.; Kaar, J.L.; Stoykovich, M.P. Photonic Crystal Kinase Biosensor. *J. Am. Chem. Soc.* **2014**, *136*, 6896–6899. [CrossRef]
11. Sharma, A.C.; Jana, T.; Kesavamoorthy, R.; Shi, L.; Virji, M.A.; Finegold, D.N.; Asher, S.A. A General Photonic Crystal Sensing Motif: Creatinine in Bodily Fluids. *J. Am. Chem. Soc.* **2004**, *126*, 2971–2977. [CrossRef] [PubMed]
12. Elsherif, M.; Hassan, M.U.; Yetisen, A.K.; Butt, H. Wearable Contact Lens Biosensors for Continuous Glucose Monitoring Using Smartphones. *ACS Nano* **2018**, *12*, 5452–5462. [CrossRef] [PubMed]
13. Yetisen, A.K.; Jiang, N.; Fallahi, A.; Montelongo, Y.; Ruiz-Esparza, G.U.; Tamayol, A.; Zhang, Y.S.; Mahmood, I.; Yang, S.A.; Kim, K.S.; et al. Glucose-Sensitive Hydrogel Optical Fibers Functionalized with Phenylboronic Acid. *Adv. Mater.* **2017**, *29*, 1606380. [CrossRef] [PubMed]
14. Weissman, J.M.; Sunkara, H.B.; Tse, A.S.; Asher, S.A. Thermally Switchable Periodicities and Diffraction from Mesoscopically Ordered Materials. *Science* **1996**, *274*, 959–963. [CrossRef] [PubMed]
15. Matsubara, K.; Watanabe, M.; Takeoka, Y. A Thermally Adjustable Multicolor Photochromic Hydrogel. *Angew. Chem. Int. Ed.* **2007**, *46*, 1688–1692. [CrossRef] [PubMed]
16. Jeon, S.-J.; Chiappelli, M.C.; Hayward, R.C. Photocrosslinkable Nanocomposite Multilayers for Responsive 1D Photonic Crystals. *Adv. Funct. Mater.* **2016**, *26*, 722–728. [CrossRef]
17. Yan, H.; Si, L.; Li, G.; Zhao, L.; Luo, W.; Ma, H.; Guan, J. Heterogeneous Thermochromic Hydrogel Film Based on Photonic Nanochains. *Nanomaterials* **2022**, *12*, 1867. [CrossRef]
18. Arsenault, A.C.; Puzzo, D.P.; Manners, I.; Ozin, G.A. Photonic-crystal full-colour displays. *Nat. Photonics* **2007**, *1*, 468–472. [CrossRef]
19. Nucara, L.; Greco, F.; Mattoli, V. Electrically responsive photonic crystals: A review. *J. Mater. Chem. C* **2015**, *3*, 8449–8467. [CrossRef]
20. Puzzo, D.P.; Arsenault, A.C.; Manners, I.; Ozin, G.A. Electroactive Inverse Opal: A Single Material for All Colors. *Angew. Chem. Int. Ed.* **2009**, *48*, 943–947. [CrossRef]
21. Ueno, K.; Matsubara, K.; Watanabe, M.; Takeoka, Y. An Electro- and Thermochromic Hydrogel as a Full-Color Indicator. *Adv. Mater.* **2007**, *19*, 2807–2812. [CrossRef]
22. Wang, J.Y.; Cao, Y.; Feng, Y.; Yin, F.; Gao, J.P. Multiresponsive Inverse-Opal Hydrogels. *Adv. Mater.* **2007**, *19*, 3865–3871. [CrossRef]

23. Mafé, S.; Manzanares, J.; English, A.; Tanaka, T. Multiple Phases in Ionic Copolymer Gels. *Phys. Rev. Lett.* **1997**, *79*, 3086–3089. [CrossRef]
24. Tanaka, T.; Fillmore, D.J. Kinetics of swelling of gels. *J. Chem. Phys.* **1979**, *70*, 1214–1218. [CrossRef]
25. Luo, W.; Cui, Q.; Fang, K.; Chen, K.; Ma, H.; Guan, J. Responsive Hydrogel-based Photonic Nanochains for Microenvironment Sensing and Imaging in Real Time and High Resolution. *Nano Lett.* **2020**, *20*, 803–811. [CrossRef] [PubMed]
26. Cai, J.; Luo, W.; Pan, J.; Li, G.; Pu, Y.; Si, L.; Shi, G.; Shao, Y.; Ma, H.; Guan, J. Glucose-Sensing Photonic Nanochain Probes with Color Change in Seconds. *Adv. Sci.* **2022**, *9*, 2105239. [CrossRef] [PubMed]
27. Li, L.; Yu, Z.; Liu, J.; Yang, M.; Shi, G.; Feng, Z.; Luo, W.; Ma, H.; Guan, J.; Mou, F. Swarming Responsive Photonic Nanorobots for Motile-Targeting Microenvironmental Mapping and Mapping-Guided Photothermal Treatment. *Nano-Micro Lett.* **2023**, *15*, 141. [CrossRef] [PubMed]
28. Weng, L.; Liang, S.; Zhang, L.; Zhang, X.; Xu, J. Transport of Glucose and Poly(ethylene glycol)s in Agarose Gels Studied by the Refractive Index Method. *Macromolecules* **2005**, *38*, 5236–5242. [CrossRef]
29. Schuszter, G.; Gehér-Herczegh, T.; Szűcs, Á.; Tóth, Á.; Horváth, D. Determination of the diffusion coefficient of hydrogen ion in hydrogels. *Phys. Chem. Chem. Phys.* **2017**, *19*, 12136–12143. [CrossRef]
30. Luo, W.; Ma, H.; Mou, F.; Zhu, M.; Yan, J.; Guan, J. Steric-Repulsion-Based Magnetically Responsive Photonic Crystals. *Adv. Mater.* **2014**, *26*, 1058–1064. [CrossRef]
31. Shin, J.; Braun, P.V.; Lee, W. Fast response photonic crystal pH sensor based on templated photo-polymerized hydrogel inverse opal. *Sens. Actuators B Chem.* **2010**, *150*, 183–190. [CrossRef]
32. Yetisen, A.K.; Butt, H.; da Cruz Vasconcellos, F.; Montelongo, Y.; Davidson, C.A.B.; Blyth, J.; Chan, L.; Carmody, J.B.; Vignolini, S.; Steiner, U.; et al. Light-Directed Writing of Chemically Tunable Narrow-Band Holographic Sensors. *Adv. Opt. Mater.* **2014**, *2*, 250–254. [CrossRef]
33. Hu, Y.; He, L.; Han, X.; Wang, M.; Yin, Y. Magnetically responsive photonic films with high tunability and stability. *Nano Res.* **2015**, *8*, 611–620. [CrossRef]

polymers

MDPI

Article

Development of Advanced Solid-State Thermochromic Materials for Responsive Smart Window Applications

Kai Zeng [1], Chang Xue [1,*], Jinbo Wu [1,2] and Weijia Wen [1,3,*]

1 Materials Genome Institute, Shanghai University, Shanghai 200444, China; zengkai-mgi@shu.edu.cn (K.Z.); jinbowu@t.shu.edu.cn (J.W.)
2 Faculty of Materials Science, Shenzhen MSU-BIT University, Shenzhen 518115, China
3 Department of Physics, The Hong Kong University of Science and Technology, Clear Water Bay, Kowloon, Hong Kong
* Correspondence: xuechang33@shu.edu.cn (C.X.); phwen@ust.hk (W.W.)

Abstract: This study introduces the synthesis and detailed characterization of a novel thermochromic material capable of reversible alterations in its thermotropic transmittance. Through an emulsion polymerization process, this newly developed material is composed of 75–85% octadecyl acrylate and 0–7% allyl methacrylate, demonstrating a pronounced discoloration effect across a narrow yet critical temperature range of 24.5–39 °C. The synthesized powder underwent a battery of tests, including differential scanning calorimetry and thermogravimetric analysis, as well as scanning electron microscopy. These comprehensive evaluations confirmed the material's exceptional thermal stability, uniform particle size distribution, and strong anchoring properties. Building upon these findings, we advanced the development of thermochromic polyvinyl butyral films and laminated glass products. By utilizing a coextrusion technique, we integrated these films into laminated glass, setting a new benchmark against existing glass technologies. Remarkably, the incorporation of thermochromic PVB films into laminated glass led to a significant reduction in solar irradiance of 20–30%, outperforming traditional double silver low-emissivity glass. This achievement demonstrates the exceptional shading and thermal insulation properties of the material. The research presented herein not only pioneers a valuable methodology for the engineering of smart materials with tunable thermotropic transmittance but also holds the key to unlocking enhanced energy efficiency across a spectrum of applications. The potential impact of this innovation on the realm of sustainable building materials is profound, promising significant strides toward energy conservation and environmental stewardship.

Keywords: smart materials; emulsion polymerization; solid-state thermochromic materials; transmittance; thermal management; environmental sustainability

1. Introduction

In the field of construction, energy utilization has emerged as a critical issue, with a staggering 40% of total energy consumption in regions such as the European Union, the United States, and other developed nations attributed to this sector [1]. In China, the architectural sector's energy consumption represented 19.1% of the country's total energy use in 2012, a figure that increased at an average annual growth rate of approximately 8.3% from 2001 to 2012 [2]. In response, researchers have explored ways to reduce energy consumption and improve conservation within the architectural domain worldwide. A variety of materials and energy management strategies have been developed, with the integration of structural and functional elements into material design leading to significant breakthroughs in solar energy management within buildings. These innovations are instrumental in reducing carbon emissions and bolstering the energy efficiency of buildings, all in the pursuit of a more sustainable future [3–5].

Scientists have harnessed a diverse array of methods to reduce energy usage in building operations. These strategies include the application of lightweight porous materials [6], thermal insulation coatings [7], solar reflective coatings [8], and cool light-color paints. Moreover, the use of solar spectral selective absorbing coatings [9] and the strategic placement of ventilation windows on rooftops are among the tactics employed to enhance thermal insulation. Despite the range of available methods, there are inherent limitations in fully exploiting solar energy throughout the seasons, which impacts the ability to maintain optimal warmth in winter and coolness in summer. Nonetheless, the advent of energy-efficient windows made from glass combined with color-changing materials offers an auspicious solution. These chromogenic materials adjust their optical properties in response to external stimuli, presenting a dynamic thermal regulation approach. Classified according to their stimuli, photochromic [10], electrochromic [11], and thermochromic [12] materials react to changes in light [13], electric fields [14], and temperature [15], respectively.

Electrochromic glass is particularly notable as a leading innovation in smart glass technology and is capable of reversible transitions between transparent and tinted states upon the application of a specific voltage. In its tinted state, it markedly diminishes visible light transmittance and the solar heat gain coefficient but retains transparency, making it an effective tool for light and heat management within buildings [16]. However, electrochromic glass has shortcomings, such as its inability to effectively reduce glare from direct sunlight, especially intense glare from east–west orientations or top surfaces during periods of strong outdoor illumination or direct sun exposure [17]. Thermochromic materials, on the other hand, are capable of altering their color or optical transmission in response to temperature changes [12,18,19]. As the surface temperature of thermochromic materials varies, so do their optical and thermal properties [20]. At the transition temperature, these materials undergo a property change. Their defining characteristic is their sensitivity to surface temperature: above the transition temperature, they absorb and reflect most of the sun's infrared rays while partially absorbing and scattering visible light.

Conversely, when the surface temperature is below the threshold, sunlight can pass directly through the material into the room. Given the desired performance characteristics, intrinsic modification of glass presents a challenge. The pure thermochromic material, VO_2 film, undergoes a phase transition from a metal to a semiconductor at a critical temperature of 68 °C, reflecting the near-infrared region with high efficiency at temperatures above this point [21]. To increase the practicality and industrial viability of VO_2 films, researchers have delved into the realm of doping, specifically incorporating metal ions such as tungsten (W) into the VO_2 lattice. This innovative approach is designed to lower the phase transition temperature of VO_2 films, making them more suitable for a broader range of everyday applications [22]. The doping process not only adjusts the temperature at which VO_2 transitions between phases but also reduces the energy threshold required for this transition, enabling it to occur at lower temperatures. For example, the introduction of only 0.4% tungsten can significantly lower the lattice transition energy from 15.2 °C to 8.3 °C, effectively reducing the phase transition temperature from 68 °C to 56 °C [23].

Chen et al. [24] developed a sophisticated bilayer VO_2 film, which is only 55 nanometers thick and includes both a VO_2 layer and a titanium dioxide antireflective layer. This groundbreaking structure led to a substantial increase in the film's visible light transmittance at lower temperatures, with a notable increase from 40.3% to 61.5% at 20 °C. Shen et al. [25] achieved a uniform dispersion of zirconium-doped VO_2 nanoparticles within an oily polyurethane solution, which, when applied to a polyethylene terephthalate substrate and dried, resulted in the successful formation of a composite film [26]. This VO_2-PU composite film displays a visible light transmittance (τvis) of 60.4% and impressive solar modulation ability, as indicated by a solar transmittance change ($\Delta\tau$solar) of 14.1% at 25 °C [27]. Long et al. [28] have also made progress in this field, developing a modified composite film with a phase transition temperature of 41.3 °C for application on the exterior surface of hollow glass. Outdoor testing facility trials have demonstrated that this modified VO_2 film can reduce the cooling energy consumption by 11.1% compared with that of standard hollow glass.

Polymer thermochromic films represent a distinct category of thermochromic films, achieving thermochromism through the integration of specific ligand metal ions, such as nickel or cobalt, into a transparent polymer matrix, such as polyvinyl butyral (PVB). These metal ions undergo color changes at various temperatures due to their ligand structures, thus enabling thermochromic properties and reversible transmittance changes across the visible and near-infrared spectra [29]. Giovannini et al. [30] and Lu et al. [31] utilized building simulation techniques to integrate visible light and near-infrared data from ligand metal ion-based thermochromic films at various temperatures into a building model. Their study revealed that employing such thermochromic glass could lead to energy savings of 3% to 10% compared with traditional glass under identical continental climate conditions, with additional potential to reduce lighting energy usage by 5% to 20%.

In a pivotal study by Lee [32], the focus was on the synthesis of reversible thermotropic materials via polymerization techniques, resulting in a novel reactive thermotropic liquid crystalline polymer. El-Deeb [33] explored characterization techniques for thermochromic materials, employing UV–Vis and Raman spectroscopy to examine changes in optical properties with temperature, whereas SEM and DSC microscopy techniques were used to investigate the microstructure and phase transitions of the materials. Busi Kumar Babu et al. [34] addressed the applications of reversible smart materials in smart windows, highlighting their ability to actively modulate light and heat transmission, thereby improving energy efficiency and indoor comfort levels and highlighting their potential environmental benefits, such as reduced carbon emissions.

In this novel study, we successfully synthesized an advanced solid-state thermochromic powder, which is highly important for the development of thermochromic materials. By integrating this powder into transparent resins, we have significantly bolstered the material's resilience and dimensional integrity, culminating in a more robust end product. The resulting smart film not only inherits the dynamic thermochromic properties of the powder but also actively regulates solar transmittance. At temperatures above a predefined threshold, the film augments solar scattering to curtail heat absorption, whereas below this threshold, it permits increased solar penetration. This research is not only a testament to our commitment to scientific advancement but also a critical step toward sustainable construction practices, offering newly developed solutions to the pressing issue of energy conservation.

2. Synthesis and Characterization

2.1. Synthesis

We commenced the synthesis by meticulously preparing the aqueous phase, which was meticulously maintained at 35 °C and composed of 0.30 g of sodium dodecyl sulfate, 0.075 g of $NaHCO_3$, and 140 g of deionized water. Concurrently, under well-ventilated conditions, 75 wt% ODA completely melted at 60 °C. The monomer blend, consisting of 10 wt% butyl acrylate, 5 wt% allyl methacrylate (ALMA), and 10 wt% styrene, was prewarmed in a water bath at 35 °C before being combined with the molten ODA under continuous magnetic stirring. This mixture was then held at 35 °C for an additional 15 min to ensure thorough blending.

In the next phase, 0.6 g of the oil-soluble initiator Dilauroyl peroxide was introduced into the monomer blend, ensuring complete dissolution through magnetic stirring. The blend was subsequently stirred for 10 min at 35 °C to achieve a preliminary dispersion, forming an o/w crude emulsion. This emulsion was then subjected to ultrasonic homogenization for 30 min, creating an o/w microemulsion while being cooled in a 25 °C water bath with strict internal temperature monitoring.

The o/w microemulsion was degassed under an argon atmosphere with stirring for 15 min, after which the inert gas flow was carefully reduced, and the mixture was heated to 70 °C. Polymerization was allowed to proceed at this temperature with consistent stirring for 120 min.

A mixed phase consisting of a monomer phase with 96 wt% methyl methacrylate and 4 wt% ethyl acrylate and an aqueous phase with 0.020 g sodium dodecyl sulfate, 0.010 g $NaHCO_3$, 0.020 g NaPDS, and 20 g deionized water was subsequently stirred vigorously into the o/w microemulsion for 30 min. The oil bath was then removed, and the particles were dispersed and solidified with 150 mL of a 0.5% magnesium sulfate solution under vigorous agitation. The solid product was vacuum-filtered, followed by five thorough washes with 250 mL of water each to eliminate any water-soluble polymeric substances, the fabrication process is shown in Figure 1.

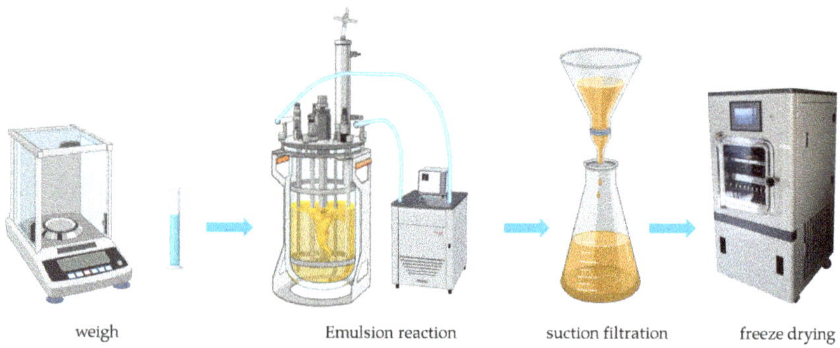

Figure 1. Fabrication process of thermochromic powder.

The thermochromic PVB film was ultimately produced via a twin-screw extruder. This process involved blending 90 parts of PVB, 10 parts of the synthesized thermochromic powder, and 33 parts of the plasticizer DBP, which were then extruded to form the final PVB film.

2.2. *Characterization*

The surface morphology and microstructure of the synthesized materials were examined via a JEOL FE–SEM 6700F scanning electron microscope from Akishima, Japan. Differential scanning calorimetry (DSC) and thermogravimetric (TG) measurements were conducted with a PerkinElmer DSC-7 instrument from the Waltham, MA, USA, using samples weighing 2.0–5.0 mg placed in aluminum pans. The DSC scans were performed at a rate of 2 °C per minute.

The fluorescence characteristics of the materials were determined via a Hitachi F-4500 fluorescence spectrophotometer from Tokyo, Japan. The optical properties of the thermochromic laminated glass were monitored over a wavelength range of 300–2500 nm via a Lambda 1050 spectrophotometer from PerkinElmer, Waltham, MA, USA equipped with a heating unit. Outdoor comparative tests assessing temperature, illumination, and irradiance were conducted via a state-of-the-art online testing device, ensuring accurate and real-time data acquisition.

2.3. *Outdoor Experimental*

The experimental setup featured a downsized physical model of a testing chamber, expertly crafted from 75 mm-thick polystyrene foam encased in color steel sandwich panels. The testing chamber was meticulously designed at a 1:4 scale. With dimensions of 1.2 m in length, 1.1 m in width, and 0.8 m in height, the chamber was equipped with an external window measuring 0.6 m by 0.5 m. The aspect ratio of the windows to the walls was carefully set at 0.34, and the windows included a sill positioned at a height of 0.2 m.

To counteract the effects of high summer temperatures on the thermal environment and the experimental equipment within the testing chamber, a protective enclosure, measuring 2.0 m × 1.9 m × 1.5 m, was strategically installed around the chamber. This configuration was designed to simulate an actual office room environment, similar to one with an air conditioning system operating from an adjacent room during the summer months.

Temperature monitoring within the testing chamber was meticulously carried out via T-type thermocouples, which were tasked with measuring both the indoor temperature and the surface temperatures of the glass—both on the interior and exterior sides. A state-of-the-art temperature data collector was used to automatically gather and log these readings at 10 min intervals.

Illuminance levels within and outside the testing chambers were accurately measured with an illuminance sensor, and in accordance with the temperature data collection, these readings were automatically recorded every 10 min via a data collector. Moreover, a solar radiation sensor was deployed to measure the intensity of the indoor solar radiation and the total solar radiation striking the vertical wall of the test chamber. These radiation data were also systematically collected and recorded at 10 min intervals.

3. Results

3.1. Influence of ODA on the Temperature and Transmittance Characteristics of Thermochromic Materials

The exploration of microemulsion polymerization with organic monomers has captured significant interest across the globe. In our study, ODA was strategically selected as the polymerization matrix for incorporating thermotropic functional groups. This choice led to the successful synthesis of a polymer shell microemulsion with a particle size of approximately 5 μm through emulsion polymerization. The integration of ODA monomers significantly enhances the thermochromic polymer's capacity to modulate the transmittance of PVB in response to temperature variations. However, the inclusion of ODA in excess quantities may result in the increased formation of homopolymers in the aqueous phase, potentially compromising the overall transmittance of the material. With increasing ODA content, there was a notable increase in system viscosity and a concomitant increase in latex particle size. When modified with PVB, the transmittance at room temperature initially increases but then decreases, mirroring a similar trend at higher temperatures.

Figure 2 elegantly depicts the relationship between visible light transmittance (τvis) and the phase transition temperature of laminated glass containing ODA monomers and thermochromic dimming materials.

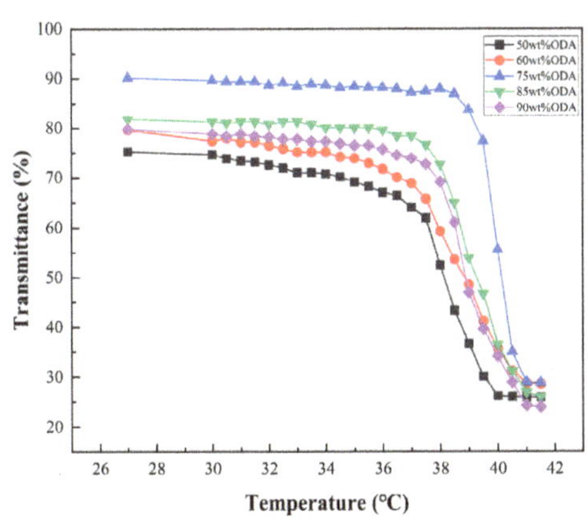

Figure 2. Visible light transmittance of laminated glass prepared with thermochromic powder synthesized with different concentrations of ODA monomers at different temperatures.

The data illustrate that increasing the ODA concentration from 75 wt% to 85 wt% marginally decreases the phase transition temperature (Ts) from 40.2 °C to 39.5 °C. Nonetheless, further increasing the ODA content from 85 wt% to 90 wt% led to a minimal reduction in the phase transition temperature of approximately 1 °C. This finding indicates that beyond a certain concentration threshold, increasing the proportion of ODA monomers has a negligible effect on diminishing the phase transition temperature of the thermochromic material.

3.2. Effect of the ALMA Monomer on the Temperature and Transmittance Characteristics of Thermochromic Materials

At an ODA monomer concentration of 75%, the phase transition temperature of the thermochromic powder can be finely adjusted from 24.5 °C to 39.5 °C by varying the ALMA content. Figure 3 clearly shows the relationship between the visible light transmittance (τvis) of laminated glass, which incorporates a thermochromic powder with a 75% ODA formulation, and various concentrations of ALMA. An increase in ALMA from 0 to 7 wt% substantially decreases the phase transition temperature (Ts) from 39.5 °C to 24.5 °C. This pronounced reduction in Ts is likely due to the structural expansion of the thermochromic powder induced by the presence of ALMA in the branched chains, which weakens the bonding forces between the ODA polymer and polyethylene adipate, facilitating the aggregation and precipitation of the polymer at lower temperatures.

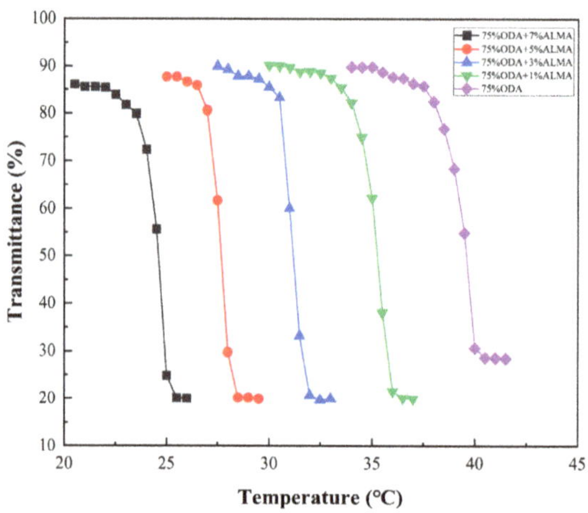

Figure 3. Visible light transmittance of laminated glass prepared with thermochromic powder synthesized at different concentrations of 75% ODA monomer with various ALMA concentrations at different temperatures.

However, further increasing the ALMA content beyond the optimal level is advised. The excessive incorporation of ALMA can cause the polymer to exhibit turbidity at or below room temperature (25 °C), potentially leading to an undesirable color shift in practical applications and adversely affecting product performance, especially in window applications where clarity is paramount.

This provides a visual representation of the visible light transmittance of laminated glass prepared with thermochromic powder synthesized at various concentrations of 75% ODA monomer alongside different ALMA concentrations at various temperatures.

3.3. SEM

Figure 4 presents a scanning electron microscope (SEM) image of a sample synthesized with a composition of 75% ODA and 1% ALMA. The image reveals a uniform array of spherical particles that are clearly free from significant agglomeration and feature a rugged surface marked by pronounced anchor-like structures. These uniformly distributed surface protrusions are anticipated to effectively interlock with PVB, thus fostering the formation of an enhanced PVB film during the twin-screw melt extrusion process.

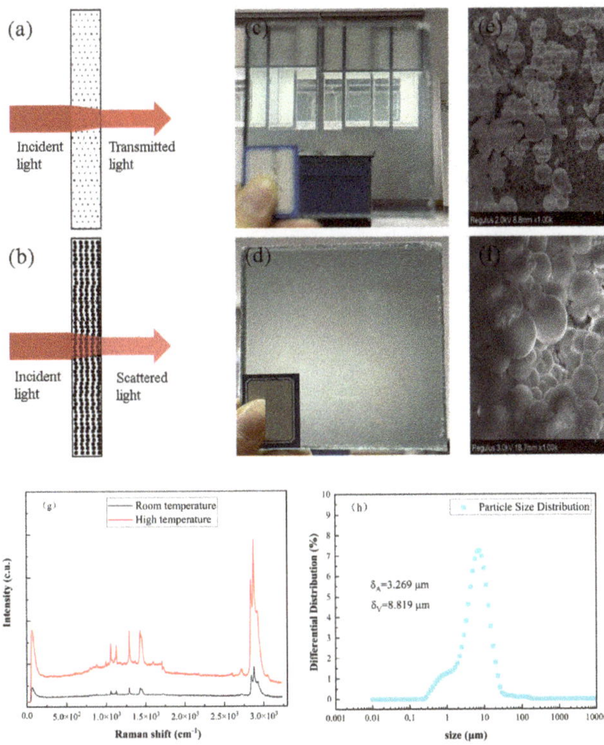

Figure 4. Synthesized thermochromic powder and PVB film for light absorption: (**a**) room temperature; (**b**) high temperature; (**c**,**d**) visibility through a polymer at room temperature; (**e**,**f**) optical micrographs of the same polymers at different temperatures; (**g**) Raman scattering spectra of the synthesized thermochromic powder; (**h**) particle size distribution of the solid-state thermochromic powder: δ_A: average particle size of the surface area; δV: volume average particle size.

SEM analysis further provides precise measurements, indicating that the surface area mean diameter of the solid-state thermochromic powder is 3.269 μm, with a volume mean diameter of 8.819 μm. An increase in particle size, which correlates with a larger surface area, may increase the light absorption capacity of thermochromic materials, potentially intensifying the thermochromic effect as the particles absorb more heat. However, larger particle sizes could also impact thermal conductivity, resulting in increased size, larger gaps, and extended pathways for heat transfer, which might lead to slower responses to temperature fluctuations and diminished stability in terms of the thermochromic effect. Moreover, larger particles can introduce haze during processing, adversely affecting the clarity and performance of the final product.

From the Raman spectroscopy analysis conducted at 25 °C and 50 °C, the characteristic peaks at 90.89 cm^{-1}, 1230 cm^{-1}, 1450 cm^{-1}, and 2750 cm^{-1} remained consistent in position, indicating that no inherent material changed across the various temperatures. However, the heightened amplitude of these peaks at 50 °C suggests temperature-induced alterations in the material's structure or order, changes that could be linked to the thermochromic properties as such effects are typically the consequence of structural transformations within the material. When subjected to heat, modifications in the molecular or lattice structure can precipitate changes in optical properties, often reflected in parameters such as the band structure and scalar coupling.

The intensity of the Raman shift peaks also acts as an indicator of the excitation level and, to a degree, the quantity of molecules present. At elevated temperatures, an increased number of molecules are excited by the laser, corroborating the influence of the DSC characteristics on the material's thermal variation. This interplay demonstrates the complex relationships among material structure, thermal properties, and their collective role in shaping the thermochromic behavior of a material.

3.4. TG–DSC

The experiments were conducted at a heating rate of 2 °C/min, and the results depicted in Figure 5 shed light on the thermal stability of the solid-state thermochromic powder. TG analysis revealed the onset of weight loss at 250 °C, with a significant reduction to 90% at 450 °C, suggesting that the material underwent a phase transition that initiated weight loss or decomposition at 250 °C. Additionally, the TG curve points to a 20% weight loss at 360 °C and 80% weight loss at 425 °C, implying potential intermediate phase transitions within this temperature range, contributing to the further decomposition of the material.

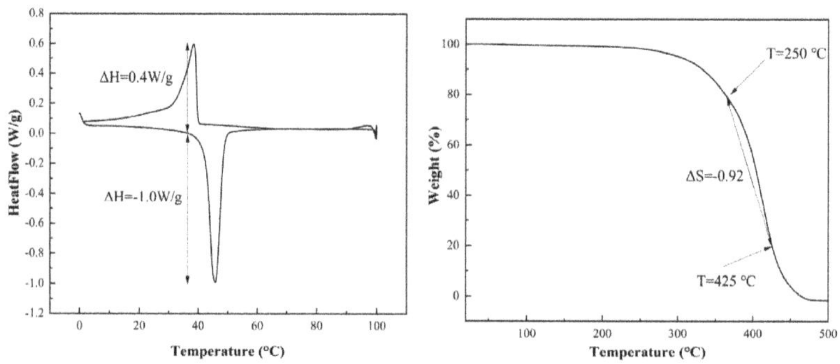

Figure 5. DSC and TG analysis of the synthesized thermochromic powder.

The DSC curve revealed noticeable heat flow changes at 40 °C during the heating cycle, with a pronounced heat absorption peak of −1.0 w/g at 42 °C. This peak likely signifies a phase transition occurring between 40 °C and 42 °C. Similarly, during the cooling cycle, the heat flow begins to change at 40 °C, reaching a peak release of 0.6 w/g at 39 °C. These heat flow variations are indicative of exothermic and endothermic transitions associated with structural and compositional changes within the thermochromic material.

These phase transitions, which are evident in both TG and DSC analyses, reflect modifications to the material's structure, composition, and crystallography. Such transformations are crucial for adjusting the crystal lattice, which subsequently influences the optical properties of the material and induces a thermochromic effect—a reversible color change with temperature. Understanding these thermal behaviors is essential for refining the thermal response of thermochromic materials for diverse applications. The synthesized thermochromic powder material, which contains hydroxyl and carboxyl groups, engages in weak physical hydrogen bonding with the hydroxyl groups in the PVB molecules. An increase in the quality of the core-shell structure and an alteration in the material's thermal transition properties are realized via the addition of methyl methacrylate, which creates protrusions that marginally lower the film's transition temperature below 42 °C, as indicated by the DSC curves.

3.5. FTIR

Figure 6 shows the FTIR (Fourier transform infrared) spectra of the synthesized thermochromic powder, revealing essential information about the molecular structure and its influence on the thermal properties. Within the spectrum, a broad peak indicative of multimolecular O–H associations is observed in the range of 3300–3600 cm^{-1}. The C-H stretching vibrations are represented by a distinct peak in the 2800–2900 cm^{-1} region, and the presence of a carbon–carbon double bond is denoted by a peak within 1620–1680 cm^{-1}.

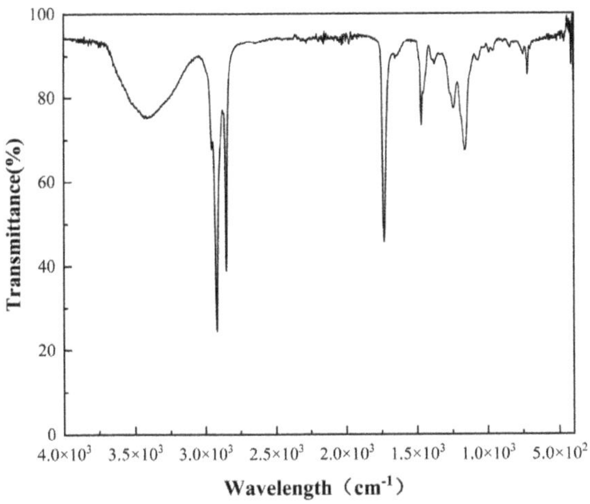

Figure 6. FTIR analysis of the synthesized thermochromic powder.

The molecular structure is a critical determinant of the thermal performance and phase transition behavior of materials. For the synthesized thermochromic powders, the inclusion of short alkyl side chains elevates the decomposition temperature, suggesting a more thermally robust material. These side chains contribute to enhanced molecular mobility, which is associated with a reduced phase transition temperature, thereby facilitating a more pronounced thermochromic effect at lower temperatures and improving the material's overall thermal stability.

Additionally, the incorporation of aromatic rings during synthesis enhances this thermal stability. Through π-electron delocalization, aromatic rings confer extra rigidity to the molecular structure, limiting rotational motion, increasing the melting point of the material, and increasing its thermal stability.

The incorporation of polar functional groups such as hydroxyl (–OH) or carbonyl (C=O) groups is also evident in the FTIR spectra. These groups promote stronger hydrogen bonding and dipole–dipole interactions, which influence the thermal behavior of the material. These interactions are not only vital for the material's structural integrity but also play a significant role in defining the strength and nature of the thermochromic response.

In summary, the FTIR spectra provide insight into the complex interplay of functional groups in the thermochromic powder. The presence and arrangement of these groups within the material's molecular architecture not only modify its inherent properties but also dictate the strength and mechanisms of the thermochromic transitions. Understanding these spectral signatures is key to tailoring the material for specific thermal responses and stability requirements in practical applications.

3.6. Mechanical Properties of the PVB Film

The stress-strain curve clearly illustrates the mechanical properties of the modified PVB film, as detailed in Figure 7 and Table 1. Compared with traditional PVB films, the modified film markedly increased the tensile strength, reaching 23.85 MPa. Upon the integration of the thermochromic powder, which is facilitated by weak hydrogen bonding, the elongation at break is elevated to 233.4%, surpassing the 200% elongation at break typical of conventional PVB. It is widely recognized that robust hydrogen bonds between PVB polymer chains can impede the processability and deformability of a film; however, the introduction of a modest amount of plasticizer is insufficient to disrupt these bonds [35].

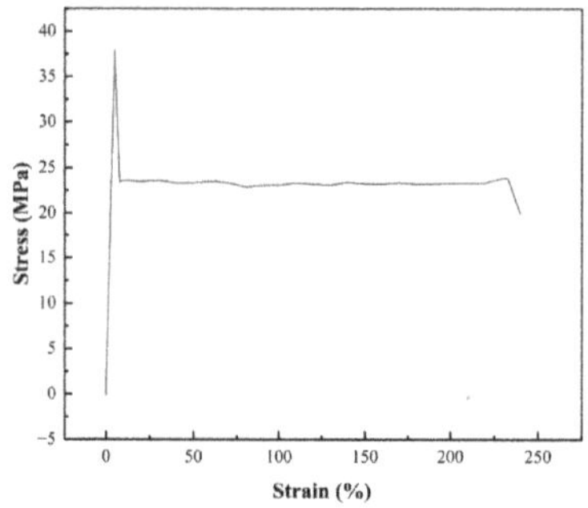

Figure 7. Stress–strain curves.

Table 1. Tensile strength, elongation at break, and breakage rate of the samples.

Performance	Thickness (mm)	Tensile Stress (MPa)	Elongation at Break (%)	Melt Index (g/10 min)	Peel Force (N/mm)
Test value (1)	0.76	23.85	233.4	1.35	35
Test value (2)	0.75	23.83	233.1	1.35	35
Test value (3)	0.76	23.86	233.4	1.35	35

Compared with its traditional counterpart, the modified PVB film has a diminished impact value of 7. This reduction in impact value is attributed to the subtle alteration in the hydrogen bond strength upon the addition of the thermochromic powder. The incorporation of this powder into the PVB matrix is posited to distribute tensile stress more effectively, thus enhancing elongation at break. This suggests that the modified thermochromic powder within the PVB contributes to a more ductile and resilient film, with improved mechanical properties that are particularly beneficial for applications requiring flexibility and impact resistance.

3.7. Outdoor Experimental Characterization

The analysis of the impact of three distinct types of external window glass on the indoor thermal environment has indicated that each glass variant significantly contributes to a reduction in the indoor temperature and an increase in the thermal comfort level. In terms of temperature control effectiveness, thermochromic dimming insulating glass emerged as the superior performer, with double silver low-emissivity (Low-E) insulating glass following runner-up and single silver low-E insulating glass taking the last position.

When examining the temperature differential across the glass surfaces, as shown in Figure 8, the single silver low-E insulating glass exhibited the smallest disparity between its inner and outer surfaces, which was lower than that of its double silver counterpart and substantially lower than the disparity observed with thermochromic dimming insulating glass. Notably, the inner surface temperature of the thermochromic dimming insulating glass remained the coolest, averaging 2.8 °C cooler than that of the single silver low-E insulating glass. This reduced surface temperature results in a decreased impact of thermal radiation on both objects and individuals within the room.

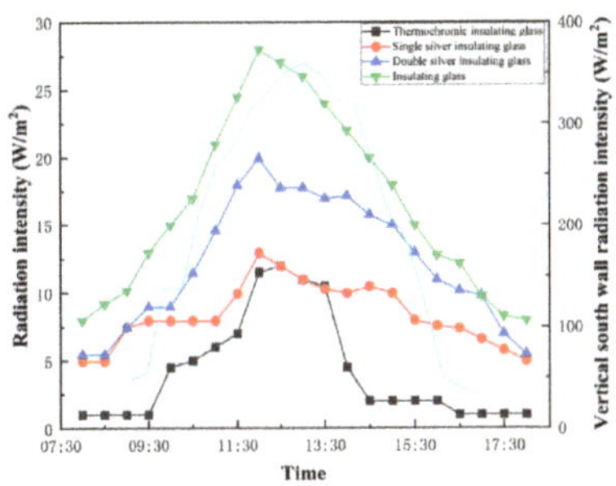

(**a**) Solar radiation intensity

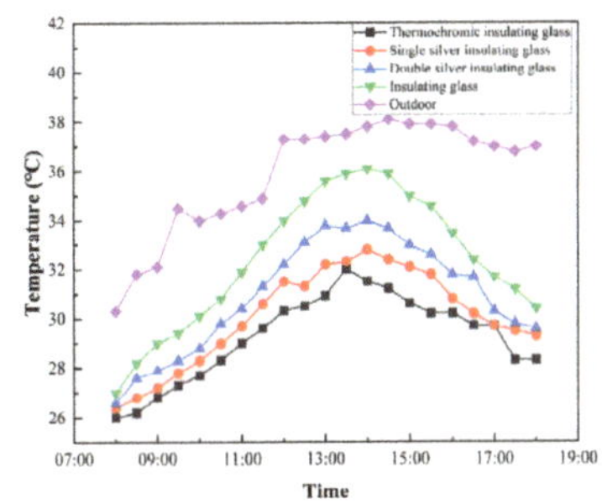

(**b**) Average indoor temperature

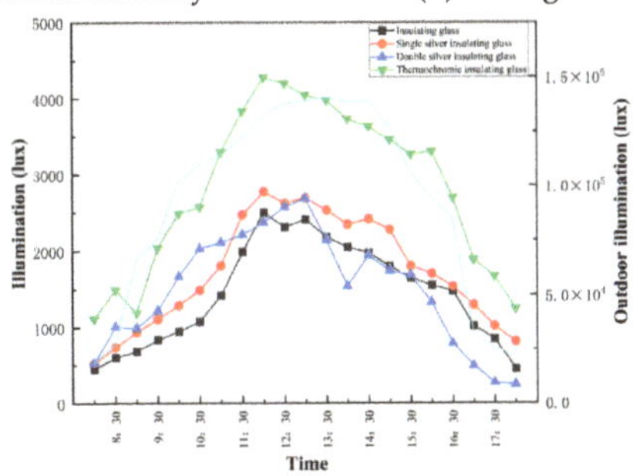

(**c**) Indoor illuminance

Figure 8. Southward comparison data under different outer windows in summer. The light blue line in (**a**) represents the irradiation intensity of the outdoor vertical facade, while the light blue line in (**c**) represents the outdoor illumination.

In terms of mitigating solar radiation intensity, the pivotal period requiring control in south-facing rooms is from 11:00 to 15:00, when solar radiation is at its zenith. The thermochromic dimming insulating glass demonstrated exceptional performance in controlling direct solar radiation through windows. This type of glass possesses the lowest solar radiation heat transfer coefficient, suggesting that its reduction in indoor solar radiation heat is primarily through managing the portion absorbed by the glass that is subsequently conveyed indoors as heat. Consequently, thermochromic dimming insulating glass significantly outperforms the other types of thermal insulation, with double silver low-E insulating glass ranking second and single silver low-E insulating glass being the least effective.

On the basis of data collected during a typical summer day, the indoor environment experiences continuous heating due to the combined effects of heat transfer from the indoor–outdoor temperature difference and solar radiation. With respect to the impact on the indoor lighting environment, thermochromic glazing ensures ample illumination even under overcast conditions during the testing period, owing to its 64% visible light transmittance. Consequently, compared with single- and double-silver low-E insulating glass, thermochromic glazing not only offers superior temperature control but also provides excellent lighting effects.

4. Conclusions

To succinctly encapsulate the research findings on the development of tunable thermochromic properties through emulsion polymerization and the newly developed application of thermochromic powders in smart film technology, the following key points encapsulate the outcomes of our rigorous experimental and investigative efforts:

(1) The incorporation of the ODA monomer in emulsion polymerization has been empirically demonstrated to produce a powder with thermally responsive discoloration properties. A specific concentration of 75% ODA within the polymer matrix facilitates a color change at 40.2 °C while maintaining over 90% visible light transmittance. Moreover, the transition temperature is adjustable within a precise range of 24.5 °C to 39.5 °C by modulating the ALMA content.
(2) Comprehensive analytical tests, including DSC-TG and SEM, confirmed the thermal characteristics and morphology of the synthesized powder. The powder comprises spherical particles with a remarkably uniform diameter of approximately 10 μm, which enhances its thermal processing capabilities and potential for effective anchoring within a matrix.
(3) Integrating thermochromic powder into insulating glass results in the creation of thermochromic smart laminated glass with adjustable shading capabilities, significantly enhancing thermal performance. Compared with single silver low-emissivity glass, this smart glass maintains the lowest inner surface temperature, with a notable temperature difference of 2.8 °C. This feature substantially reduces thermal radiation, thereby improving occupant comfort and protecting interior objects from excessive heat exposure. Furthermore, compared with double silver low-E glass, it achieves a 20–30% reduction in solar irradiance within the test rooms.
(4) In terms of indoor lighting conditions, thermochromic laminated glass offers effective control over direct sunlight. Its adjustable shading feature allows for the entry of diffused light, which can reduce the reliance on artificial illumination within the interior. By diminishing the need for artificial lighting or active cooling systems, this capability has the potential to lead to significant energy savings.

In the future, further testing is needed for thermochromic PVB, including plans for year-round performance assessments to evaluate the service life and aging characteristics under various environmental conditions. It is essential to verify the long-term stability and efficiency of our material, further optimize the formulation and synthesis process of the thermochromic materials and assess the full life cycle of the finished products. These innovations are expected to play a pivotal role in enhancing building energy efficiency and reducing the carbon footprint. Their applications are anticipated to extend across various sectors, including large-scale shading smart windows, daylight-responsive roofing systems, and glass greenhouses, thereby contributing to the advancement of sustainable architectural practices.

Author Contributions: Conceptualization, K.Z. and W.W.; methodology, K.Z.; software, K.Z.; validation, K.Z., C.X. and. J.W.; formal analysis, K.Z.; investigation, K.Z.; resources, J.W.; data curation, K.Z.; writing—original draft preparation, K.Z.; writing—review and editing, K.Z.; visualization, K.Z.; supervision, J.W.; project administration, W.W.; funding acquisition, W.W. All authors have read and agreed to the published version of the manuscript.

Funding: This work was supported in part by the Project of the Hetao Shenzhen-Hong Kong Science and Technology Innovation Cooperation Zone (HZQB-KCZYB-2020083).

Institutional Review Board Statement: Not applicable.

Data Availability Statement: Data are contained within the article.

Conflicts of Interest: The authors declare that they have no conflicts of interest.

References

1. Gan, V.J.L.; Lo, I.M.C.; Ma, J.; Tse, K.T.; Cheng, J.C.P.; Chan, C.M. Simulation optimisation towards energy efficient green buildings: Current status and future trends. *J. Clean. Prod.* **2020**, *254*, 120012. [CrossRef]
2. Chen, H.; Lee, W.L.; Wang, X. Energy assessment of office buildings in China using China building energy codes and LEED 2.2. *Energy Build.* **2015**, *86*, 514–524. [CrossRef]
3. Nussbaumer, R.J.; Halter, M.; Tervoort, T.; Caseri, W.R.; Smith, P. A simple method for the determination of refractive indices of (rough) transparent solids. *J. Mater. Sci.* **2005**, *40*, 575–582. [CrossRef]

4. Xu, X.; Wu, X.; Zhao, C.; Wang, J.; Ge, X. Simulation and improvement of energy consumption on intelligent glasses in typical cities of China. *Sci. China Technol. Sci.* **2012**, *55*, 1999–2005. [CrossRef]
5. Lee, E.S.; Pang, X.; Hoffmann, S.; Goudey, H.; Thanachareonkit, A. An empirical study of a full-scale polymer thermochromic window and its implications on material science development objectives. *Sol. Energy Mater. Sol. Cells* **2013**, *116*, 14–26. [CrossRef]
6. Sun, Y.; Chu, Y.; Wu, W.; Xiao, H. Nanocellulose-based lightweight porous materials: A review. *Carbohydr. Polym.* **2021**, *255*, 117489. [CrossRef]
7. Khitab, A.; Abdin, Z.U.; Ahmed, I.; Karim, T. 8—Thermal insulation of buildings through classical materials and nanomaterials. In *Recent Advances in Nano-Tailored Multi-Functional Cementitious Composites*; Şahmaran, M., Shaikh, F., Yıldırım, G., Eds.; Woodhead Publishing: Sawston, UK, 2022; pp. 277–303.
8. Maharjan, S.; Liao, K.-S.; Wang, A.J.; Curran, S.A. Highly effective hydrophobic solar reflective coating for building materials: Increasing total solar reflectance via functionalized anatase immobilization in an organosiloxane matrix. *Constr. Build. Mater.* **2020**, *243*, 118189. [CrossRef]
9. Elyasi, R.; Nowee, S.M.; Namaghi, H.A.; Ramezanian, N. Preparation and characterization of absorbing tubes with spectrally selective coatings using economical methods for low- to mid-temperature solar thermal collectors. *Sol. Energy Mater. Sol. Cells* **2015**, *141*, 57–70. [CrossRef]
10. Gulati, L.K.; Gulati, G.K.; Kumar, S. Photochromic materials as a photosensitizer in reversible reactive singlet oxygen generation. *Dye. Pigment.* **2022**, *199*, 110104. [CrossRef]
11. Baetens, R.; Jelle, B.; Gustavsen, A. Properties, requirements and possibilities of smart windows for dynamic daylight and solar energy control in buildings: A state-of-the-art review. *Sol. Energy Mater. Sol. Cells* **2010**, *94*, 87–105. [CrossRef]
12. Gong, X.; Li, J.; Chen, S.; Wen, W. Copolymer solution-based "smart window". *Appl. Phys. Lett.* **2009**, *95*, 251907. [CrossRef]
13. Khaled, K.; Berardi, U. A cross-climate assessment of the visual and energy performance of flexible photochromic films for in situ window retrofits. *Energy Build.* **2023**, *300*, 113664. [CrossRef]
14. Roostaei Firouzabad, M.; Razi Astaraei, F. Energetical effect of electrochromic glazing on the double-skin façade of the building in different climates. *Energy Build.* **2024**, *316*, 114344. [CrossRef]
15. Sun, M.; Sun, H.; Wei, R.; Li, W.; Lai, J.; Tian, Y.; Li, M. Energy-Efficient Smart Window Based on a Thermochromic Hydrogel with Adjustable Critical Response Temperature and High Solar Modulation Ability. *Gels* **2024**, *10*, 494. [CrossRef]
16. Cheng, Y.; Zhang, X.; Fang, C.; Chen, J.; Wang, Z. Discoloration mechanism, structures and recent applications of thermochromic materials via different methods: A review. *J. Mater. Sci. Technol.* **2018**, *34*, 2225–2234. [CrossRef]
17. Piccolo, A. Thermal performance of an electrochromic smart window tested in an environmental test cell. *Energy Build.* **2010**, *42*, 1409–1417. [CrossRef]
18. Sadineni, S.B.; Madala, S.; Boehm, R.F. Passive building energy savings: A review of building envelope components. *Renew. Sustain. Energy Rev.* **2011**, *15*, 3617–3631. [CrossRef]
19. Tan, Y.; Chen, R.; Xiao, Y.; Wang, C.; Zhou, C.; Chen, D.; Li, S.; Wang, W. Temperature-responsive 'cloud' with controllable self-assembled particle size for smart window application. *Appl. Mater. Today* **2021**, *25*, 101248. [CrossRef]
20. Zhang, Z.; Liangmiao, Z.; Zhou, Y.; Cui, Y.; Chen, Z.; Liu, Y.; Li, J.; Long, Y.; Gao, Y. Thermochromic Energy Efficient Windows: Fundamentals, Recent Advances, and Perspectives. *Chem. Rev.* **2023**, *123*, 7025–7080. [CrossRef]
21. Seyfouri, M.M.; Binions, R. Sol-gel approaches to thermochromic vanadium dioxide coating for smart glazing application. *Sol. Energy Mater. Sol. Cells* **2017**, *159*, 52–65. [CrossRef]
22. Hoffmann, S.; Lee, E.S.; Clavero, C. Examination of the technical potential of near-infrared switching thermochromic windows for commercial building applications. *Sol. Energy Mater. Sol. Cells* **2014**, *123*, 65–80. [CrossRef]
23. Xiao, X.; Zhang, H.; Chai, G.; Sun, Y.; Yang, T.; Cheng, H.; Chen, L.; Miao, L.; Xu, G. A cost-effective process to prepare VO2 (M) powder and films with superior thermochromic properties. *Mater. Res. Bull.* **2014**, *51*, 6–12. [CrossRef]
24. Chen, Z.; Gao, Y.; Kang, L.; Du, J.; Zhang, Z.; Luo, H.; Miao, H.; Tan, G. VO2-based double-layered films for smart windows: Optical design, all-solution preparation and improved properties. *Sol. Energy Mater. Sol. Cells* **2011**, *95*, 2677–2684. [CrossRef]
25. Shen, N.; Chen, S.; Chen, Z.; Liu, X.; Cao, C.; Dong, B.; Luo, H.; Liu, J.; Gao, Y. The synthesis and performance of Zr-doped and W-Zr-codoped VO2 nanoparticles and derived flexible foils. *J. Mater. Chem. A* **2014**, *2*, 15087–15093. [CrossRef]
26. Shafaghat, A.; Keyvanfar, A. Dynamic façades design typologies, technologies, measurement techniques, and physical performances across thermal, optical, ventilation, and electricity generation outlooks. *Renew. Sustain. Energy Rev.* **2022**, *167*, 112647. [CrossRef]
27. Cui, Y.; Ke, Y.; Liu, C.; Chen, Z.; Wang, N.; Zhang, L.; Zhou, Y.; Wang, S.; Gao, Y.; Long, Y. Thermochromic VO2 for Energy-Efficient Smart Windows. *Joule* **2018**, *2*, 1707–1746. [CrossRef]
28. Long, L.; Ye, H.; Zhang, H.; Gao, Y. Performance demonstration and simulation of thermochromic double glazing in building applications. *Sol. Energy* **2015**, *120*, 55–64. [CrossRef]
29. Seeboth, A.; Ruhmann, R.; Mühling, O. Thermotropic and Thermochromic Polymer Based Materials for Adaptive Solar Control. *Materials* **2010**, *3*, 5143. [CrossRef]
30. Giovannini, L.; Favoino, F.; Pellegrino, A.; Lo Verso, V.R.M.; Serra, V.; Zinzi, M. Thermochromic glazing performance: From component experimental characterization to whole building performance evaluation. *Appl. Energy* **2019**, *251*, 113335. [CrossRef]
31. Lu, Y.-C.; Hsueh, C.-H. Simulations and experimental studies of thermochromic performances of multilayered VO2 nanocone array for energy-saving smart window applications. *Appl. Surf. Sci.* **2024**, *661*, 160083. [CrossRef]

32. Lee, W.J.; Kwac, L.K.; Kim, H.G.; Chang, J.-H. Thermotropic liquid crystalline copolyester fibers according to various heat treatment conditions. *Sci. Rep.* **2021**, *11*, 11654. [CrossRef] [PubMed]
33. El-Deeb, A.S.; Abd elkader, M.M.; Ismail, A.M.; Hassaan, M.Y. Optical, FTIR, electrical and dielectrical properties of a glass system for smart windows applications. *Optik* **2020**, *221*, 165358. [CrossRef]
34. Kumar Babu, B.; Ghosh, S.; Chakrabortty, S. Chapter 10—Recent developments in smart window engineering: From antibacterial activity to self-cleaning behavior. In *Energy Saving Coating Materials*; Dalapati, G.K., Sharma, M., Eds.; Elsevier: Amsterdam, The Netherlands, 2020; pp. 227–263.
35. Liu, X.; Gallavardin, T.; Burel, F.; Vuluga, D. Influence of quadruple hydrogen bonding on polyvinyl butyral resin properties. *Polym. Degrad. Stab.* **2023**, *208*, 110243. [CrossRef]

polymers

MDPI

Article

Facile Synthesis of Thermoresponsive Alternating Copolymers with Tunable Phase-Transition Temperatures

Zichen Huang [1], Fan Chen [1], Qi Wang [2], Dingxiang Zhang [1], Hongdong Wang [2,*] and Xiacong Zhang [1,*]

[1] Department of Polymer Materials, School of Materials Science and Engineering, Shanghai University, Shanghai 200444, China; zb933@shu.edu.cn (F.C.)
[2] School of Mechatronic Engineering and Automation, Shanghai University, Shanghai 200444, China
* Correspondence: whd20@shu.edu.cn (H.W.); xczhang@t.shu.edu.cn (X.Z.)

Abstract: A series of novel amphiphilic alternating CPEG copolymers were synthesized through an amine–epoxy click reaction comprising aliphatic amine and polyethylene glycol diglycidyl ether (PEGDE). These polymers were characterized in detail via nuclear magnetic resonance (NMR), gel permeation chromatography (GPC), Fourier-transform infrared spectroscopy (FTIR), and thermogravimetric analysis (TGA) to confirm the successful synthesis. Due to their amphiphilic structure, these polymers display thermoresponsiveness, with tunable cloud points (Tcps) that are adjustable from 20.8 °C to 46.8 °C by altering the side-chain length of the aliphatic amine, varying the mixing ratios of copolymers, the solution's pH, and salt additions. This tunable thermoresponsive behavior positions CPEG copolymers as promising candidates for a range of functional material applications.

Keywords: thermoresponsive polymers; amphiphilic copolymers; borate ester; polyethylene glycol

1. Instruction

Stimuli-responsive polymers are able to respond to changes in the external environment, such as temperature [1–3], light [4–6], pH [7,8], mechanical forces [9], and chemical stimuli. They can change their physical and/or chemical properties reversibly or irreversibly. These polymers undergo reversible or irreversible changes in their physical and/or chemical properties, making them invaluable for applications in drug delivery [10,11], biosensors [12], diagnostics, and separation technologies [13]. Among them, thermoresponsive polymers, which exhibit reversible phase-transitions triggered by temperature changes, form one of the most prominent types due to their unique properties and broad biomedical applications, including bioimaging [14], drug delivery [15,16], injectables [17,18], smart surfaces [19–22], adhesives [23], and tissue engineering [24–26]. The most common thermoresponsive polymers, such as poly(*N*-alkylacrylamides), are poly(*N*-isopropylacrylamide) (PNIPAM) and the polyethylene glycol (PEG) derivative OEGMA [27]. PNIPAM, for instance, undergoes a pronounced phase-transition upon heating while showing significant hysteresis during cooling due to both intramolecular and intermolecular hydrogen bonding. Lutz [28] investigated the lower critical solution temperature (LCST) of poly(2-(2-methoxyethoxy) ethylmethacrylate-*co*-(ethylene glycol) methacrylate) (P(MEO_2MA-*co*-$OEGMA_{475}$)) copolymers, which exhibited similar or superior temperature-sensitive transition properties in comparison to PNIPAM, with added benefits for biological applications due to reduced toxicity. Polymers with LCSTs or cloud point temperatures (Tcps) in aqueous solutions are of particular interest. When heated above their Tcps, these polymers undergo entropy-driven dehydration, causing polymer chain collapse and aggregation into mesoglobules. One critical characteristic of thermoresponsive polymers is the tunability of their Tcps. Tcp is primarily governed by the hydrophilic–hydrophobic balance within the polymer structure, with higher hydrophilicity generally increasing Tcp. The thermoresponsive behavior of block copolymers is typically determined by their thermoresponsive blocks, while random copolymers' Tcps depend on the overall hydrophobicity of the entire polymer. By adjusting the ratio of hydrophilic and hydrophobic units, the Tcps of copolymers can be tuned accordingly. However, most thermoresponsive polymers are covalently linked, meaning that making post-synthesis structural adjustments to finely tune their LCSTs without breaking the polymer's covalent bonds is challenging. Therefore, it remains an important challenge to develop novel strategies such as supramolecular interactions [29] and dynamic covalent chemistry [30] for preparing thermoresponsive polymers with truly tunable Tcps.

Polyethylene glycol diglycidyl ether (PEGDE), with its α, ω-epoxy terminal groups, is a highly reactive cross-linker commonly used in hydrogel fabrication due to its ability to undergo crosslinking reactions. For that reason, PEGDE is always used as a cross-linker in hydrogel fabrication. It has been proven that when this polymer is subcutaneously injected at lower doses, it is non-poisonous [31]. Meanwhile, injectable hyaluronic acid (HA)-based hydrogels mainly comprising PEGDE are presently used as the new standard for soft-tissue fillers [32]. The formation of a PEGDE polymer network can be used effectively for improving the mechanical strength of solid electrolytes in Li-ion batteries [33]. PEGDE can also be used to produce microelectrode biosensors [34]. Therefore, PEGDE demonstrates significant potential across various fields, from biomedical applications to energy storage applications.

In our previous study, a series of hydrogels were synthesized through a cross-linking process involving CPEG polymers and alginate grafted with phenylboronic acid (Alg-PBA) [35]. CPEG polymers, obtained by reacting PEGDE with pentylamine, exhibited amphiphilic structures and demonstrated thermoresponsive behavior in aqueous solutions. Building on this foundation, the current study aims to investigate the thermoresponsive properties of a series of alternating copolymers with tunable hydrophilic and hydrophobic segments. These copolymers were synthesized via condensation polymerization between different primary amines and PEGDE. Their thermoresponsive behavior was characterized via UV–vis spectroscopy in order to determine their Tcps under different conditions, including variations in the aliphatic amine side-chain length, mixing ratios of copolymers, solution pH, and salt concentration. In addition, their self-assembly behavior was thoroughly examined using dynamic light scattering (DLS), allowing for detailed exploration of the mechanisms driving their thermoresponsiveness. Notably, the reactive diethanolamine (DEA) units in the polymer backbone enable the formation of borate esters upon reaction with boronic acid, and this is stabilized via intramolecular N-B coordination. The introduction of phenylboronic acid derivatives further allowed the fine-tuning of Tcp, expanding the versatility of these materials. This research advanced our understanding of CPEG's thermoresponsive behavior and expanded their potential applications in responsive material design.

2. Materials and Methods

2.1. Materials

Poly(ethylene glycol) diglycidyl ether (PEGDE; average molecular weight = 500) was purchased from Aladdin (Shanghai, China). The *n*-pentylamine (C5-NH_2, >98%) and *n*-hexylamine (C6-NH_2 > 98%) were purchased from TCI (Tokyo, Japan). The *n*-butylamine (C4-NH_2, >98%) was purchased from Adamas-beta (Shanghai, China). All other chemicals were used without further purification.

2.2. Synthesis of CPEG

C5PEG was synthesized from C5-NH_2 and PEGDE as monomers via the amine–epoxy click reaction. Generally, PEGDE (2 g, 3.8 mmol) and C_5-NH_2 (480 μL, 3.8 mmol) were added into a 10 mL reaction tube. The super-dry reagent of DMF was also added as a solvent. The air in the tube should be replaced by N_2. After stirring for 12 h at 60 °C, dimethylformamide (DMF) was removed by rotary evaporation. The polymer was dissolved in dichloromethane (DCM) and extracted with a 10 wt% NaOH aqueous solution. Using rotary evaporation to remove DCM, the colorless viscous liquid was obtained. C4PEG and C6PEG were synthesized in a similar manner. The only difference was that C_4-NH_2 and C_6-NH_2 were applied as monomers. C4-C6PEG was synthesized from C4-NH_2 (240 μL, 1.9 mmol), C6-NH_2 (240 μL, 1.9 mmol), and PEGDE (2 g, 3.8 mmol) as monomers via the same solution and processes as C5PEG.

2.3. Characterization of Synthesized Polymers

^{1}H NMR spectra of C6PEG, C5PEG, and C4PEG were recorded on an NMR spectrometer (AV-500; Bruker, Zurich, Switzerland) using $CDCl_3$ as the solvent. Fourier-transform infrared (FTIR) spectroscopies of C6PEG, C5PEG, and C4PEG were obtained using an FTIR spectrometer (Nicolet Is 50; Thermo Fisher Scientific, Waltham, MA, USA). The number-average molecular weight (Mn), weight-average molecular weight (Mw), and polydispersity index (Đs) of the synthesized C6PEG, C5PEG, and C4PEG were characterized via gel permeation chromatography (GPC) using a gel permeation chromatography system (Breeze; Waters Corporation, Milford, CT, USA) at a test concentration of 2 mg·mL^{-1} with DMF as the eluent. Thermogravimetric analysis (TGA) was carried out using a PerkinElmer Q5000IR thermobalance, and nitrogen was used as the purging gas at a heating rate of 20 °C min^{-1}. The liquid samples were around 5 mg. The DSC measurement was carried out on a Thermal Advantage DSC (DSC 25; TA Instruments, New Castle, DE, USA) equipped with a refrigerator cooling system. The polymer (10 mg) was placed in aluminum pans (non-hermetic) and scanned at a heating rate of 10 °C min^{-1} from −80 to 100 °C under a dry nitrogen atmosphere.

2.4. Characterization of the Self-Assembly Process of Polymer Micelles

The aggregate sizes of C6PEG, C5PEG, and C4PEG micelles were measured using a dynamic light scattering (DLS) instrument (DynaPro NanoStar M3300; Wyatt, Santa Barbara, CA, USA). Prior to measurement, the solutions were filtered through a 0.45 μm filter to remove dust. The hydrodynamic radii of aggregates were calculated following the Stokes–Einstein equation.

The critical micelle concentrations (CMC) of C6PEG, C5PEG, and C4PEG in water were measured using 1,6-diphenyl-1,3,5-hexatriene (DPH) as a hydrophobic dye probe. Solutions of C6PEG, C5PEG, and C4PEG at varying concentrations (0.01, 0.1, 0.5, 1, 3, and 5 wt%) were prepared, and 4 μM DPH was added into each solution. These solutions were equilibrated in the dark for 4 h before their absorbance from 300 to 450 nm was measured via a UV–vis spectrophotometer (V750; JASCO, Hachioji, Tokyo, Japan). The inflection points of the absorbance difference between 377 and 400 nm against the logarithm concentration of the copolymer solution were equal to the CMC value [35].

2.5. Characterization of Factors of Polymer Phase-Transition Temperature

A UV–vis spectrophotometer equipped with a thermo-controlled bath (V750; JASCO, Hachioji, Tokyo, Japan) was utilized to analyze the transmittance variations of CPEG solutions with different salts, pHs, and phenylboronic acid derivatives. The heating rate was set at 0.5 °C·min^{-1}. Measurements were performed using a light-source wavelength of 700 nm, and absorbance was recorded every 5 s.

3. Result and Discussion

3.1. Synthesis and Characterizations of Copolymers

The amphiphilic alternating copolymers, CPEG, were synthesized via amine–epoxy condensation copolymerization between the PEGDE and alkylamines. The structural formulas are presented in Figure 1. The copolymerization process involved the reaction of primary amine groups with the epoxy groups under DMF as the solvent at 60 °C. This amine–epoxy condensation resulted in the formation of secondary amines as intermediates. These secondary amines could further react with additional epoxy groups, facilitating the growth of the polymer chain via condensation polymerization. Three different alkylamines—*n*-butylamine, *n*-pentylamine, and *n*-hexylamine—were used to synthesize the copolymers C4PEG, C5PEG, and C6PEG, respectively. The successful polymerization resulted in transparent, viscous liquid products once the solvent was removed, with yields exceeding 82% for all three polymers.

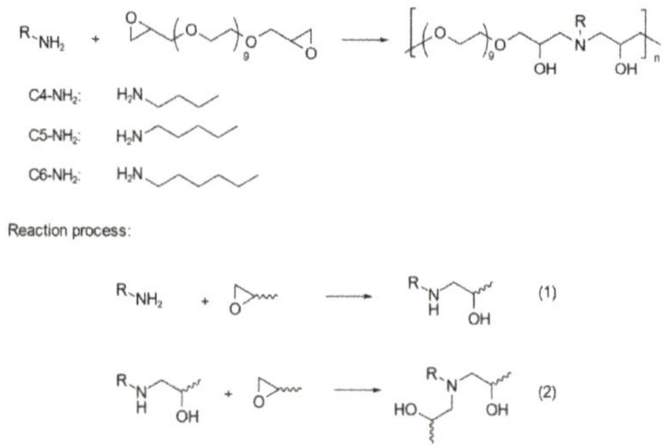

Figure 1. Synthesis scheme of C4PEG, C5PEG, and C6PEG polymers.

The chemical structures of the CPEG copolymers were characterized via ^{1}HNMR, as shown in Figure 2. Peak a at 0.88 ppm represented the terminal methyl group of the alkyl side chain. The chemical shifts of methylene groups located between the initial and terminal methyl groups of the alkyl side chain were in the range of 1.3–1.4 ppm (b). The difference in Peak b exhibited the difference in alkyl side chains. Peak c represented the initial methylene of the alkyl side chain, which was linked with the nitrogen atom, and their chemical shifts changed to 2.5 ppm after the amine–epoxy click reaction due to nitrogen effects. Peak f belonged to the polyethylene glycol (PEG) chain. The disappearance of the peaks in PEGDE refers to the epoxy groups, which were originally at 2.5–2.8 (Peak 1) and 3.3 ppm (Peak 2), respectively. The appearance of peaks at 3.5 ppm (d) and 2.5 ppm (c), along with the disappearance of Peaks 1 and 2, indicates that the reaction between the epoxy and the primary amine was successful.

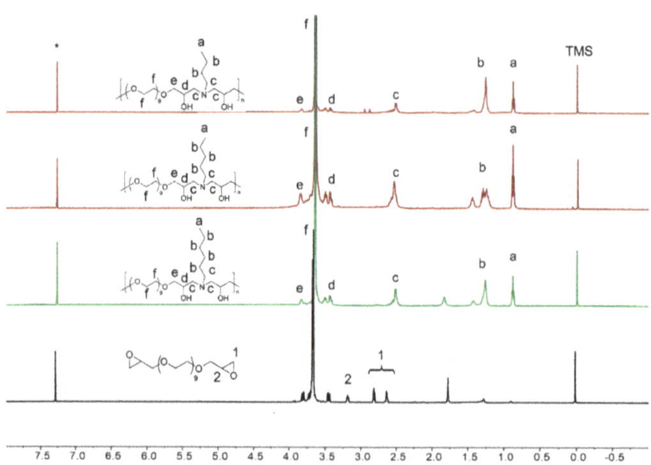

Figure 2. ^{1}H NMR spectra of C4PEG, C5PEG, C6PEG, and PEGDE in $CDCl_3$. Chemical shifts are reported as δ values (ppm) relative to internal Me_4Si (TMS). Solvent signals are marked with *.

The molecular structures of CPEG were further confirmed via attenuated total reflection Fourier-transform infrared (ATR-FTIR) spectroscopy, as presented in Figure 3. The broad absorption band observed between 3300 and 3600 cm^{-1} corresponded to hydroxyl (O−H) stretching vibrations, which are indicative of the presence of hydroxy groups in the polymer chains. Additionally, a broad peak located at approximately 2860 cm^{-1} was associated with the stretching vibration of alkyl (−C−H) bonds, which are typical of the polymer backbone. The intense band appearing at around 1098 cm^{-1} was attributed to C−O stretching vibrations, confirming the presence of ether groups in the molecular structure. These characteristic peaks suggested the successful synthesis of CPEG variants, with consistent molecular structures across different chain lengths (C4PEG, C5PEG, and C6PEG).

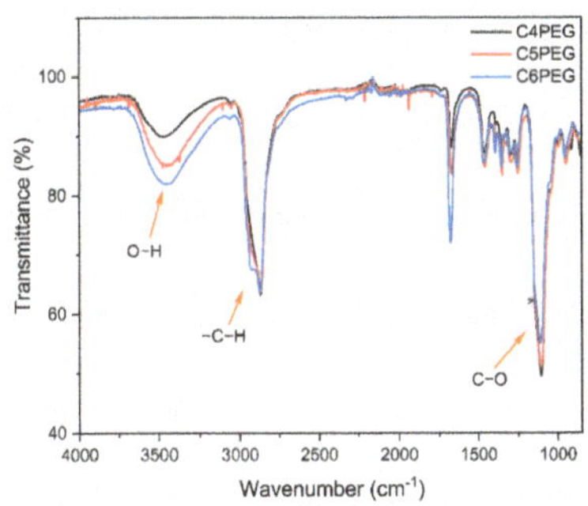

Figure 3. ATR-FTIR spectra of C4PEG, C5PEG, and C6PEG.

The average molecular weights of the CPEG polymers were measured using gel permeation chromatography (GPC), and the resulting GPC curves are presented in Figure S1. The presence of shoulders in the GPC analysis is due to oligomers. The Đs for the three CPEG samples were approximately 1.3, indicating relatively narrow molecular weight distributions. This was due to the low degree of polymerization with respect to CPEG removal during the alkaline washing process. The molecular weights of the CPEG polymers were found to be quite similar, with number-average molecular weights (Mn) ranging from 6.6 $\times 10^3$ to 7.2 $\times 10^3$ and weight-average molecular weights (Mw) ranging from 8.3 $\times 10^3$ to 9.2 $\times 10^3$. However, the degree of polymerization for the CPEGs synthesized using this method was relatively low, at approximately 16–18. This lower degree of polymerization could be attributed to the reduced reactivity of the secondary amine in the reaction, as well as potential inaccuracies in the proportion of raw materials used during the synthesis. To increase the degree of polymerization, an alternative approach was explored by synthesizing C5PEG without solvents. Since both PEGDE and the amine were liquids, bulk polymerization was considered feasible, resulting in an increase in monomer concentration. As observed in the data shown in Table 1, this solvent-free method successfully increased the average molecular weight of the C5PEG polymer. The Mn increased to 13.0 $\times 10^3$, and the Mw increased to 26.4 $\times 10^3$. However, this method also resulted in an increase in polydispersity, which rose to 2.0.

Table 1. Conditions for and results of the polymerization.

Sample	Aliphatic Amine	Mn (Da) ($\times 10^{-3}$)	Mw (Da) ($\times 10^{-3}$)	Đ	Tcp (°C)
C4PEG	*n*-butylamine	6.6	8.3	1.2	46.8
C5PEG	*n*-pentylamine	7.2	9.2	1.3	29.7
C6PEG	*n*-hexylamine	7.1	9.2	1.3	20.8
C5PEG (without solvent)	*n*-pentylamine	13.0	26.4	2.0	-

The thermal properties of CPEG copolymers were evaluated via TGA and DSC, as shown in Figures 4 and 5. The results demonstrated that these series of copolymers exhibit similar thermal stability. The initial decomposition temperature of C4PEG was 295.0 °C, while C5PEG and C6PEG display decomposition temperatures of 329.7 and 312.0 °C, respectively. The glass transition temperatures of C4PEG, C5PEG, and C6PEG were −64.84 °C −60.46 °C, and −65.24 °C, respectively. Interestingly, thermal stability and glass transition temperature generally increased with the length of the side group, with C5PEG being an exception to this trend. This anomaly could be attributed to the distinct arrangement of the odd-numbered carbon chains in C5PEG compared to the even-numbered carbon chains in C4PEG and C6PEG, which likely affected molecular packing and thermal behavior. The even number for the carbon chain was associated with larger average molecular interchain spacings. The increase in the average molecular interchain spacing value resulted in a decrease in thermal conductivity. This was the reason why C5PEG exhibited a better thermal stability and glass transition temperature [36].

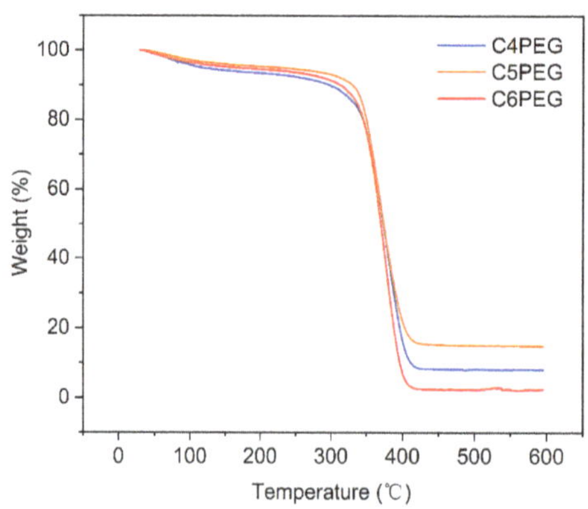

Figure 4. The TGA curves of C4PEG, C5PEG, and C6PEG.

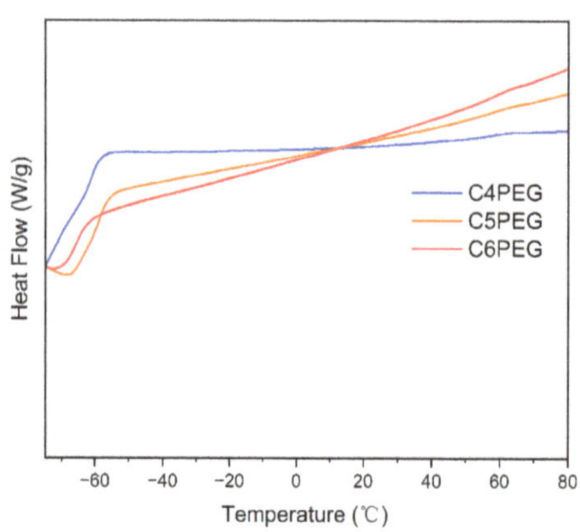

Figure 5. The DSC curves of C4PEG, C5PEG, and C6PEG.

3.2. Thermoresponsive Behaviors of Copolymers

The structural characteristics of the polymers were determined by alternating the arrangement of PEGDE and alkylamine units. The variation in alkyl chain length from butyl (C4) and amyl (C5) to hexyl (C6) introduced hydrophobicity gradients along the polymer chain. This amphiphilic nature of the copolymers was expected to influence their self-assembly and thermoresponsive behavior. The thermoresponsiveness of the CPEG aqueous solution is shown in Figure 6a. When the temperature was below the transition temperature, the C5PEG aqueous solution remained transparent. It was found that the thermoresponsiveness of all samples was reversible, and the solutions returned from turbid to transparent upon the cooling of the systems. Temperature-varied UV−vis spectroscopy was used to measure Tcps. Tcp is defined as the temperature corresponding to the 50% transmittance point. The transmittance of each CPEG solution decreased rapidly after its temperature reached a point above the transition temperature, as shown in Figure 6b. The Tcps of C4PEG, C5PEG, and C6PEG were 46.8, 29.7, and 20.8 °C, respectively, demonstrating that their Tcps would decrease as the lengths of side chain increased. This was attributed to the enhanced hydrophobicity of the CPEG polymer, which was triggered by the longer side alkyl chains. The increase in the hydrophobicity of the alkyl units might have an influence on the self-assembly and aggregation of polymers in water. Accordingly, C6PEG in the aqueous solution was turbid at room temperature, while C4PEG in the aqueous solution with higher hydrophilicity was transparent at the same temperature.

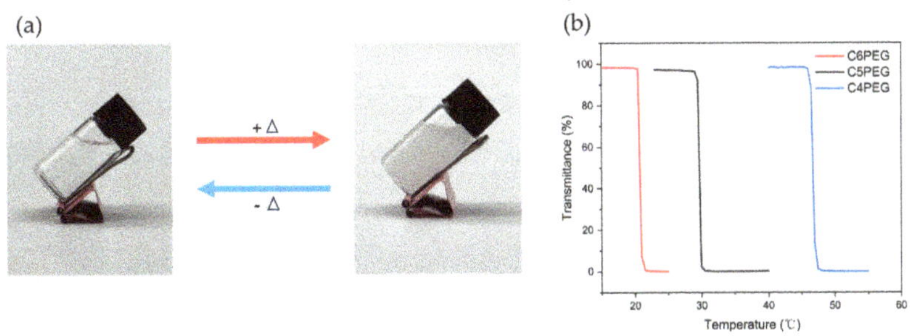

Figure 6. (**a**) Photographs of an aqueous solution of C5PEG at 20 °C (**left**) and 50 °C (**right**). (**b**) The curves of transmittance versus temperature for C4PEG, C5PEG, and C6PEG in the 5 wt% aqueous solution.

C6PEG and C4PEG were chosen for further study; these had the lowest Tcp and highest Tcp among the CPEGs. The, in general, thermally induced aggregation behaviors of CPEGs are shown in the substantial differences in their Tcps. The thermally induced phase transition was thus studied via dynamic light scattering (DLS), and the aggregate sizes of the CPEG polymers were determined. The results are plotted in Figure 7 from sample C6PEG (a), which comprises a mixture of C4PEG and C6PEG (b), and C4PEG (c) with a diluted concentration (0.5 wt%). Lower concentrations were chosen to minimize multiple light scattering effects, ensuring that size measurements were more accurate [37]. As the temperature increased, the size of the copolymer particles increased from approximately 2 nm to above 200 nm, indicating a characteristic thermoresponsive aggregation behavior.

For the further study of the thermally induced aggregation behaviors of CPEGs, the critical micelle concentrations (CMCs) of the CPEGs were measured. To measure CMC, DPH hydrophobic dye was added to the polymer solution. The absorbance of DPH increased after the micelles were formed. The CMC of different polymers depended on the differences in absorbance between 377 and 400 nm against the logarithm of copolymer concentrations. The absorbance curves between 300 and 400 nm in the C4PEG and C6PEG solutions are shown in Figures S2 and S3, respectively. As shown in Figure S4, the CMC values of C6PEG, C5PEG, and C4PEG were 0.22 wt%, 0.40 wt%, and 0.65 wt%, respectively [35]. This variation in CMC values was attributed to differences in the hydrophobic alkyl chain length in the aliphatic amine segments of the copolymers. Longer alkyl chains, such as those in C6PEG, resulted in lower CMC values, while shorter chains, such as in C4PEG, resulted in higher CMC values. The CMC of C4PEG (0.65 wt%) was higher than the concentration used in the DLS experiments, indicating that C4PEG possibly dissolved single-chain conformations in water with a temperature of 20 °C. In contrast, the concentration of C6PEG in the DLS tests exceeded its CMC, providing evidence that 0.5 wt% C6PEG formed micelles at 20 °C. However, the particle size of C6PEG at 20 °C was similar to that of C4PEG, at approximately 2 nm, suggesting that C6PEG might have formed pre-assembled aggregates rather than fully developed micelles. The PDI values of C4PEG, the mixture of C4PEG and C6PEG, and C6PEG were 0.266, 0.436, and 0.565 under 20 °C, respectively. For C4PEG, the mixture of C4PEG and C6PEG, and C6PEG at 20 °C, the values were 0.266, 0.436, and 0.565, respectively. The higher PDI value of the mixture compared to individual copolymers suggested a broader size distribution, indicating the coexistence of different micelle populations. At 50 °C, the particle sizes of the mixture of C4PEG and C6PEG were comparable to those of the individual copolymers, further supporting the hypothesis that the two polymers formed micelles together. Additionally, both the PDI and particle size in the mixed solution at 50 °C were intermediate between the values for the pure copolymers, further confirming the co-assembly of C4PEG and C6PEG into micelles.

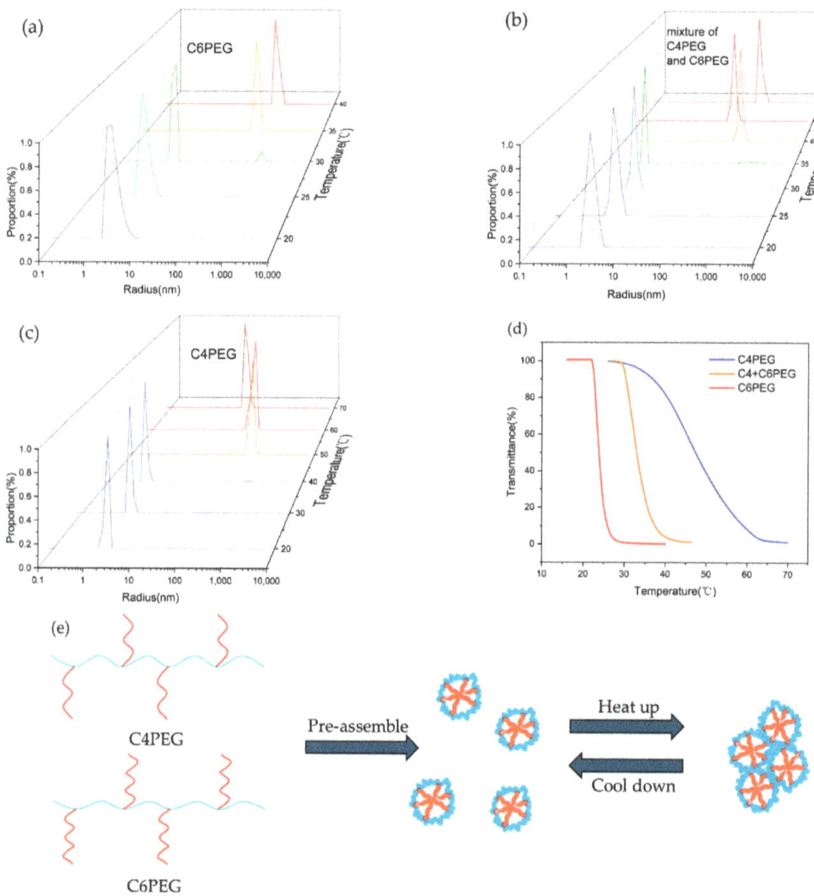

Figure 7. Dependence of the hydrodynamic radius (R_h) of (**a**) C6PEG, (**b**) C4PEG, and (**c**) the mixture of C4PEG and C6PEG in aqueous solutions on temperature. (**d**) Transition temperatures of C4PEG, C6PEG, and the mixture of C4PEG and C6PEG at 0.5 wt% concentration. (**e**) Schematic diagram of the thermally induced aggregation behaviors of the CPEG mixture.

For comparison, the Tcp of C4PEG, C6PEG, and the mixture of C4PEG and C6PEG was tested at 0.5 wt% concentration via temperature-varied UV−vis spectroscopy. The Tcp of the C4PEG and C6PEG mixture was found to be intermediate relative to the individual transition temperatures of C4PEG and C6PEG, as shown in Figure 7d. This shift in Tcp, supported by dynamic light scattering (DLS) data, confirmed that C4PEG and C6PEG co-assembled into micellar structures. The hydrophobic alkyl chains of both polymers influenced the Tcp of the mixture. The thermally induced self-assembly process is illustrated in Figure 7e. At low temperatures, the polymer chains in the aqueous solution self-assembled into micelles, with hydrophilic PEG chains forming the exterior and hydrophobic alkyl side chains enclosed within the micelle core. As the temperature increased, micelles began to coalesce, driven by the dehydration of the hydrophilic domains. During this process, the polymer chains experienced a reduction in solvation, causing them to aggregate on the micelle surfaces. With a further increase in temperature, the micelles continued to grow, ultimately transitioning into larger particles at a radius of approximately 250 nm—nearly 200 times the original micelle size. This size increase resulted in visible aggregation in the aqueous solution, resulting in a sharp decrease in transmittance. This growth in particle size was characteristic of thermally induced aggregation, closely resembling behaviors observed in PNIPAM systems [38]. The hydrophobic interactions between the alkyl chains of C4PEG and C6PEG were crucial to this transition, as particle growth was directly linked to the temperature-induced collapse of hydrophilic domains and the subsequent aggregation of hydrophobic segments. These findings indicate that the Tcp of the CPEG polymer could be tuned by varying the mixing ratios of copolymers with different hydrophilicity conditions.

3.3. Methods for Tuning the Phase-Transition Temperature

To further explore tunability with respect to phase-transition temperatures (Tcp), the effects of C6PEG and C4PEG with varying mix ratios were studied at a total copolymer concentration of 2.5 wt% using temperature-dependent UV−vis spectroscopy. By examining different C4PEG and C6PEG ratios, the aim was to broaden the application potential of the system. Figure 8 illustrates the cloud point changes for mixtures with C4PEG contents at 0% (20.8 °C), 25% (26.9 °C), 50% (28.5 °C), 75% (31.9 °C), and 100% (46.8 °C). The results demonstrated, in a clear trend, that Tcp could be easily adjusted by altering the relative proportions of the two copolymers. The data indicated a decrease in Tcp by approximately 1.5 °C when the C4PEG content increased from 50% to 75%, with a smaller shift observed for pure C4PEG. This tunable Tcp range, spanning from 20.8 °C to 46.8 °C, encompassed both room temperature (25 °C) and body temperature (37 °C) contexts, and this holds significant promise for medical applications, including temperature-sensitive indicators and drug delivery systems.

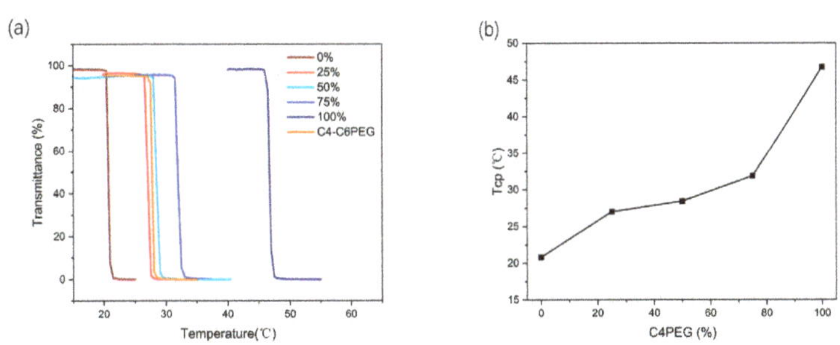

Figure 8. (**a**) Effects of different mixing ratios in C6PEG and C4PEG mixtures (copolymer concentration 2.5 wt%) on the change in cloud points; the proportion of C4PEG is shown in the figure. Comparison of the cloud points of C4PEG and C6PEG copolymers and the C4PEG and C6PEG mixture. (**b**) Tcp-dependence of the C4PEG and C6PEG mixture on the content of C4PEG.

For comparative analysis, a copolymer designated as C4-C6PEG was synthesized directly from poly(ethylene glycol) diglycidyl ether (PEGDE) and an equimolar mixture of n-butylamine and n-hexylamine. The Tcp of C4-C6PEG was found to be 27.7 °C, closely matching the Tcp of a 1:1 C4PEG and C6PEG mixture (28.5 °C). This similarity further underscored the robustness of the Tcp tuning achievable by blending, as comparable phase-transition temperatures could be obtained by either copolymerizing mixed amines or by physically blending distinct copolymers. These findings demonstrate that fine-tuning the phase-transition temperature through simple mixing strategies enables customized thermoresponsive behavior, making these copolymer systems adaptable for a range of biomedical and environmental applications.

With the exception of the changing proportions of the CPEG mixture, salt addition and pH regulation are common methods for changing phase-transition temperatures [39]. Given its moderate Tcp, C5PEG was selected as a model system to assess Tcp variations in response to different salts and pH. Previous studies have shown that salt can significantly influence the phase-transition behavior of thermoresponsive polymers, an effect often described via the Hofmeister (HS) series. While cations generally exert a minor influence on Tcp, according to the HS series, anions play a primary role. Therefore, sodium was used as the fixed cation, and anion types were varied among different salts. The effects of NaCl, Na_2SO_4, and $NaPF_6$ on the thermally induced phase transition of C5PEG were investigated by adding these salts (50 mmol·L^{-1}) to 2.5 wt% C5PEG solutions. The observed Tcp variations largely aligned with the HS series. For example, the addition of "salt-out" anions, such as Cl^- and SO_4^{2-}, lowered Tcp from 29.1 °C to 26.7 °C and 22.2 °C, respectively. In contrast, the "salt-in" anion PF_6^- increased Tcp from 29.1 °C to 32.1 °C. Anions with lower lyotropic numbers tend to dehydrate the polymer chains by displacing water molecules, thus strengthening hydrophobic interactions and promoting polymer collapse into particle states. Conversely, anions with higher lyotropic numbers interfere with the structure of bulk water, enhancing polymer hydrophilicity [40].

Beyond salt addition, the solution's pH also affected Tcp. As shown in Figure 9b, C5PEG solubility in water increased with increasing acidity. Tcp values at pH 11 and pH 9 were 28.8 °C and 30.3 °C, respectively, but increased to 40.2 °C and 45.2 °C at pH 8 and pH 7, respectively; moreover, it exceeded 80 °C at pH 5. In basic conditions, C5PEG remains largely insoluble due to the hydrophobic nature of its tertiary amine groups. However, in acidic conditions, the presence of H^+ ions protonates the tertiary amine, converting it into a hydrophilic quaternary ammonium salt, which increases Tcp significantly [41]. Despite its effectiveness, pH adjustments show limitations as a control strategy. As shown in Figure 9b, Tcp increased by roughly 15 °C when the pH value decreased from pH 9 to pH 7. Such a sharp and uncontrollable shift is impractical for applications requiring fine-tuned thermoresponsiveness.

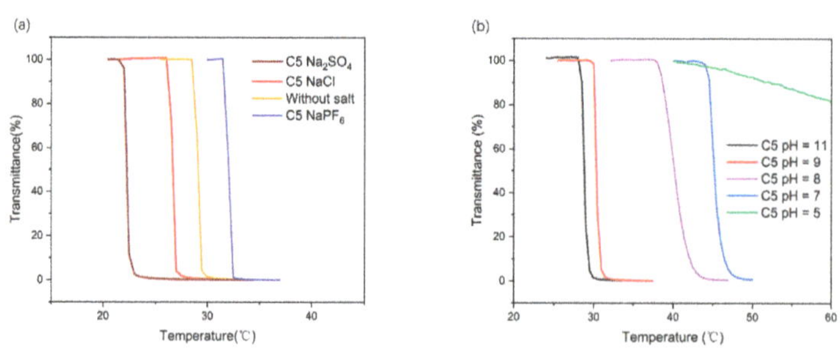

Figure 9. (**a**) Effects of different salt additions on the cloud point of the 2.5 wt% C5PEG aqueous solution. (**b**) Effects of different pH values on the turbidity point of the 2.5 wt% C5PEG aqueous solution.

Phenylboronic acid derivatives can react with the diethanolamine (DEA) unit in the main chain of CPEG to form a borate ester and realize boron–nitrogen (B–N) coordination with two five-member rings. This unique coordination enables hydrophobic groups to be anchored to the polymer's main chain, changing its overall hydrophilicity.

Specifically, as shown in Figure 10, the addition of phenylboronic acids with hydrophobic substituents—such as nitro and butyl groups—resulted in notable Tcp reductions, with Tcp values observed at 32.4 °C and 35 °C, respectively. This decrease in Tcp demonstrates that incorporating hydrophobic groups via phenylboronic acid modification effectively enhances polymer hydrophobicity and lowers the transition temperature. These findings indicate that modifying thermoresponsive polymers with different phenylboronic acid derivatives provides a viable strategy for controlling Tcp.

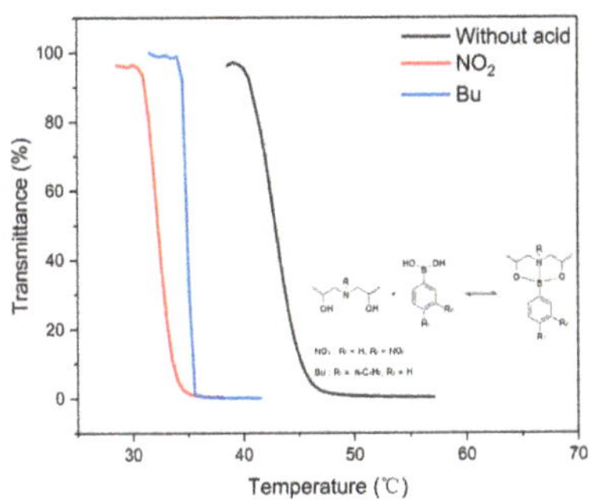

Figure 10. Effects of 3-nitrophenylboronic acid and 4-butylphenylboronic acid additions on the cloud point of C5PEG and the formation of borate esters upon reaction with boric acid, stabilized via intramolecular N-B coordination.

4. Conclusions

In this study, a series of novel amphiphilic CPEG copolymers were facilely synthesized through an amine–epoxy click reaction between PEGDE and various primary amines. The chemical structures of CPEGs were confirmed using NMR, ATR-FTIR, and GPC. Detailed characterization confirmed their amphiphilic structures and thermoresponsive behavior via DLS. The transition temperature (Tcp) of the CPEG copolymers was shown to be highly tunable, with a range from 20.8 °C to 46.8 °C achieved by varying side-chain lengths, mixing ratios, salt additions, and phenylboronic acid modifications. While pH adjustments also influenced Tcp, they offered less precision for fine control. These findings highlight the versatility of this CPEG copolymer series for phase-transition temperature regulation, enabling post-synthesis adjustments to Tcp, in which Tcp can be tailored post-synthesis. The wide range of transition temperatures renders them suitable for drug delivery and application in other thermoresponsive medical materials.

Supplementary Materials: The following supporting information can be downloaded at https://www.mdpi.com/article/10.3390/polym16243470/s1, Figure S1. (a) GPC curve of C4PEG. (b) GPC curve of C5GPC. (c) GPC curve of C5PEG (without solvent). (d) GPC curve of C6PEG; Figure S2. Absorbance curves of C4PEG solutions at different concentrations across 300–450 nm; Figure S3. Absorbance curves of C6PEG solutions at different concentrations across 300–450 nm; Figure S4. Absorption of the hydrophobic probe DPH in C4PEG and C6PEG copolymer solutions. CMC was determined via extrapolation of the difference of absorbance at 377 and 400 nm. The measurements were carried out at 25 °C; Figure S5. ATR-FTIR spectra of C6PEG and C4PEG across 4000–2500 cm^{-1} and 1900–850 cm^{-1}; Figure S6. The impact of different polymer concentrations in solution on Tcp; Figure S7. Repetitive experiments in 25%C4PEG and 75%C4PEG solution.

Author Contributions: H.W. and X.Z. supervised and guided the experiments; Z.H. finished the main manuscript and experiments; Z.H. and F.C. conducted characterizations. All authors contributed to the analysis and discussion of the data and reviewed the manuscript. All authors have read and agreed to the published version of the manuscript.

Funding: This study was supported by the National Natural Science Foundation of China (52275203).

Institutional Review Board Statement: Not applicable.

Data Availability Statement: Data are contained within the article or Supplementary Materials.

Conflicts of Interest: The authors declare no conflicts with respect to financial and personal interests.

References

1. Yue, H.; Yuan, L.; Zhang, W.; Zhang, S.; Wei, W.; Ma, G. Macrophage Responses to the Physical Burden of Cell-Sized Particles. *J. Mater. Chem. B* **2018**, *6*, 393–400. [CrossRef] [PubMed]
2. Cook, M.T.; Haddow, P.; Kirton, S.B.; McAuley, W.J. Polymers Exhibiting Lower Critical Solution Temperatures as a Route to Thermoreversible Gelators for Healthcare. *Adv. Funct. Mater.* **2021**, *31*, 2008123. [CrossRef]
3. Kim, Y.-J. Thermo-Responsive Polymers and Their Application as Smart Biomaterials. *J. Mater. Chem. B* **2017**, *5*, 4307–4321. [CrossRef] [PubMed]
4. Takashima, Y.; Hatanaka, S.; Otsubo, M.; Nakahata, M.; Kakuta, T.; Hashidzume, A.; Yamaguchi, H.; Harada, A. Expansion–Contraction of Photoresponsive Artificial Muscle Regulated by Host–Guest Interactions. *Nat. Commun.* **2012**, *3*, 1270. [CrossRef]
5. Pantuso, E.; De Filpo, G.; Nicoletta, F.P. Light-Responsive Polymer Membranes. *Adv. Opt. Mater.* **2019**, *7*, 1900252. [CrossRef]
6. Bertrand, O.; Gohy, J.-F. Photo-Responsive Polymers: Synthesis and Applications. *Polym. Chem.* **2017**, *8*, 52–73. [CrossRef]
7. Nese, A.; Lebedeva, N.V.; Sherwood, G.; Averick, S.; Li, Y.; Gao, H.; Peteanu, L.; Sheiko, S.S.; Matyjaszewski, K. pH-Responsive Fluorescent Molecular Bottlebrushes Prepared by Atom Transfer Radical Polymerization. *Macromolecules* **2011**, *44*, 5905–5910. [CrossRef]
8. Kocak, G.; Tuncer, C.; Bütün, V. pH-Responsive Polymers. *Polym. Chem.* **2017**, *8*, 144–176. [CrossRef]
9. Li, J.; Nagamani, C.; Moore, J.S. Polymer Mechanochemistry: From Destructive to Productive. *Acc. Chem. Res.* **2015**, *48*, 2181–2190. [CrossRef]
10. Xia, D.; Wang, F.; Pan, S.; Yuan, S.; Liu, Y.; Xu, Y. Redox/pH-Responsive Biodegradable Thiol-Hyaluronic Acid/Chitosan Charge-Reversal Nanocarriers for Triggered Drug Release. *Polymers* **2021**, *13*, 3785. [CrossRef]
11. Reyes-Ortega, F.; Delgado, Á.; Schneider, E.; Checa Fernández, B.; Iglesias, G. Magnetic Nanoparticles Coated with a Thermosensitive Polymer with Hyperthermia Properties. *Polymers* **2017**, *10*, 10. [CrossRef] [PubMed]
12. Moral-Zamorano, M.; Quijada-Garrido, I.; San-Miguel, V.; Serrano, B.; Baselga, J.; Hashmi, S.; Stadler, F.J.; García-Peñas, A. Concentration Effect over Thermoresponse Derived from Organometallic Compounds of Functionalized Poly(N-Isopropylacrylamide-Co-Dopamine Methacrylamide). *Polymers* **2021**, *13*, 3921. [CrossRef]
13. Ji, T.; Ji, Y.; Meng, X.; Wang, Q. Temperature-Responsive Separation Membrane with High Antifouling Performance for Efficient Separation. *Polymers* **2024**, *16*, 416. [CrossRef] [PubMed]
14. Chen, X.; Zhang, J.; Wu, K.; Wu, X.; Tang, J.; Cui, S.; Cao, D.; Liu, R.; Peng, C.; Yu, L.; et al. Visualizing the In Vivo Evolution of an Injectable and Thermosensitive Hydrogel Using Tri-Modal Bioimaging. *Small Methods* **2020**, *4*, 2000310. [CrossRef]
15. Kim, S.; Lee, K.; Cha, C. Refined Control of Thermoresponsive Swelling/Deswelling and Drug Release Properties of Poly(N-Isopropylacrylamide) Hydrogels Using Hydrophilic Polymer Crosslinkers. *J. Biomater. Sci. Polym. Ed.* **2016**, *27*, 1698–1711. [CrossRef]
16. Gheysoori, P.; Paydayesh, A.; Jafari, M.; Peidayesh, H. Thermoresponsive Nanocomposite Hydrogels Based on Gelatin/Poly (*N*–Isopropylacrylamide) (PNIPAM) for Controlled Drug Delivery. *Eur. Polym. J.* **2023**, *186*, 111846. [CrossRef]
17. Alexander, A.; Ajazuddin; Khan, J.; Saraf, S.; Saraf, S. Polyethylene Glycol (PEG)–Poly(N-Isopropylacrylamide) (PNIPAAm) Based Thermosensitive Injectable Hydrogels for Biomedical Applications. *Eur. J. Pharm. Biopharm.* **2014**, *88*, 575–585. [CrossRef] [PubMed]
18. Alexander, A.; Ajazuddin; Khan, J.; Saraf, S.; Saraf, S. Poly(Ethylene Glycol)–Poly(Lactic-Co-Glycolic Acid) Based Thermosensitive Injectable Hydrogels for Biomedical Applications. *J. Control. Release* **2013**, *172*, 715–729. [CrossRef]
19. Koriyama, T.; Takayama, Y.; Hisatsune, C.; Asoh, T.-A.; Kikuchi, A. Interaction of Bioactive Compounds on Capillary Inner Surfaces Bearing a Dense Thermoresponsive Polymer Brush. *J. Biomater. Sci. Polym. Ed.* **2017**, *28*, 900–912. [CrossRef]
20. Kobayashi, J.; Yamato, M.; Okano, T. On-off Affinity Binding Modulation on Thermoresponsive Polymer-Grafted Surfaces for Capture and Release of Proteins and Cells. *J. Biomater. Sci. Polym. Ed.* **2017**, *28*, 939–957. [CrossRef]
21. Nagase, K.; Yamato, M.; Kanazawa, H.; Okano, T. Poly(*N*-Isopropylacrylamide)-Based Thermoresponsive Surfaces Provide New Types of Biomedical Applications. *Biomaterials* **2018**, *153*, 27–48. [CrossRef] [PubMed]
22. Sponchioni, M.; Manfredini, N.; Zanoni, A.; Scibona, E.; Morbidelli, M.; Moscatelli, D. Readily Adsorbable Thermoresponsive Polymers for the Preparation of Smart Cell-Culturing Surfaces on Site. *ACS Biomater. Sci. Eng.* **2020**, *6*, 5337–5345. [CrossRef] [PubMed]
23. Forg, S.; Karbacher, A.; Ye, Z.; Guo, X.; von Klitzing, R. Copolymerization Kinetics of Dopamine Methacrylamide during PNIPAM Microgel Synthesis for Increased Adhesive Properties. *Langmuir* **2022**, *38*, 5275–5285. [CrossRef] [PubMed]
24. Hogan, K.J.; Mikos, A.G. Biodegradable Thermoresponsive Polymers: Applications in Drug Delivery and Tissue Engineering. *Polymer* **2020**, *211*, 123063. [CrossRef]
25. Bhuiyan, M.H.; Clarkson, A.N.; Ali, M.A. Optimization of Thermoresponsive Chitosan/β-Glycerophosphate Hydrogels for Injectable Neural Tissue Engineering Application. *Colloids Surf. B* **2023**, *224*, 113193. [CrossRef]
26. Sponchioni, M.; Capasso Palmiero, U.; Moscatelli, D. Thermo-Responsive Polymers: Applications of Smart Materials in Drug Delivery and Tissue Engineering. *Mater. Sci. Eng. C* **2019**, *102*, 589–605. [CrossRef]

27. Shymborska, Y.; Stetsyshyn, Y.; Awsiuk, K.; Raczkowska, J.; Bernasik, A.; Janiszewska, N.; Dąbczyński, P.; Kostruba, A.; Budkowski, A. Temperature- and pH-Responsive Schizophrenic Copolymer Brush Coatings with Enhanced Temperature Response in Pure Water. *ACS Appl. Mater. Interfaces* **2023**, *15*, 8676–8690. [CrossRef]
28. Lutz, J.-F. Thermo-Switchable Materials Prepared Using the OEGMA-Platform. *Adv. Mater.* **2011**, *23*, 2237–2243. [CrossRef]
29. Yan, J.; Zhang, X.; Li, W.; Zhang, X.; Liu, K.; Wu, P.; Zhang, A. Thermoresponsive Supramolecular Dendronized Copolymers with Tunable Phase Transition Temperatures. *Soft Matter* **2012**, *8*, 6371–6377. [CrossRef]
30. Zhang, X.; Yan, J.; Li, W.; Zhang, A. Thermoresponsive Dynamic Covalent Dendronized Polymers. *RSC Adv.* **2016**, *6*, 44216–44223. [CrossRef]
31. Kim, D.-H.; Han, J.-H.; Kwon, H.-C.; Lim, S.-J.; Han, S.-G.; Jung, H.-S.; Lee, K.-H.; Kang, J.-H.; Han, S.-G. Toxicity Assessment of a Single Dose of Poly(Ethylene Glycol) Diglycidyl Ether (PEGDE) Administered Subcutaneously in Mice. *Toxics* **2021**, *9*, 354. [CrossRef] [PubMed]
32. Zerbinati, N.; Sommatis, S.; Maccario, C.; Capillo, M.C.; Grimaldi, G.; Alonci, G.; Protasoni, M.; Rauso, R.; Mocchi, R. Toward Physicochemical and Rheological Characterization of Different Injectable Hyaluronic Acid Dermal Fillers Cross-Linked with Polyethylene Glycol Diglycidyl Ether. *Polymers* **2021**, *13*, 948. [CrossRef] [PubMed]
33. Cui, Y.; Liang, X.; Chai, J.; Cui, Z.; Wang, Q.; He, W.; Liu, X.; Liu, Z.; Cui, G.; Feng, J. High Performance Solid Polymer Electrolytes for Rechargeable Batteries: A Self-Catalyzed Strategy toward Facile Synthesis. *Adv. Sci.* **2017**, *4*, 1700174. [CrossRef] [PubMed]
34. Vasylieva, N.; Barnych, B.; Meiller, A.; Maucler, C.; Pollegioni, L.; Lin, J.-S.; Barbier, D.; Marinesco, S. Covalent Enzyme Immobilization by Poly(Ethylene Glycol) Diglycidyl Ether (PEGDE) for Microelectrode Biosensor Preparation. *Biosens. Bioelectron.* **2011**, *26*, 3993–4000. [CrossRef] [PubMed]
35. Wang, H.; Wang, Q.; Wang, K.; Wang, J.; Zhang, X.; Liu, Y.; Zhang, J. A Self-Healing Thermogelling Polymer with Tunable Transparency Based on Biomolecule Alginate Grafting Phenylboronic Acid. *Int. J. Biol. Macromol.* **2024**, *274*, 133485. [CrossRef] [PubMed]
36. Purushothaman, M.; Krishnan, P.S.G.; Nayak, S.K. Poly(Alkyl Lactate Acrylate)s Having Tunable Hydrophilicity. *J. Appl. Polym. Sci.* **2014**, *131*, app.40962. [CrossRef]
37. Ashizawa, K. Nanosize particle analysis by dynamic light scattering (DLS). *Yakugaku Zasshi* **2019**, *139*, 237–248. [CrossRef]
38. Schild, H.G. Poly(N-Isopropylacrylamide): Experiment, Theory and Application. *Prog. Polym. Sci.* **1992**, *17*, 163–249. [CrossRef]
39. Roy, D.; Brooks, W.L.A.; Sumerlin, B.S. New Directions in Thermoresponsive Polymers. *Chem. Soc. Rev.* **2013**, *42*, 7214–7243. [CrossRef]
40. Maeda, Y.; Higuchi, T.; Ikeda, I. FTIR Spectroscopic and Calorimetric Studies of the Phase Transitions of *N*-Isopropylacrylamide Copolymers in Water. *Langmuir* **2001**, *17*, 7535–7539. [CrossRef]
41. Guo, S.; Zhang, H.; Lu, X.; Xiao, H.; Zhao, Y. Sensing Carbon Dioxide through Solution Transparency Change of Gas-Responsive Polymers. *J. Mater. Chem. C* **2019**, *7*, 4049–4056. [CrossRef]

polymers

MDPI

Article

Mechanisms of the Photomechanical Response in Thin-Film Dye-Doped Glassy Polymers

Zoya Ghorbanishiadeh, Ankita Bhuyan, Bojun Zhou, Morteza Sheibani Karkhaneh and Mark G. Kuzyk *

Department of Physics, Washington State University, Pullman, WA 99163, USA;
z.ghorbanishiadeh@wsu.edu (Z.G.); ankitaccg@gmail.com (A.B.); bojun.zhou@hotmail.com (B.Z.);
morteza.sheibani.2016@gmail.com (M.S.K.)
* Correspondence: kuz@wsu.edu

Abstract: This work aims to determine the mechanism of the photomechanical response of poly(Methyl methacrylate) polymer doped with the photo-isomerizable dye Disperse Red 1 using the non-isomerizable dye Disperse Orange 11 as a control to isolate photoisomerization. Samples are free-standing thin films with thickness that is small compared with the optical skin depth to assure uniform illumination and photomechanical response throughout their volume, which differentiates these studies from most others. Polarization-dependent measurements of the photomechanical stress response are used to deconvolute the contributions of angular hole burning, molecular reorientation and photothermal heating. While photo-isomerization of dopant molecules is commonly observed in dye-doped polymers, the shape changes of a molecule might not couple strongly to the host polymer through steric mechanical interactions, thus not contributing substantially to a macroscopic shape change. To gain insights into the effectiveness of such mechanical coupling, we directly probe the dopant molecules using dichroism measurements simultaneously while measuring the photomechanical response and find mechanical coupling to be small enough to make photothermal heating—mediated by the transfer of optical energy as heat to the polymer—the dominant mechanism. We also predict the fraction of light energy converted to mechanical energy using a model whose parameters are thermodynamic material properties that are measured with independent experiments. We find that in the thin-film geometry, these dye-doped glassy polymers are as efficient as any other material but their large Young's modulus relative to other organic materials, such as liquid crystal elastomers, makes them suitable in applications that require mechanically strong materials. The mechanical properties and the photomechanical response of thin films are observed to be significantly different than in fibers, suggesting that the geometry of the material and surface effects might play an important role.

Keywords: dye-doped polymers; rheology; thermal expansion; photomechanics; photothermal heating; orientational hole burning; molecular reorientation; dichroism; thin films; Young's modulus; photomechanical efficiency

1. Introduction

Photoisomerization is a reversible process where a molecule's conformation changes upon absorbing a photon. Underlying processes in azobenzene chromophores include trans–cis conversion and an open–closed ring transition [1]. Such molecules, when doped into a polymer, will induce a bulk photomechanical response. Figure 1 shows the Disperse Red 1 (DR1) molecule, which is a commercially available azobenzene dye that we use in our studies.

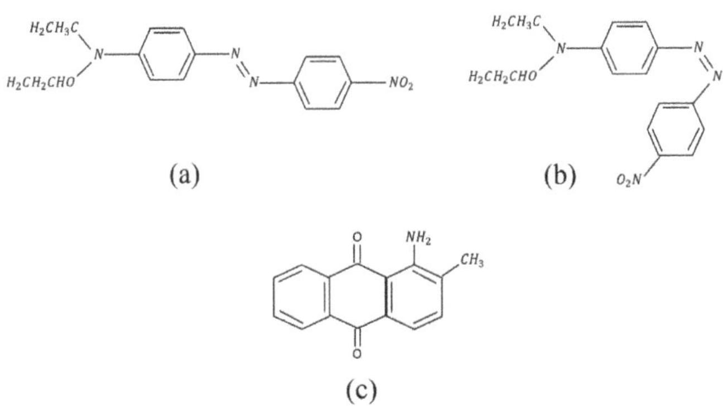

Figure 1. Molecular structure of (**a**) the trans state of Disperse Red 1 (DR1), (**b**) the cis form, and (**c**) Disperse Orange 11 (DO11) [2].

The trans form, being of lower energy, is dominant at low temperatures and in the dark. Certain wavelengths of light can excite the trans molecule into the cis state, which can subsequently relax nonradiatively to the trans state or the de-excitation can be induced by absorbing a photon [1,3]. This specific switchable property of azo dyes has attracted considerable attention for their promise in novel applications such as optical actuation, refs. [4–8]; optical data storage, refs. [9–11]; all-optical switching, ref. [12]; holography, ref. [13]; grating fabrication, refs. [14,15]; and nonlinear optical effects [16–18].

Photomechanical materials convert light energy into mechanical work [1,3,4]. Two dominant mechanisms include photothermal heating, where the light energy absorbed by the dopant molecules is transferred as heat to the polymer, hence increasing the material's temperature and inducing thermal expansion, and photoisomerization, where through steric interaction dye dopants stress the surrounding polymer when changing shape or reorienting [1,4,19]. Common methods to study photoisomerization and photo-(re)orientation include photo-induced birefringence and dichroism [10,20–23]. However, changes in molecular shape and orientation might not result in a photomechanical effect of the bulk polymer. An understanding of the mechanical coupling efficiency of the molecule to the host polymer thus requires experiments that both probe the dopant molecules and measure the bulk photomechanical response. Past studies using second harmonic generation to probe molecular reorientation of a dye molecule in a polymer matrix suggested weak mechanical coupling based on the high-level of molecular mobility [24]. However, there have been no direct measurements of the effect on the photomechanical response.

Kawabe and Okoshi separated the contribution of angular hole burning and molecular reorientation using dichroism and birefringence in DR1-doped PMMA thin films as a function of dye concentration. They measured the absorption spectra before, during, and after illuminating the sample with a polarized pump in the presence of a probe beam. They also probed the sample's birefringence and found that the light absorbance along and perpendicular to the pump polarization decreases with the perpendicular polarization dropping more than the parallel one. Kawabe and Okoshi concluded that angular hole burning is a fast process that dominates dichroism at short exposure times, and molecular reorientation becomes notable only after a longer exposure time and is stronger in samples with higher dye concentration [23]. Rodriguez et al., on the other hand, found that the birefringence and absorption coefficient decreased significantly parallel to the pump polarization, while the reduction was negligible in the perpendicular direction, ref. [22], indicative of hole burning. In contrast to these studies, our work uses thin freestanding films whose optical skin depth is much larger than the thickness, yielding uniform illumination for a more accurate determination of the population's orientational dynamics and a more uniform photomechanical force. In addition, our experiments also simultaneously measure the photomechanical response so that molecular reorientation can be correlated with changes in the orientation distribution function and relative isomer populations.

Silva et al. studied V-shaped azo-chromophores doped into PMMA thin films as a function of dye concentration to investigate their optical storage performance. They observed a decrease in the absorption spectra when the probe light is polarized parallel to the pump and an increase when the polarization is perpendicular to the pump. They conclude that molecular reorientation dominates angular hole burning in their materials. The long-lasting residual persistence of the angular hole due to reorientation was investigated for its potential in optical memory. The high quantum efficiency of photo-induced birefringence is related to angular hole burning due to a change in isomer populations [10]. These processes at the molecular level can lead to a bulk photomechanical response if the molecules couple to the polymer host, so understanding the molecular process is an important step in optimizing materials for optical actuation applications [1,4,25].

The present work combines (1) new theory development, (2) a unique experimental design that simultaneously measures the photomechanical tensor response while probing with optical dichroism as a function of time during optical pumping and (3) a free-standing thin-film sample that can be uniformly illuminated without the constraints of a substrate. Each of these alone differentiates our approach from most others. The goal of the work is to determine the relative contributions of photo-heating, photoisomerization, molecular reorientation and angular hole burning to the photomechanical stress response. Measurements include light-induced dichroism and stress as a function of time, intensity and temperature in DR1-doped PMMA (DR1/PMMA) thin films using Disperse Orange 11-doped PMMA (DO11/PMMA) thin films as a control. DO11 is the ideal control because it does not photoisomerize and its optical absorption spectrum is similar to DR1, making it possible to match the optical absorbance of the two dyes at the measurement wavelength by controlling the relative concentrations. The wealth of data obtained, along with a measurement of Young's modulus, allows us to determine the photomechanical figure of merit, which quantifies the efficiency of converting a photon's energy to mechanical work. The understanding gained from studies of the various mechanisms will inform the development of more efficient materials for applications that require the direct conversion of light to work.

2. Theory

This section develops the models that will be used in the analysis of the data. There are many separate and independent theoretical parts to this work. We start with an overview here to prepare the reader with a broader overview.

The start of Section 2.1 defines the constitutive relations of the photomechanical response. The important result is that the photomechanical stress is a function of the pump intensity that illuminates the material, which is well approximated by a power series in the field. Secondly, the stress response varies in time after the pump light is turned on, which we assume takes on an exponential form because it fits the data well.

Section 2.1.1 focuses solely on a model for photothermal heating, where light energy absorbed by the dye molecules is transferred to the host polymer as heat. The central idea is that the heating response is independent of the heat source, be it light or hot air in an oven. First, a phenomenological model for the stress as a function of temperature is introduced that fits the data for a film whose length is constrained. Then, the section ends with determining the photomechanical coefficients due to heat by relating the temperature increase to the pump light's intensity.

Section 5.1.2 relates the stress induced by two orthogonal pump polarizations relative to the principal sample axis to the two tensor components of the photomechanical stress tensor. The various mechanisms have distinct photomechanical stress tensor ratios, so a measure of the stress induced by two orthogonal pump polarizations is a metric of the mechanisms.

The theoretical results will be used to fit and interpret the data. The reader who is interested in the experimental results can skip the theory section.

2.1. Photomechanical Response

The stress response of a material to light is a function of the intensity. For small enough intensities, most functions can be approximated by a power series expansion of the form

$$\sigma(t \to \infty) = -\sum_{n=1}^{\infty} \kappa_\sigma^{(n)} I^n \approx -(\kappa_\sigma^{(1)} I + \kappa_\sigma^{(2)} I^2), \tag{1}$$

where σ is the stress, $\kappa_\sigma^{(n)}$ the nth-order photomechanical coefficient and I the absorbed intensity [1,4]. The first two terms in the expansion are found to model typical materials in the intensity ranges measured. If the material is mechanically stressed, there would also be a term corresponding to $n = 0$, but this would not be a material property but rather an experimental constraint. Equation (1) thus describes a material in its relaxed state. The photomechanical coefficients will be different in an initially stressed material, but we will assume that all pre-stress due to sample mounting will be small enough to ignore this effect.

When $\kappa_\sigma^{(n)} > 0$, light causes the sample's equilibrium length to increase. When the sample length is held constant by the clamps, the material becomes compressed if it is not to buckle. By definition, a compressional force that opposes expansion is negative, so $\sigma < 0$. The experiment measures this stress—defined to be along the z-direction—as a function of time after the pump light (measured separately for each pump polarization) is turned on.

The photomechanical stress response as a function of time for fixed pump intensity is assumed to follow a bi-exponential function of the form

$$\sigma_{\text{on}}(t) = \sigma_0 + \sigma_1(1 - e^{(-t/t_1)}) + \sigma_2(1 - e^{(-t/t_2)}), \tag{2}$$

and

$$\begin{aligned} \sigma_{\text{off}}(t) = \sigma_0 &+ \sigma_1(1 - e^{(-t_0/t_1)})e^{-(t-t_0)/t_3} \\ &+ \sigma_2(1 - e^{(-t_0/t_2)})e^{-(t-t_0)/t_4}. \end{aligned} \tag{3}$$

The parameters t_n are time constants, σ_n the amplitudes and t_0 is the time offset between pump turn-on and turn-off [4]. Figure 2 shows typical data and fits to a bi-exponential as given by Equations (2) and (3).

2.1.1. Heating

The photothermal heating mechanism is a two-step process that converts the absorbed light energy into heat, which increases the material's temperature and induces thermal expansion. Direct heating in an oven should yield the same amount of expansion for the same amount of deposited energy; the source of heat is irrelevant.

We assume that the temperature dependence of stress is of the form

$$\sigma(T) = \sigma_A + \frac{\sigma_B}{(1 + (\frac{T}{T_0})^n)}. \tag{4}$$

Here, σ_A is the pre-stress due to sample mounting and is not related to the material property, n an exponent, T_0 a critical temperature and σ_B the strength of the temperature-dependent response. Note that these parameters may have physical meaning, but this function and set of parameters are just a guess and are based on the fact that they fit the data well and will thus allow for an analytical analysis of the data using one simple equation. Figure 3 shows a fit to the data as an example, which will be described in greater detail later. For example, we can determine the change in stress per change in temperature simply by taking the derivative of this analytical form and using it to determine the thermal photomechanical stress response coefficients at any temperature. Any functional form that fits the data can be used as the starting point.

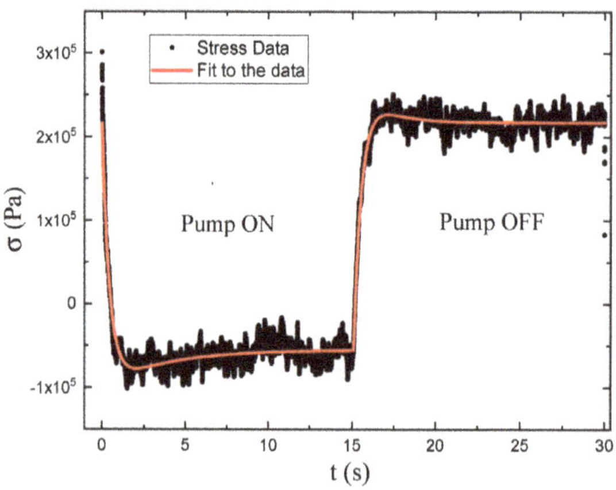

Figure 2. Typical photomechanical stress response as observed in a film as a function of time and a fit to the theory given by Equations (2) and (3).

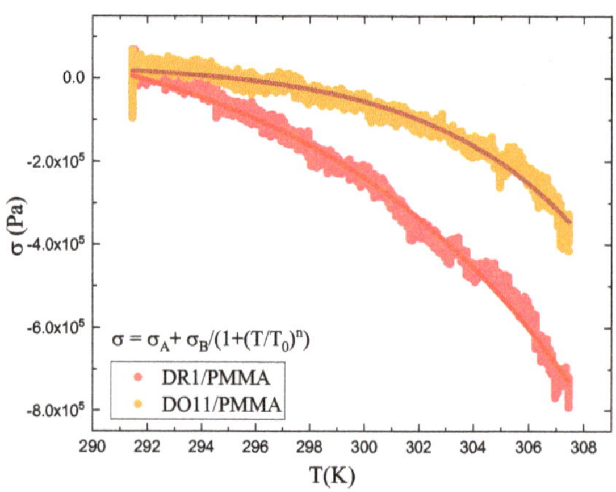

Figure 3. Measured temperature-dependent stress at constant sample length and with no light applied of DR1 and DO11 thin films (points) and a fit of the data (curves) to Equation (4).

The connection of the temperature-dependent stress to the thermal photomechanical response originates in the temperate change ΔT induced by the heat deposited by the light of intensity I, given by [2]

$$\Delta T = \frac{\tau}{c\rho t} I, \tag{5}$$

where τ is the time it takes to reach steady state after the pump light is turned on, where the energy absorbed from the light balances the heat radiated from the sample, ρ is the density, c the specific heat and t the sample thickness.

The approach is to evaluate Equation (4) at $T = T + \Delta T$ where ΔT is given by Equation (5) and then expand the function for small intensity such that $\Delta T \ll T$. The stress as a function of intensity is then given by the series expansion

$$\sigma(T,I) = \sigma\left(T, \frac{\tau}{c\rho t} I\right) = \sigma(T,0) + \left.\frac{\partial\sigma}{\partial T}\right|_{T,I=0} \frac{\tau}{c\rho t} I + \frac{1}{2}\left.\frac{\partial^2\sigma}{\partial T^2}\right|_{T,I=0} \left(\frac{\tau}{c\rho t}\right)^2 I^2 + \ldots, \tag{6}$$

where we have expanded the stress as a series of I under the assumption that the temperature change due to the stress is small compared with the ambient temperature.

The change in stress with temperature, given by the derivative of Equation (4), yields

$$\frac{d\sigma}{dT} = -\frac{\frac{n\sigma_B}{T}\left(\frac{T}{T_0}\right)^n}{\left(1+\left(\frac{T}{T_0}\right)^n\right)^2} \tag{7}$$

and the second derivative is given by

$$\frac{d^2\sigma}{dT^2} = -\frac{n\sigma_B}{T^2}\left(\frac{T}{T_0}\right)^{2n} \cdot \frac{(n-1)-(n+1)\left(\frac{T}{T_0}\right)^n}{\left(1+\left(\frac{T}{T_0}\right)^n\right)^3}. \tag{8}$$

Substituting Equations (7) and (8) into Equation (6), we can read off the photomechanical stress response coefficients, yielding

$$\kappa_\sigma^{(1)}(T) = \sigma_B \frac{\frac{n\tau}{c\rho t T}\left(\frac{T}{T_0}\right)^n}{\left(1+\left(\frac{T}{T_0}\right)^n\right)^2} \tag{9}$$

and

$$\kappa_\sigma^{(2)}(T) = 2n\sigma_B\left(\frac{\tau}{c\rho t}\right)^2 \frac{1}{T^2}\left(\frac{T}{T_0}\right)^{2n} \cdot \frac{(n-1)-(n+1)\left(\frac{T}{T_0}\right)^n}{\left(1+\left(\frac{T}{T_0}\right)^n\right)^3}. \tag{10}$$

When the photomechanical response is well approximated as a quadratic function of the intensity, we can define an effective intensity-dependent photomechanical stress response coefficient $\kappa_\sigma^{(1)}(I)$ given by

$$\kappa_\sigma^{(1)}(I,T) = \kappa_\sigma^{(1)}(T) + \kappa_\sigma^{(2)}(T) I$$

$$= \sigma_B \frac{\frac{n\tau}{c\rho t T}\left(\frac{T}{T_0}\right)^n}{\left(1+\left(\frac{T}{T_0}\right)^n\right)^2} + 2n\sigma_B\left(\frac{\tau}{c\rho t}\right)^2 \frac{1}{T^2}\left(\frac{T}{T_0}\right)^{2n} \cdot \frac{(n-1)-(n+1)\left(\frac{T}{T_0}\right)^n}{\left(1+\left(\frac{T}{T_0}\right)^n\right)^3} I. \tag{11}$$

Here, we pause to stress a point of potential confusion regarding the signs. Recall that the negative sign in Equation (1) is introduced because of the convention that the sample length increases when the photomechanical constant is positive, which leads to a decrease in the stress when the sample is in the clamped configuration. In the clamped experiment, where the material is pre-stressed, the magnitude of the measured force will decrease when the sample expands. Hence, a negative slope of the stress versus temperature is an indicator of thermal expansion. To summarize, the negative sign in Equation (1) requires that the sign of Equation (9) be opposite to the sign of Equation (7).

Alternatively, we can eliminate the parameter σ_B in Equation (9) using Equation (4), yielding the semi-empirical model [2]

$$\kappa_\sigma^{(1)}(I) = \frac{\tau}{\rho c t} \cdot \frac{n(\sigma - \sigma_A)}{T\left(1 + \left(\frac{T_0}{T}\right)^n\right)}. \tag{12}$$

This model is semi-empirical in that $\sigma(I)$, the stress in the limit of infinite time after the pump of intensity I is turned on, is an experimentally determined value. So, we emphasize that in the semi-empirical approach, we are using the measured stress induced by the light rather than by a temperature change because of the one-to-one correspondence between temperature change and the intensity, which is the source of the deposited heat. The pre-stress, called σ_0 in the photomechanical measurements and σ_A in the temperature-dependent stress, is the initial stress from the sample being mounted under tension. **Thus, the empirically determined intensity-dependent time constant $\tau(I)$ and intensity-dependent $\sigma(I)$ together are responsible for the intensity dependence of $\kappa_\sigma^{(1)}(I)$.**

In principle, the time constant and stress for the heating mechanism should be independent of the intensity. However, a substantial increase in the sample's temperature can change its elastic properties and thus lead to a nonlinearity that yields an intensity-dependent photomechanical constant. **Therefore, a discrepancy between the zero-intensity photomechanical constant and the heating theory is an indicator of other mechanisms at play.** Below, we start with comparisons of the temperature derivatives of the stress as a proxy for photothermal heating. Later in the paper, we will fit the time-dependent stress to a bi-exponential and apply the heating theory to the time constants and stresses so determined to isolate the process that corresponds to photothermal heating.

Since the samples are prepared to be isotropic, and below we will see this to be approximately true, the heating rate should be independent of the polarization of the light and thermal expansion should also be isotropic. If the photomechanical response is due solely to heating, we expect the linear photomechanical stress response to be given by Equation (9) or (12), which assumes that the absorbed heat leads to a temperature change that drives thermal expansion [4].

The heating theory presented here is a zeroth-order approximation of a more comprehensive theory. For example, a material that undergoes photo-isomerization will result in a change in the amount of light absorbed due to hole burning even if hole burning does not directly lead to a photomechanical effect. Hole burning will, however, change the amount of heat transferred to the material, which is not accounted for in this theory. Other issues include the following: (1) This is a low-temperature theory that only applies below the phase transition temperature. (2) The presence of dopant dyes with two or more conformations affect the heat capacity and thermal expansion coefficient by adding additional degrees of freedom that need to be taken into account. This will be evident by the large difference in the thermally induced stress between DO11 and DR1 dye doping as described later. The present theory will be sufficient to interpret our data, and the more refined theory of the contribution of photothermal heating is under development and will be the topic of a future paper.

2.1.2. Hole Burning and Photo-Stress Anisotropy

Light-induced angular hole burning, which is commonly observed in azo-dye-doped polymers such as PMMA/DR1, induces an anisotropic photomechanical stress response [1,2]. Hole burning generally refers to the process where an originally isotropic material, represented by a spherical orientational distribution, develops an anisotropy like the dimples on an apple's north and south poles. These dimples represent a depletion of molecules orientated in that direction and therefore called an orientational hole. There are many processes that can lead to such a hole, such as the molecules aligned along the light's polarization axis being excited and molecules diffusing away from the light's polarization axis. The former is usually referred to as angular hole burning, while the latter is called molecular reorientation.

Assuming that the photomechanical response is a linear function of the intensity, as observed in Figure 6, the ratio of stress for two orthogonal pump polarizations is given by [1,2]

$$\frac{\sigma_{//}}{\sigma_{\perp}} = \frac{-\kappa_{//}^{(1)} I}{-\kappa_{\perp}^{(1)} I} = \frac{\kappa_{//}^{(1)}}{\kappa_{\perp}^{(1)}} = \frac{\kappa_{//}^{(\text{heat})} + \kappa_{//}^{(\text{hole})}}{\kappa_{\perp}^{(\text{heat})} + \kappa_{\perp}^{(\text{hole})}}, \tag{13}$$

where the stress response from the two mechanisms is assumed to be additive. Note that the parallel stress $\sigma_{//}$ is defined to be the stress induced for a pump beam that is polarized along the direction of the stress measurement while $\sigma_{\perp}$ is perpendicular to it. Note that the superscript "hole" refers to both angular hole burning and molecular reorientation. Once the heating contribution is isolated, these two mechanisms can be separated.

The photothermal heating contribution for an isotropic material is independent of polarization, so

$$\kappa_{//}^{(\text{heat})} = \kappa_{\perp}^{(\text{heat})} \equiv \kappa^{(\text{heat})}(I). \tag{14}$$

Polarized light acting on randomly oriented azo dyes doped into a polymer induces reorientational diffusion away from the light's polarization. This depletion of molecules oriented along the light's polarization results in a decreased length along the light's polarization and an increase perpendicular to it. If the orientational distribution function retains azimuthal symmetry, conservation of volume demands that

$$\kappa_{//}^{(\text{hole})} = -2\kappa_{\perp}^{(\text{hole})} \equiv \kappa^{(\text{hole})}(I). \tag{15}$$

Using Equations (14) and (15), Equation (13) becomes

$$\frac{\sigma_{//}}{\sigma_{\perp}} = \frac{\kappa_{//}^{(1)}}{\kappa_{\perp}^{(1)}} = \frac{\kappa^{(\text{heat})}(I) + \kappa^{(\text{hole})}(I)}{\kappa^{(\text{heat})}(I) - \frac{1}{2}\kappa^{(\text{hole})}(I)} = \frac{1 + \frac{\kappa^{(\text{hole})}(I)}{\kappa^{(\text{heat})}(I)}}{1 - \frac{1}{2}\frac{\kappa^{(\text{hole})}(I)}{\kappa^{(\text{heat})}(I)}}. \tag{16}$$

Since we seek to determine the relative contribution of the two mechanisms, we define their ratio as R and invert Equation (16) to obtain

$$R \equiv \frac{\kappa^{(\text{hole})}(I)}{\kappa^{(\text{heat})}(I)} = \frac{\frac{\kappa_{//}^{(1)}(I)}{\kappa_{\perp}^{(1)}(I)} - 1}{\frac{1}{2}\frac{\kappa_{//}^{(1)}(I)}{\kappa_{\perp}^{(1)}(I)} + 1}. \tag{17}$$

If the only two acting mechanisms are photothermal heating and angular hole burning, which includes both changes in the isomer populations and molecular reorientation, all of the photomechanical properties of the material can be expressed in terms of two parameters. As such, two measurements (e.g., the photo-induced stress measured for two orthogonal pump polarizations) are sufficient to determine all others. For example, Equations (15) and (17) yield

$$\kappa_{//}^{(\text{hole})} = R\kappa^{(\text{heat})}(I), \tag{18}$$

Therefore, for stress induced parallel to the light's polarization, the linear photomechanical constant is given by

$$\kappa_{//}^{(1)} = \kappa_{//}^{(\text{heat})} + \kappa_{//}^{(\text{hole})} = (1+R)\kappa^{(\text{heat})}(I) \tag{19}$$

So, in this example, R and $\kappa^{(\text{heat})}(I)$ are the independent parameters.

Conversely, Equations (18) and (19) can be inverted to determine the relative strength of the mechanisms, yielding

$$\kappa^{(\text{heat})}(I) = \frac{\kappa_{//}^{(1)}}{1+R}, \tag{20}$$

and

$$\kappa^{(\text{hole})}(I) = \frac{R\kappa_{//}^{(1)}}{1+R}. \tag{21}$$

Similarly, it is trivial to show that

$$\kappa_{//}^{(\text{heat})} = \kappa_{\perp}^{(\text{heat})} = \kappa^{(\text{heat})}(I) = \frac{\kappa_{//}^{(1)}}{1+R}, \tag{22}$$

$$\kappa_{//}^{(\text{hole})} = \frac{R\kappa_{//}^{(1)}}{1+R}, \tag{23}$$

and

$$\kappa_{\perp}^{(\text{hole})} = -\frac{1}{2}\frac{R\kappa_{//}^{(1)}}{1+R}. \tag{24}$$

For the second order, Equation (17) can be generalized to obtain the ratio R as a function of intensity from

$$\frac{\sigma_{//}(I)}{\sigma_{\perp}(I)} = \frac{\kappa_{//}(I)}{\kappa_{\perp}(I)} = \frac{\kappa_{//}^{(1)} I + \kappa_{//}^{(2)} I^2}{\kappa_{\perp}^{(1)} I + \kappa_{\perp}^{(2)} I^2}, \tag{25}$$

which yields

$$R = \frac{\kappa^{(\text{hole})}(I)}{\kappa^{(\text{heat})}(I)} = \frac{\dfrac{\kappa_{//}^{(1)} + \kappa_{//}^{(2)} I}{\kappa_{\perp}^{(1)} + \kappa_{\perp}^{(2)} I} - 1}{\dfrac{1}{2}\dfrac{\kappa_{//}^{(1)} + \kappa_{//}^{(2)} I}{\kappa_{\perp}^{(1)} + \kappa_{\perp}^{(2)} I} + 1}. \tag{26}$$

3. Approach

The Disperse Red 1 (DR1) and Disperse Orange 11 (DO11) dyes used in our studies are shown in Figure 1. The azo bond (–N=N–) changes conformation upon heating [1,3] or absorbing light [1], while DO11 remains mostly unchanged. The sole mechanism in DO11 should be a photothermal response (aside from photo tautomerization [26,27]—which is not associated with a detectable photomechanical effect [1]). As such, DO11 is used as a control to determine the relative contributions of photoisomerization and photothermal heating in DR1 by comparing samples of the two materials with dye concentrations that absorb the same amount of light.

The smallest photomechanical part of a material whose properties are the same as the material is called a photomorphon [1,4]. The photomorphon, in turn, is made of the photomechanical unit (usually the active dye molecule) along with the passive material surrounding it. In a dye-doped polymer, a single dopant molecule is the photomechanical unit and with the surrounding polymer makes the photomorphon. The photomechanical unit can be placed in series and/or in parallel with the photomechanical unit, depending on the structure of the polymer.

In an isotropic material, the amount of heat deposited is independent of the light's polarization. Thus, the measured light-induced stress from thermal expansion along the direction the sample is clamped is independent of the polarization of the light. Angular hole burning refers to the process by which molecules that are aligned along the light's polarization are converted to a more compact and more-nearly spherical isomer, thus inducing anisotropy in the material. This anisotropy leads to both an anisotropic refractive index/absorbance and stress. "Angular hole burning" is a relatively fast process compared with the other mechanisms. The slower molecular reorientational process, which also leads to anisotropy, results from a depletion of molecules oriented along the light's polarization axis and an increase perpendicular to it, yielding anisotropy in the stress and refractive index/absorbance [1,2,4], which builds over longer time scales. While this is also technically "angular hole" burning due to the depletion of molecules oriented along the pump polarization, it is called "molecular reorientation" to differentiate it from the faster process and to stress that the molecules are reorienting away from the light's polarization. As described later, the characteristics of these two types of anisotropies can be used to deconvolute the relative contributions of photo-thermal heating, angular hole burning and reorientation.

In materials with azo-dye molecules, molecular reorientation away from the light's polarization is mediated by the preferential absorption of light by trans molecules that are aligned with the light's polarization, followed by rotational diffusion of the more compact and more mobile cis form. The higher-energy cis isomers that are formed, upon de-excitation to the trans form, result in a random distribution of orientation, leaving a depleted trans population along the light's polarization axis [1,2,4,28–30]. Generally, molecular reorientation results in both the length of the sample and the absorbance along the light's polarization axis to decrease. They both increase in the plane perpendicular to the polarization axis.

In addition to studying the mechanisms of the photomechanical response, we also evaluate the figure of merit (FOM), which quantifies the efficiency of a material in converting light energy to mechanical work. This gives an estimate of the efficiency of photoisomerization (DR1) and photothermal heating (DO11). The thin films used in our experiments are specifically tailored to absorb light uniformly so that the photomechanical stress response κ_σ can be determined accurately. Along with Young's modulus E, the material figure of merit is given by [1,2,4]

$$FOM = \frac{\left(\kappa_\sigma^{(1)}\right)^2}{E}. \tag{27}$$

We note that Equation (27) assumes that Young's modulus is unaffected by light to simplify the required set of measurements to obtain the FOM.

4. Experiment

4.1. Sample Preparation

Dye-doped polymeric thin-film samples are formed by spin coating, starting with solutions of Propylene Glycol Methyl Ether Acetate (PGMEA) and γ-Butyrolactone co-solvents and PMMA with either DR1 or DO11 dye solutes. The samples' optical absorbances at λ = 488 nm are matched using 2.76 times more DO11 than DR1 dye in PMMA polymer by weight [4]. We use 0.5 weight % DR1 and 1.38 weight % DO11 as a percentage of PMMA. The DR1 solutions are composed of 7.33 mL of PGMEA, 0.5 W% DR1, 3.1 mL γ-Buterolactone and 1.8565 g PMMA. For DO11 solutions, we use the same solvents with 1.38 W% DO11 and 1.8491 g PMMA. The solutions are stirred overnight to fully dissolve the solids. Subsequently, the solution is transferred to a new clean bottle using a filtering syringe to remove any remaining residues. The filtered solution is refrigerated overnight before spin coating.

A vacuum chuck securely holds a clean glass substrate horizontal. The spin speed and spin time are set and the substrate flooded with a solution with a sterile syringe. The samples are spun for a fixed time and speed (typically 1000 to 10,000 RPM for 30 s), forming films in the thickness range of 1 to 10 micron. The coated substrates are placed in an oven at 90 °C for 30 min to insure that all of the solvent is removed. After being removed from the oven, the sample is submerged in distilled water to delaminate the film from the substrate.

4.2. Setup

Figure 4 shows a schematic diagram of our experiment, which simultaneously measures the light-induced stress and monitors the transmittance of two orthogonal probe beams to determine the evolution of the population upon irradiation. The λ = 488 nm wavelength light from an argon laser passes through a half-wave plate and a polarizing beam splitter (PBS), then illuminates the sample. This beam acts as the pump, which induces a photomechanical response. The polarizer (P) is set to select the polarization of the pump light and the preceding half-wave plate is rotated to adjust its intensity. In a typical experiment, a run consists of the polarizer set to one angle, followed by a run where it is set perpendicular to the original orientation to obtain the polarization dependence of the response due to the pump along the two principal axes of the material.

Two orthogonally polarized probe beams are derived from the same laser. One of them is diverted from the rejected polarization by the PBS, while the other one is picked off by the glass slide G downstream from the PBS. Both probe beams and the pump beam pass through the same lens and fully overlap at the lens' focus at the same point in the sample. The two probe beams' polarizations are orthogonal to each other and set along the principal axes of the material, which are vertical and horizontal in our experiment. The pump polarization aligns with one or the other probe beam's polarization.

The pump light induces a physical change in the sample, which results in a stress that is detected with a force sensor placed in series with the sample and attached to the upper clamp. The probe beam powers are much lower than the pump power, so the probe beams induce a negligible force. They remain on at all times and are monitored with photo detectors, which are interfaced with an Arduino and custom circuitry [31]. The pump beam is unblocked/blocked every 15 s with an automatic shutter (Sh). The transmitted pump light is also monitored with a photodetector. All intensities and the force are simultaneously recorded.

The pump light is first polarized vertically for a sequence of runs over time for several intensities so that the intensity-dependent photomechanical response and light-induced dichroism can be simultaneously determined. The same experiments are then run for the horizontal pump. This set of data determines the evolution of angular hole burning as determined from the probe beams and correlated with the photomechanical response. DR1-doped PMMA thin films are compared with a DO11 thin film as a control due to it not having multiple conformations. All samples are pre-stressed in the force-sensing clamp before starting the experiments to prevent buckling during sample expansion.

The same setup measures Young's modulus by translating the upper clamp to stretch the sample. The setup records the strain and force to determine Young's modulus. The in situ oven enables the temperature-dependent stress to be measured, which can be used to determine the effect of heating on the sample's properties. Figure 5 shows the force measurement setup used to hold the sample and measure the Young's modulus and the stress.

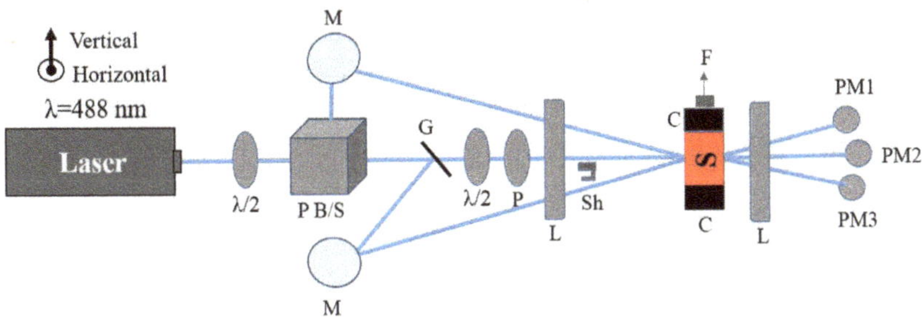

Figure 4. The apparatus used for the intensity and polarization-dependent response measurement. M: Mirror, P B/S: Polarized beam-splitter, G: Glass slide, P: Polarizer, $\lambda/2$: Half-wave plate, L: Lens, Sh: Shutter, C: Clamp, S: Sample, F: Force sensor, PM1, PM2, PM3: Power meters.

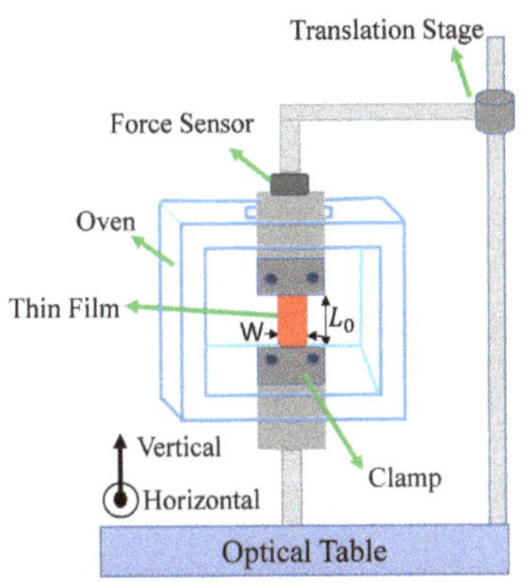

Figure 5. The force measurement setup and sample, which is placed at the intersection of the three beams as diagrammed in Figure 4.

4.3. Experimental Procedure

The $\lambda = 488$ nm line of an argon ion laser is used as the light source from which all the other beams are derived. For DR1-doped PMMA samples: 0.5 wt% DR1, $L_0 = 5.3$ mm, thickness $t = 10$ µm, width $w = 4.8$ mm and area $A = tw = 4.76 \times 10^{-8}$ m^2 in the Young's modulus experiment and $L_0 = 8.28$ mm, thickness $t = 10$ µm, width $w = 4.73$ mm and area $A = tw = 4.73 \times 10^{-8}$ m^2 in the photomechanical experiments. For D011-doped PMMA samples: 1.38 wt% DO11, $L_0 = 5.3$ mm, thickness $t = 10$ µm, width $w = 4.7$ mm and area $A = tw = 4.74 \times 10^{-8}$ m^2 in the Young's modulus experiment and $L_0 = 8.28$ mm, thickness $t = 10$ µm, width $w = 4.6$ mm and area $A = tw = 4.56 \times 10^{-8}$ m^2 in the photomechanical experiments.

The experiments seek to determine the parameters required to determine the contributions of angular hole burning and photothermal heating to the photomechanical response κ from temperature-induced stress measurements, Young's modulus measured at room temperature, and simultaneous pump/probe/stress measurements. We use the data to determine the efficiency of the material in converting light energy to mechanical force. This efficiency provides a metric for comparing materials.

5. Results and Discussion

This section brings together all of the pieces, where the theory is applied to the data and the results interpreted.

Section 5.1 starts with introducing two hypotheses based on the observed long-term data. Section 5.1.1 summarizes the parameters determined form the temperature-dependent stress experiment, which will be used to model the photothermal heating contribution. Section 5.1.2 applies the theory of hole burning and photo-stress anisotropy to the polarization-dependent photomechanical response to determine the long-time contributions of the mechanisms. Section 5.2 uses dichroism to determine how the orientational order of the material evolves over long times. The conclusion is that molecular reorientation and orientational hole burning are taking place, but these effects do not couple strongly to the polymer, and so they do not result in an appreciable photomechanical response. Also found are small permanent changes in the material order. Section 5.3 quantifies photomechanical efficiencies by a figure of merit. Section 5.4 separates the amplitudes of the photomechanical response at the two characteristic response times in an effort to associate individual mechanisms with each time scale, using predictions of the photothermal heating theory to determine which timescale corresponds to heating. The results found are partially consistent with expectations, but unexpected results bring up new questions that will require further research to answer.

5.1. Photomechanical Response

Comparisons of light-induced stress and absorbance in DR1- and DO11-doped PMMA thin films can be used to deduce the mechanisms of the photomechanical response in photoisomerizable samples. The entire area of the sample is illuminated by the pump in a 15 s on and 15 s off cycle. After multiple repetitions, the intensity is increased and the protocol repeated. Typically, the experiment cycles through fifteen different intensities.

As shown in Figure 2, the photomechanical stress response as a function of time for fixed pump intensity follows a bi-exponential function of the form given by Equations (2) and (3). The data are fit to the model with a time offset of $t_0 = 15\,\mathrm{s}$ to determine the time constants and amplitudes.

One parameter of interest is the long-time stress response $\sigma_{eq}(t \to \infty) = \sigma_1(I) + \sigma_2(I)$, where $\sigma_1(I)$ and $\sigma_2(I)$ come from the fits to the data to Equations (2) and (3) at each intensity. These are plotted in Figure 6. The long-term stress response of DR1 and DO11 is the same within experimental uncertainties at each intensity. **However, this does not mean that $\sigma_1(I)$ of DR1 is the same as $\sigma_1(I)$ of DO11 and that $\sigma_2(I)$ of DR1 is the same as $\sigma_2(I)$ of DO11, just that the sum $\sigma_2(I) + \sigma_2(I)$ is the same.** As such, the heating contribution to the photomechanical response may be drastically different in the two materials even though their photomechanical response is the same.

The data for each polarization are fitted to a quadratic polynomial which is in the form of Equation (1), and it shows that $\kappa^{(1)}I$ is large compared to $\kappa^{(2)}I^2$ over the data range, so the stress response is linear within fractional deviations of 0.02 and 0.13 for DR1 and 0.05 and 0.12 for DO11 with horizontal and vertical polarizations of light, respectively, where the fractional deviation is given by $(\kappa^{(2)}I^2)/(\kappa^{(1)}I + \kappa^{(2)}I^2)$ at the highest measured intensity.

The linear photomechanical constant is the same within experimental uncertainty for both pump polarizations, i.e., $\kappa_{\perp}^{(1)} = \kappa_{//}^{(1)}$, in DR1 and in DO11. This observation leads to two possible hypotheses.

Hypothesis 1. *This observation is as expected in DO11, which has only one conformation and so it should be dominated by photothermal heating alone, a process that is polarization-independent. However, since DR1 is known to photoisomerize [1,4], which leads to angular hole burning [1,4,28–30], the data suggest that the light-induced change in the molecular orientational distribution function does not translate into a large enough photomechanical response to be detectable within experimental uncertainties relative to the dominating photothermal heating mechanism. In addition, the magnitudes* $\kappa_{\perp}^{(1)}$ *and* $\kappa_{//}^{(1)}$ *of DO11 and DR1 agree with each other within experimental uncertainty, strengthening the conclusion that the mechanism of the response is the same in both materials. However, as seen in Figure 6, the horizontal polarization data for DR1 dye is systematically lower than the vertical polarization data and at the edge of experimental uncertainty, allowing for a small contribution from a mechanism that is not found in DO11. This could originate from photoisomerization.*

Hypothesis 2. *Heating and other mechanisms are contributing to the DR1-doped polymers and the magnitude of the heating effect is different in each material so that the individual mechanics contribute differently but in aggregate give the same total response observed in Figure 6. Testing these hypotheses requires that each mechanism be isolated.*

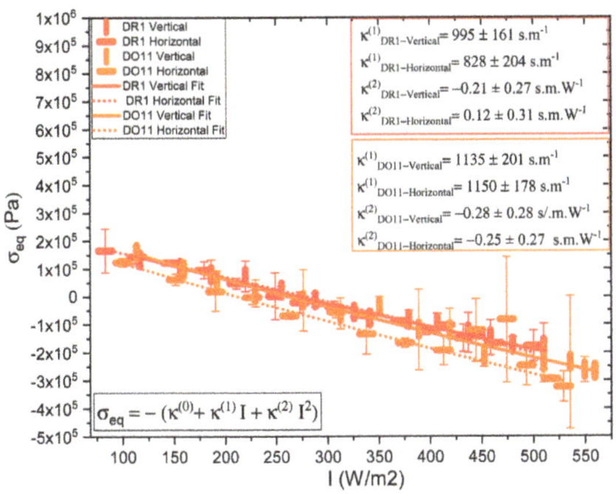

Figure 6. Photomechanical stress response of DR1 and DO11 thin films for horizontal and perpendicular polarization of light as a function of intensity at infinite time, when all processes have come into equilibrium.

The sections below describe how one isolates each mechanism by controlling individual experimental parameters such as temperature and intensity while monitoring the stress and transmitted probe intensity.

5.1.1. Heating

Thermally induced stress of clamped DR1 and DO11 thin films is studied by placing the samples in an oven with no light applied and measuring the stress as a function of temperature from 291.15 K to 307.15 K over a time span of 3 min. The result is shown in Figure 3. The data are fit to Equation (4) and the fitting parameters are shown in Table 1.

Equation (9) has the parameters n, T_0 and σ_B, which are determined from fits to Equation (4) to temperature-dependent stress data as shown in Figure 3. T is the temperature of the sample prior to illumination and τ is the time constant of heating, which can be determined theoretically from the geometry of the material or empirically from the measured stress as a function of time [32]. **The time constant τ can depend on the intensity of the pump beam, which gives $\kappa_{\sigma}^{(1)}(I)$ its dependence on intensity.**

The large difference between DR11- and DO11-doped films in Figure 3 implies that, for the same amount of absorbed light energy, the photothermal mechanism in DR1-doped polymers will be larger than in DO11-doped polymers and that the other mechanisms, when added to heating, makes the total response similar in the two materials. This supports the second hypothesis.

Determining the cause for this large difference is beyond the scope of this paper but is a worthy goal for future studies. There are potentially multiple causes for this difference including the number of internal degrees of freedom added by the dopant (isomerization in DR1 and tautomerization in DO11), which can be modeled with statistical mechanical approaches, the efficiency of heat transfer to the polymer relative to the other competing processes such as molecular shape changes and plasticization due to the dyes. Combinations of these processes working together could magnify the effects of each.

Table 1. The parameters determined by fitting the temperature-dependent stress data to Equation (4) for DR1- and DO11-doped PMMA thin films.

Sample	σ_A (MPa)	σ_B (MPa)	T_0 (°K)	n
DR1/PMMA thin film	−300 (±0)	300 (±0.002)	367.88 (±0.28)	32.40 (±0.14)
DO11/PMMA thin film	−300 (±0)	300 (±0.0004)	344.95 (±0.18)	57.97 (±0.24)

5.1.2. Hole Burning and Photo-Stress Anisotropy

Table 2 summarizes the parameters determined from fits to the data shown in Figure 6 for thin films. For comparison, also shown are the values from the literature for dye-doped fibers [2]. The effect of dye-doping and annealing of the photomechanical effect in fibers has also been studied [33]. The positive sign of $\kappa^{(\mathrm{heat})}$ in both materials and in both thin films and fibers indicates that photothermal heating drives expansion. The ratio of the hole burning to the heating response, R, vanishes within experimental uncertainty, confirming that photothermal heating dominates in thin films and fibers. In DO11 $\kappa_{//}^{(\mathrm{hole})}$ and $\kappa_{\perp}^{(\mathrm{hole})}$ are expected to be zero, because DO11 is nonisomerizable, and the results show this to be true with a high degree of confidence. So, we conclude that angular hole burning and molecular reorientation are absent in DO11. DR1 shows a larger contribution from angular hole burning, and the signs of the coefficients are what is expected but vanish within experimental uncertainty. The sign of $\kappa_{//}^{(\mathrm{hole})}$ indicates that the length of the sample along the light's polarization decreases and the positive sign of $\kappa_{\perp}^{(\mathrm{hole})}$ shows that the length of the sample increases along its long axis when it is illuminated by light perpendicular to it, refs. [1,2], as would be expected. But again, the values are statistically insignificant, so angular reorientation and hole burning are negligible.

Table 2. A summary of each mechanism's contribution to the photomechanical response in dye-doped PMMA thin films. Fiber data are from the literature, refs. [2,33], where "CTA" refers to the concentration of charge transfer agent, which affects the polymer's chain length.

Sample	$\kappa_{//}^{(1)}/\kappa_{\perp}^{(1)}$	R	$\kappa_{//}^{(\mathrm{heat})} = \kappa_{\perp}^{(\mathrm{heat})}$ (s/m)	$\kappa_{//}^{(\mathrm{hole})}$ (s/m)	$\kappa_{\perp}^{(\mathrm{hole})}$ (s/m)
DR1 Thin Film	0.92 ± 0.12	−0.06 ± 0.080	961.35 ± 124.91	−54.61 ± 77.86	27.31 ± 38.88
DO11 Thin Film	0.97 ± 0.17	−0.02 ± 0.12	841.07± 106.23	−18.21 ± 98.43	9.11 ± 49.20
DR1 Fiber 1.33 × 10^{-3} CTA	0.96 ± 0.28	−0.02 ± 0.19	246 ± 80	−6 ± 46	3 ± 23
DR1 Fiber 2 × 10^{-3} CTA	0.91 ± 0.08	−0.06 ± 0.05	259 ± 22	−15 ± 14	8 ± 7
DR1 Fiber 2.66 × 10^{-3} CTA	0.94 ± 0.23	−0.04 ± 0.16	231 ± 49	−10 ± 36	5 ± 18

To make sure that the response is linear and we are in the low-intensity regime, the ratio R as a function of intensity is also calculated for both samples. R is found using Equation (26). Figure 7 shows a plot of the relative reorientational fraction R as a function of intensity, where we have evaluated $\kappa_{//}^{(1)} + \kappa_{//}^{(2)} I$ and $\kappa_{\perp}^{(1)} + \kappa_{\perp}^{(2)} I$ in Equation (26) using the values of $\kappa_{\perp}^{(1)}, \kappa_{\perp}^{(2)}, \kappa_{//}^{(1)}$, and $\kappa_{//}^{(2)}$ determined from the slopes in Figure 6 to obtain the points at each intensity I. The error bars are propagated from the measured uncertainties in $\kappa_{\perp}^{(1)}, \kappa_{\perp}^{(2)}, \kappa_{//}^{(1)}$, and $\kappa_{//}^{(2)}$. The ratio varies more over the intensity range measured for DR1 than for DO11. The ratio R for DR1 appears to change sign at low intensity, but this sign change is in the noise given the uncertainties. The increase in R for DR1 is statistically significant, suggesting that hole burning becomes a significant fraction of the response mechanisms at the highest intensities.

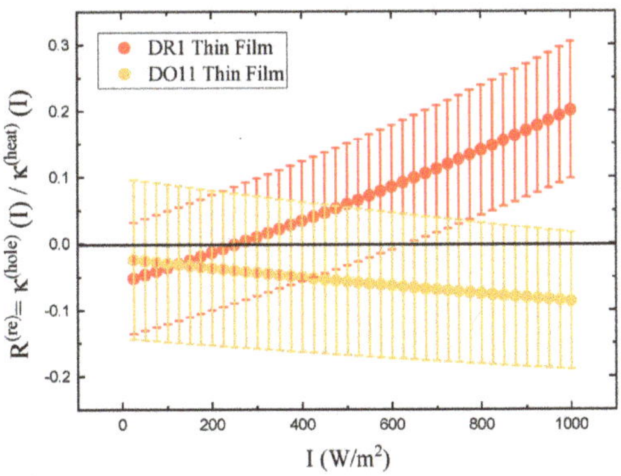

Figure 7. The ratio of the photomechanical stress due to angular hole burning and heating.

While the behavior of the thin films and fibers is similar, the linear photomechanical constants are a factor of three to four times larger in the thin films.

5.2. Evolution of Molecular Orientation

Section 5.1 describes how the photomechanical effect depends on the various mechanisms. Here we focus on measurements that directly determine the orientational order of the dopant molecules in the PMMA polymer host, which can be correlated with the photomechanical response to assess if the orientational order couples mechanically to the polymer and that it contributes to photomechanical actuation.

The transmitted and incident intensities are recorded using a power meter with and without a sample. From these, the optical absorption coefficient of the sample is determined for the two orthogonal polarizations using the Beer–Lambert law [1,2]

$$I = I_0 e^{-\alpha d}, \tag{28}$$

where

$$\alpha_{//} = -\frac{1}{d} \ln\left(\frac{I_{//}}{I_0}\right), \tag{29}$$

and

$$\alpha_{\perp} = -\frac{1}{d} \ln\left(\frac{I_{\perp}}{I_0}\right). \tag{30}$$

$\alpha_{//}$ and $\alpha_{\perp}$ are the absorbance parallel and perpendicular to the polarization of the pump beam, respectively. Figure 8 (top) shows the absorbance for vertically and horizontally polarized probe beams in DR1- and DO11-doped PMMA thin films. Figure 8 (bottom) shows a close-up view of one cycle in a region where transients from the probe lasers being turned on have subsided. For DR1, the vertically polarized pump induces a fast increase in the intensities of both probe polarizations as shown by the decreased absorbance. At later times with the pump on, the intensity of the horizontally polarized probe beam decreases (increased absorbance), while the vertically polarized probe beam shows little change.

This fast behavior can be understood as follows. Irradiating DR1 molecules with vertically polarized light preferentially excites the vertically oriented chromatophores. In the argument here, we assume that DR1 absorbs less light in its excited state and the molecules are one-dimensional, which has been shown to be a good approximation [34]. The probability of exciting a chromophore that is oriented an angle θ relative to the pump polarization is proportional to $\cos^2\theta$, making it most probably to excite vertically oriented molecules at $\theta = 0$ and improbably in the horizontal direction, where $\theta = \pi/2$. The probability of a molecule absorbing vertically polarized probe light is proportional to $\cos^2\theta$, making the probability of absorbing vertically polarized probe light in the presence of the vertically polarized pump beam is proportional to $\cos^4\theta$. Similarly, since the probability of absorbing horizontally polarized light is proportional to $\sin^2\theta$, the probability of absorbing horizontally polarized probe light in the presence of a vertically polarized pump is proportional to $\sin^2\theta\cos^2\theta$.

Finally, assuming that the sample is isotropic, which should be based on the material's processing conditions, the ratio of the change in absorbance for the two polarizations for an initially isotropic sample is given by the orientational average ratio

$$\frac{\alpha_{\perp}}{\alpha_{//}} = \frac{\langle\sin^2\theta\cos^2\theta\rangle}{\langle\cos^4\theta\rangle} = \frac{2}{3}. \quad (31)$$

Equation (31) implies that the drop in the vertically polarized probe is greater than the horizontal one and that they both should drop [1,2]. Figure 8 (bottom) shows this to be true for DR1. DO11, on the other hand, also shows a drop in the absorbance for both polarizations, but the one-dimensional approximation does not hold as well for DO11. In both cases, we see a drop in the absorbance due to the pump light exciting the molecules. This effect is labeled "hole burning" because the orientational distribution function takes on a shape like the dimple in an apple. Note that the probe beam intensities are much lower than the pump, so their affect on the excited populations should be minimal.

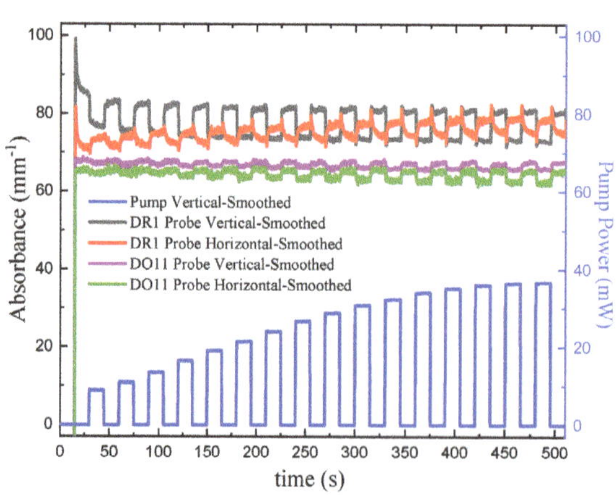

Figure 8. *Cont.*

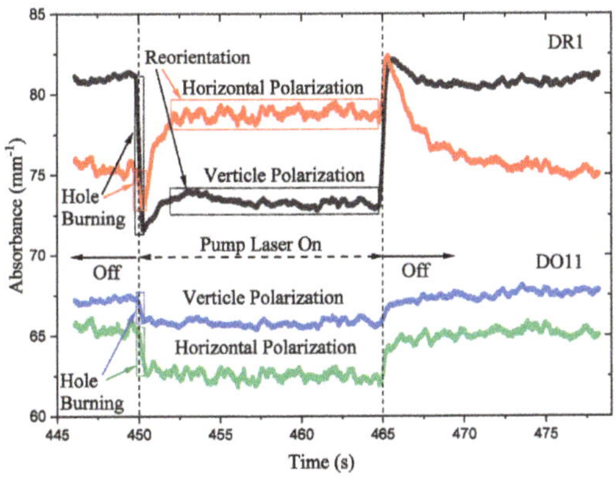

Figure 8. (**top**) Absorbance parallel and perpendicular to the pump beam polarizations in DR1- and DO11-doped PMMA thin films as a function of time with increasing pump intensity. (**bottom**) A close-up of the data for one on–off cycle showing a fast process associated with angular hole burning and two slow processes, which includes molecular reorientation.

Another fast process is photothermal heating, which increases the temperature of the material as the absorbed light is converted to heat. Since the trans isomer is of lower energy than the cis isomer, the trans population drops and the cis population increases at higher temperatures, leading to an isotropic decrease in the absorbance of light by virtue of the cis molecule's slow absorption cross-section. Thus isotropic heating will lower the absorbance of both polarizations equally. The fast component will thus be a combination of both heating and hole burning.

At longer time scales, molecules that undergo photoisomerization will reorient into the horizontal direction, as we would expect; when molecules aligned along the light's polarization turn into the smaller cis state, they reorient randomly due to diffusion and on average end up in the perpendicular orientation once they decay back to the trans state [2,4]. This decreases the vertical absorbance because fewer molecules are aligned with the pump beam and increases the horizontal absorbance where there are more molecules. The horizontal probe sees an increase in absorbance as we would expect. Similarly, the long-time vertical probe sees a decrease in the absorbance, as expected. At an intermediate time between the fast hole-burning process and the slow molecular reorientation process, another feature is observed where both probe polarizations show an increase in absorbance. The region's time scale includes the photothermal heating mechanism, which can change the relative populations between the cis and trans isomers. Separating the mechanisms to this level of detail is beyond the scope of this work, but it is worth mentioning that the three mechanisms can couple to produce complex behavior, which is difficult to model.

DO11 thin films show a decrease in absorption for both polarizations in response to the pump and an increase when the pump is turned off. The DO11 molecule is known to transition from the keto to the enol state when excited by a photon. Optical absorption of the enol state is weaker, so the absorbance decreases when the pump is on. The DO11 molecule relaxes back to the keto state when the pump is turned off [26,35], leading to an increase in the absorbance.

In both DR1 and DO11 molecules, the time dependence of the absorbance when light is turned off is the reverse of the process when the light is turned on. In the case of DR1, where there are multiple exponential processes that contribute amplitudes of opposite signs, the process is left–right anti-asymmetric, while the DO11 process is left–right symmetric.

Note that for both cases, flipping the pump-off absorbance data about the horizontal axis and shifting them by 15 s gives the same qualitative shape.

The difference in the absorbance measured by the two orthogonal probe lasers can be used to find the axial orientational order parameter $\langle P_2 \rangle$. For molecules that are approximately one-dimensional, the order parameter is given by [26,30]

$$\langle P_2 \rangle = \frac{\alpha_{//} - \alpha_{\perp}}{\alpha_{//} + 2\alpha_{\perp}}. \tag{32}$$

The order parameter $\langle P_2 \rangle = 1$ if the long axes of the molecules are fully aligned and $\langle P_2 \rangle = 0$ when they are randomly aligned. The order parameter can be negative when the molecules lie flat in the plane perpendicular to the symmetry axis.

The order parameter relative to the sample's long axis, mounted vertically, as a function of time and intensity is shown in Figure 9 for vertically and horizontally polarized pump light. The top plot references the order parameter to the long axis of the sample while the bottom plot is relative to the pump axis.

Order parameter relative to long axis of sample: We first consider the top plot of Figure 9, which references the order parameter to the long axis of the thin-film sample. The long axis is always mounted vertically. Turning the pump on and off for both polarizations does not affect the order parameter in DO11—as expected when photo-isomerization is absent. In contrast, for DR1, a *vertically polarized pump* decreases the order parameter along the film's long axis (vertical) and turning the pump off results in an increase in this order parameter back to its pre-pumped state. A *horizontally polarized pump* increases the order parameter, which decreases reversibly when the pump is turned off. This behavior is expected if the pump light that is polarized along the trans molecule's long axis excites the cis state, making a "hole" along the light's polarization direction by depleting those trans molecules. Subsequently, the number of molecules in the perpendicular orientation increases as the cis molecules relax back to the trans state and take on random orientation but leave a net depletion along the pump's polarization.

Order parameter relative to pump polarization axis: The bottom part of Figure 9 shows the order parameter relative to the pump's polarization axis. For an isotropic sample, the order parameter in a dark sample should vanish and the order parameter in the presence of pump light should be the same for both polarizations if their intensities are the same. The data show that the order parameters are not the same for the two polarizations, so the samples are not isotropic. There are several possible sources of this anisotropy, including the fact that the samples are slightly stretched in the vertical direction when mounted in the sample holder. This is consistent with the observation that a vertical pump polarization has a slight positive order parameter, while the horizontal one has a negative order parameter. In both cases, the presence of the pump leads to a decrease in the order parameter, as expected for molecular reorientation.

Figure 10 shows the same data as in Figure 9, but as a function of pump power. Because the samples are slightly anisotropic, the behavior is made clearer by considering the change in the order parameter relative to the dark state. This change in the order parameter $\Delta\langle P_2 \rangle$ is given by

$$\Delta\langle P_2 \rangle(I) = \langle P_2 \rangle_{on}(I) - \langle P_2 \rangle_{off}(0), \tag{33}$$

where $\langle P_2 \rangle_{on}(I)$ is the order parameter when the pump intensity is I and $\langle P_2 \rangle_{off}(0)$ is the order parameter of the material prior to the start of the experiment. We call $\langle P_2 \rangle_{on}(I)$ the total order parameter change because it includes the change in order parameter due to the instantaneous pump power, as well as the accumulated order parameter from all past exposure due to the fact that the dye molecules have not had the chance to relax to the isotropic state during the 14 s period that the pump light is turned off. The low levels of probe light might also add to the cumulative effect. The values of $\langle P_2 \rangle_{on}(I)$ at $I = 0$ include all the values determined when the pump light is off between measurements where the intensity is changed. If there were no cumulative effects of time as the pump intensity is increased, there would be only one point for each polarization and each material.

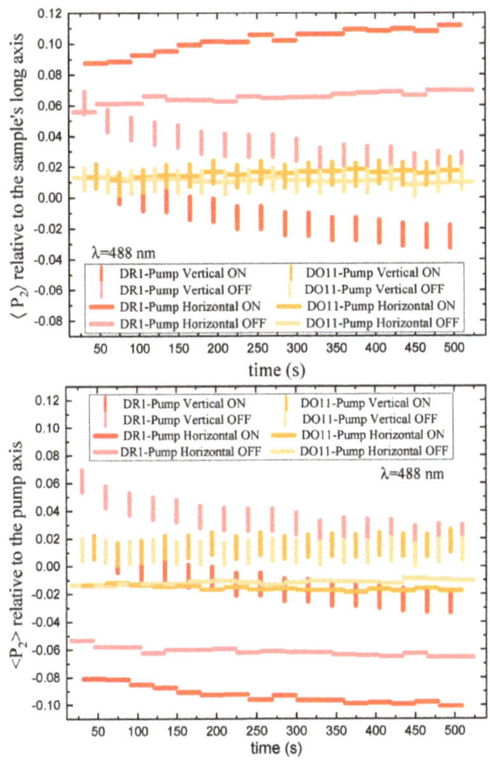

Figure 9. Measured order parameter of DR1- and DO11-doped PMMA thin films as a function of time. Time is a proxy for intensity as shown in the pump power plots on the bottom part of the top graph of Figure 8. (**top**) Order parameter relative to the sample's long axis and (**bottom**) order parameter relative to the pump beam's polarization axis.

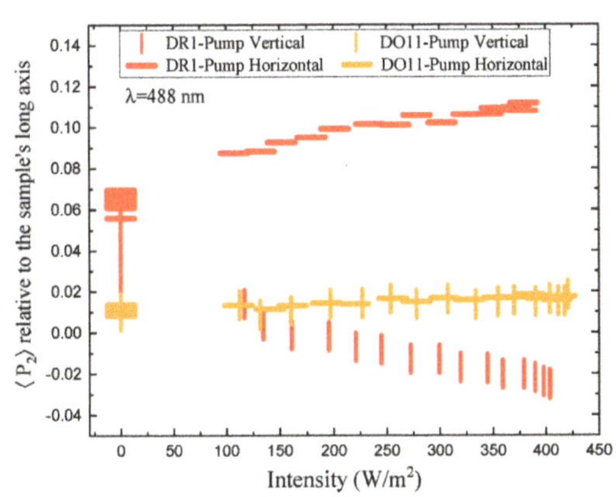

Figure 10. *Cont.*

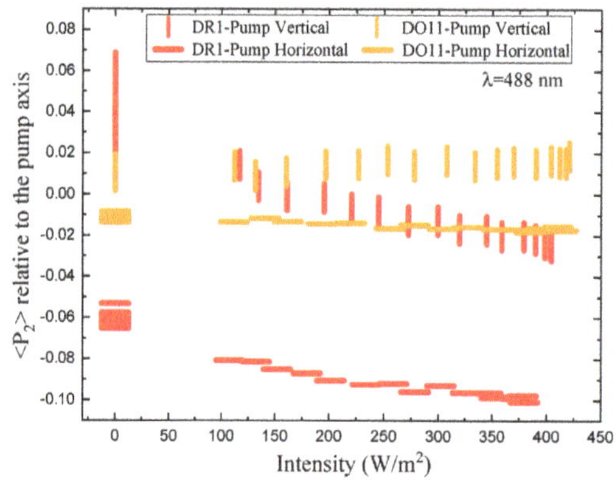

Figure 10. Order parameter of DR1- and DO11-doped PMMA thin films as a function of intensity. (**top**) Order parameter relative to the sample's long axis and (**bottom**) order parameter relative to the pump's polarization axis for vertically (vertical lines) and horizontally polarized (horizontal lines) pump. The data markers at $I = 0$ include the initial measurement before the start of the experiment, as well as all the order parameters measured when the pump light is blocked prior to increasing intensity (the $I = 0$ wells of the pump power are plotted in the upper portion of Figure 8).

Figure 11 shows the order parameter change $\Delta\langle P_2\rangle$ as a function of absorbed intensity relative to the sample's long axis (top) and the pump axis (bottom). For DO11-doped PMMA, the order parameter referenced to the long axis of the sample (top plot) shows no statistically significant change as a function of intensity, while DR1-doped PMMA shows a dramatic effect with the horizontal pump inducing an increase in order parameter and the vertical pump inducing a decrease. The decrease in order parameter for the vertical pump is of slightly larger magnitude than the increase induced by the horizontal pump. These observations are consistent with molecules reorientating away from the light's polarization and ending up in the plane perpendicular to the pump.

The order parameter referenced to the pump axis (bottom plot) shows a larger difference between the two pump polarizations for DO11 and a similar difference for DR1. But, the trend is the same; the order parameter drops in the direction of the pump polarization axis. The observation of a pump polarization-dependent change in the parameter suggests that the sample might be initially anisotropic. The change in the order parameter $\Delta\langle P_2\rangle$ appears to saturate at higher intensities, which one would expect for a fixed population of molecules.

Next, we seek to understand the source of the anisotropy. The two obvious candidates are (1) anisotropy due to the slight stretching of the sample along its long axis from mounting strain and (2) long-time molecular reorientation that builds over time if the dyes in the polymer do not have enough time to relax over the 15 s dark cycle when the pump is off. There may also be an effect due to the excitation of molecules due to the ever-present weak probe beams, whose small effects may accumulate over time.

First, we need to determine the change in order parameter due to the current pump cycle. This is best characterized by the short-time order parameter change, defined as the fractional change in the order parameter, which is normalized to the order parameter just before the pump is turned on during each cycle, $\langle P_2\rangle_{off}(\text{n})$, given by

$$\frac{\Delta\langle P_2\rangle}{\langle P_2\rangle}^{(fast)}(n) = \frac{\langle P_2\rangle_{on}(n) - \langle P_2\rangle_{off}(n)}{\langle P_2\rangle_{off}(n)}, \tag{34}$$

where n labels the cycle number.

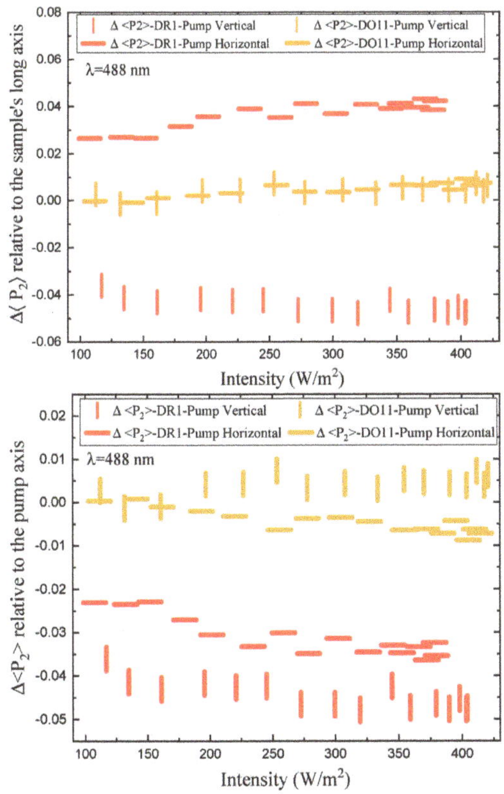

Figure 11. Change in the order parameter in DR1- and DO11-doped PMMA thin films relative to the order parameter of the pristine sample as a function of the pump intensity. This includes the order parameter change induced by the light and the accumulated change in the order parameter from past exposures. (**top**) Order parameter relative to the long axis of the sample and (**bottom**) order parameter relative to the pump beam's polarization axis.

Figure 12 shows the short-time fractional order parameter change as a function of intensity relative to the sample's long axis (top) and relative to the pump's polarization axis for both vertically and horizontally polarized pumps. DO11 shows the same fractional order parameter change for both pump polarizations, which is consistent with the absence of molecular reorientation. DR1, on the other hand, is consistent with molecular reorientation in that the order parameter change relative to the long axis for a vertically polarized pump is greater than for a horizontally polarized pump and of opposite sign. In the low-intensity regime, the ratio is about $-2/3$, as predicted by Equation (31). The ratio becomes smaller at higher intensities, which is an indication that the sample is changing over time due to pump cycling and the effects of the probe beams. This does not preclude an intensity-dependent effect that originates in nonlinearities in the underlying mechanisms. Note that extrapolating the plots to zero intensity suggests clustering around a vanishing fractional order parameter change.

The long-term effects of the light on the sample can be assessed by comparing its properties at a given cycle with the sample prior to the first cycle. To this end, we define the long-term fractional change in the order parameter in the nth cycle by referencing it to the order parameter of the sample at the start of the measurement. We start by determining the fractional change in the order parameter when the pump is off after the nth cycle, given by

$$\frac{\Delta \langle P_2 \rangle}{\langle P_2 \rangle}^{(slow,dark)}(n) = \frac{\langle P_2 \rangle_{off}(n) - \langle P_2 \rangle_{off}(first)}{\langle P_2 \rangle_{off}(first)}. \tag{35}$$

This will determine the change in the dark state of the sample in the presence of only the weak probe lasers. Doing the same for when the pump is on during the nth cycle, we obtain

$$\frac{\Delta \langle P_2 \rangle}{\langle P_2 \rangle}^{(slow,pump)}(n) = \frac{\langle P_2 \rangle_{on}(n) - \langle P_2 \rangle_{off}(first)}{\langle P_2 \rangle_{off}(first)}. \tag{36}$$

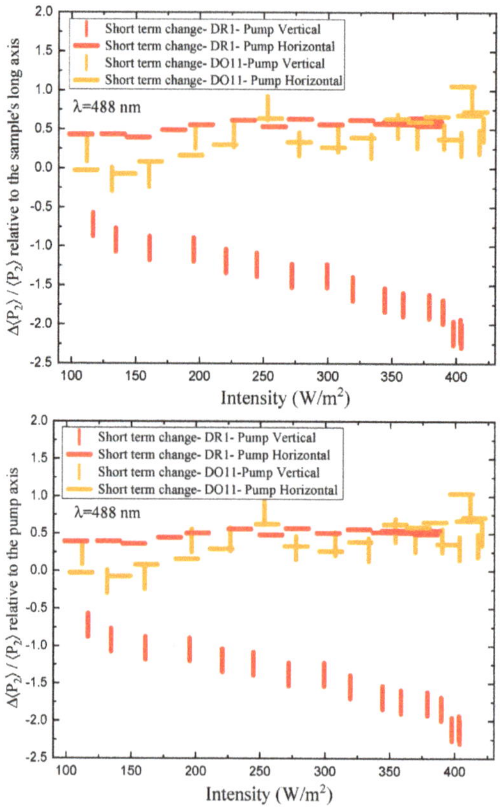

Figure 12. Fractional change in the order parameter relative to the order parameter just prior to the pump being turned on as a function of pump intensity in DR1- and DO11-doped PMMA thin films. (**top**) Fractional order parameter change relative to the sample's long axis and (**bottom**) relative to the pump axis for both vertically and horizontally polarized pump light.

The long-term fractional change in the order parameter as a function of intensity when the pump is off and the pump is on relative to the pristine sample are shown in Figures 13 and 14. The data clearly show that the dark state of the material (when pump is off, but the probes are on), over the long-term span of the experiment, evolves over time. The fractional change in the vertically referenced order parameter in DR1 decreases over time (and with increased pump intensity). This change is most likely a combination of the sample's properties changing over time under the influence of the pump and probe beams, as well as the increased pump intensities, which might produce a longer-lasting change in the material properties. Similarly, when the pump is on, the vertically referenced fractional order parameters decreases over time, but by a greater factor than the dark state. This is attributable to the fact that in the pumped state, the fractional change in the order parameter is influenced by both the long-term changes in the sample and the short-term effects of the pump.

Consider that the fast response shows an increase in the fractional order parameter change over time/intensity from about −0.75 to −2.25, the slow dark state changes from −0.3 to −0.65 and the slow pumped state changes from −0.75 to −1.3. At the lowest intensity and earliest times, there is already evidence of the dark state being anisotropic. At the highest intensity, the fast response shows the largest fractional change in the order parameter, while the long-term response is somewhat lower, but by a statistically significant amount. The long-term change in the fractional order parameter in the dark state is greater than the initial value but still significantly smaller than the pumped state. Thus, the changes in the sample are enough to have a statistically significant effect on the measurements of the pumped state at high intensity, though not a large difference. Changes in the material from long-term light exposure might explain most of the small variations in DO11-doped PMMA thin films. However, the fractional change in the order parameter in DO11 is always similar for both pump polarizations, confirming that angular hole burning and reorientation are insubstantial.

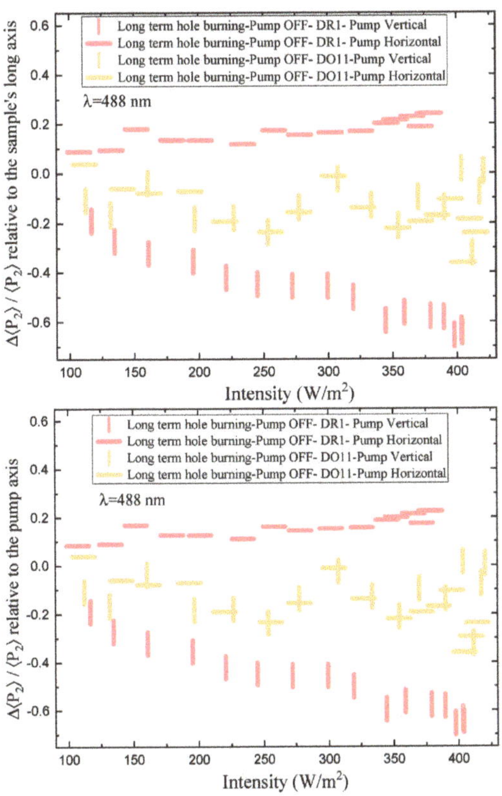

Figure 13. Long-term time evolution of the fractional order parameter change *when the pump is off* relative to the order parameter of the pristine sample as a function of the intensity of the pump prior to the initial measurement at $t = 0$ in DR1- and DO11-doped PMMA thin films. (**top**) Fractional order parameter change relative to the sample's long axis and (**bottom**) relative to the pump axis for both vertically and horizontally polarized pump light.

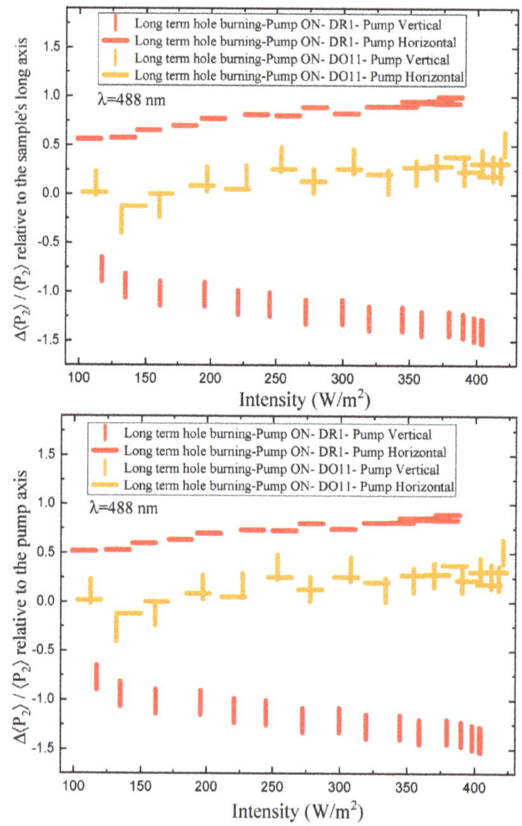

Figure 14. Long-term time evolution of the fractional order parameter change as a function of the pump intensity in DR1- and DO11-doped PMMA thin films. (**top**) Fractional order parameter change relative to the sample's long axis and (**bottom**) relative to the pump axis for both vertically and horizontally polarized pump light.

5.3. Photomechanical Efficiency

The photomechanical efficiency is given by Equation (27). It requires a measurement of both the photomechanical response and Young's modulus. The photomechanical response is described above. Here, we determine Young's modulus of DR1 and DO11 at room temperature by pulling on one end of the sample with a stepper motor running at constant speed while measuring the stress. For each sample, the experiment is repeated five times and the average of the five runs is used to decrease experimental uncertainties. Figure 15 shows the five runs for each sample. The initial flat portion up to a strain of 0.002 is due to backlash in the stepper motor and translation stage used to strain the thin film and also due to the fact that the sample is not rigid and so it is initially in a bucked state when first mounted. At each cycle, the stepper motor brings the thin film back to the same point. Young's modulus is determined from the slope above a strain of 0.004 where the sample is stretched beyond the buckling point along its long axis. Both DR1- and DO11-doped PMMA give a Young's modulus of 0.89 ± 0.01 GPa. The stress for all measurements is well below the yield point and far below the stress required to fracture the polymer.

Table 3 summarizes the thin-film results, including the efficiency in converting light energy to mechanical work using the figure of merit. For comparison, results from the literature of DR1-doped PMMA fibers are also presented [2]. For each sample, the figure of merit (FOM) is the same within experimental uncertainty for each of the two pump polarizations in the two dye-doped PMMA samples. Similarly, the thin-film figure of merit is the same within experimental uncertainty for the two materials at the same polarization. One must keep in mind that these are 1-σ error bars.

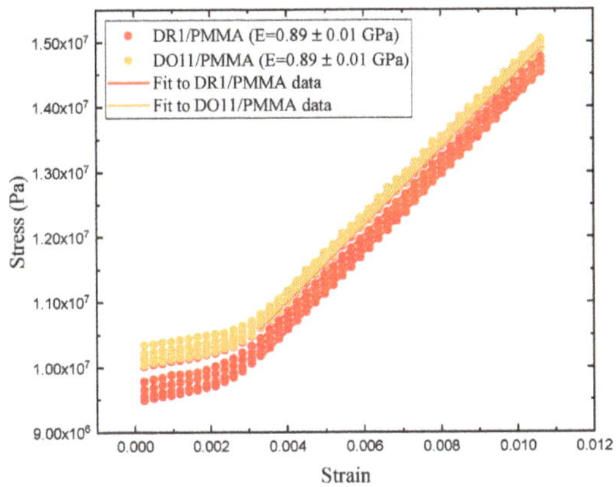

Figure 15. The measured stress versus strain in DR1- and DO11-doped PMMA thin films at room temperature. The plot shows 5 separate runs for each material. The initial slope below a strain of 0.003 is due to system backlash. Young's modulus is determined from the slope of the average of the 5 runs above a strain of 0.004.

However, there is a slight statistically significant difference between the two polarizations in DR1. This could be due to the small contribution of reorientation, which is of the same order of magnitude as the error bars. Similarly, the difference between the two different materials is different at the edge of the error bars for the horizontally polarized pump light, where we expect the effects of molecular reorientation to be the largest. But, given that these effects are at the edge of the error band, these differences are insignificant but of the correct polarization-dependence to be due to molecular reorientation.

The FOM for DR1-doped PMMA fibers is about two orders of magnitude lower than for thin films. The FOM is even lower when the fibers are annealed, as described in the literature [33]. Furthermore, the Young's modulus in the fiber geometry is more than a factor of three larger than in the thin-film geometry. The advantage of the thin films over fibers originates in both parameters; the smaller Young's modulus provides less material resistance to the photomechanical force which is leveraged by the larger photomechanical constant. The larger photomechanical constant may be due to more efficient heating due to the lower heat capacity of the thin film and the uniform illumination. The lower Young's modulus of the thin film might be due to surface effects, where the skin—of the lower modulus—contributes more to the response because it contains a larger fraction of the total sample volume. The question of volume versus surface effects is an interesting one which could be studied characterizing samples of the same length but varying cross-sectional aspect ratio.

Table 3. A summary of the results for both thin films. Fiber data are from the literature [2].

Material	Polarization	$\kappa_\sigma^{(1)}$ (s/m)	$\kappa_\sigma^{(2)}$ (s·m/W^2)	E (Gpa)	FOM (10^{-4} s^2/N)
DR1 Thin Film	Vertical	995 ± 161	-0.21 ± 0.27	0.89 ± 0.01	11.12 ± 1.80
DR1 Thin Film	Horizontal	828 ± 204	0.12 ± 0.31	0.89 ± 0.01	7.7 ± 1.809
DO11 Thin Film	Vertical	1135 ± 201	-0.28 ± 0.28	0.89 ± 0.01	14.50 ± 2.57
DO11 Thin Film	Horizontal	1150 ± 178	-0.25 ± 0.27	0.89 ± 0.01	14.90 ± 1.02
DR1 Fiber 1.33×10^{-3} CTA	Vertical	240 ± 63	-5 ± 8	3.12 ± 0.05	0.19 ± 0.05
DR1 Fiber 1.33×10^{-3} CTA	Horizontal	249 ± 35	-2 ± 0.03	3.12 ± 0.05	0.20 ± 0.03
DR1 Fiber 2×10^{-3} CTA	Vertical	243 ± 15	-5 ± 3	3.02 ± 0.04	0.20 ± 0.01
DR1 Fiber 2×10^{-3} CTA	Horizontal	266 ± 16	-10 ± 2	3.02 ± 0.04	0.23 ± 0.01
DR1 Fiber 2.66×10^{-3} CTA	Vertical	221 ± 30	-2 ± 4	2.87 ± 0.02	0.17 ± 0.02
DR1 Fiber 2.66×10^{-3} CTA	Horizontal	235 ± 48	-13 ± 4	2.87 ± 0.02	0.19 ± 0.04
DR1 Fiber 2.66×10^{-3} CTA	Vertical	263 ± 42	6 ± 4	2.87 ± 0.02	0.24 ± 0.08
DR1 Fiber 2.66×10^{-3} CTA	Horizontal	266 ± 20	1.3 ± 3	2.87 ± 0.02	0.25 ± 0.04

5.4. Photomechanical Time Response

We determine the photomechanical response times by fitting the time-dependent stress as shown in Figure 2 to a double exponential when the pump is on and a separate double exponential when the pump is off. As a function of intensity, these fits give the intensity dependence of the magnitude and characteristic photomechanical time constants. Figures A1 and A2 show the time constants for the photomechanical response, while Figures A3 and A4 show the relaxation time constants when the light is turned off for both DR1- and DO11-doped PMMA for both pump polarizations. The insets show the average time constant over the full intensity range for each material at each polarization.

While the fluctuations in the data exceed the error bar range, there is no obvious trend in the intensity dependence. The average of each run for a given sample and pump polarization shows that both the photomechanical response and its relaxation have two characteristic time constants that are about the same for each material. The fast time constant is about 0.5 s for both materials and for both the photomechanical response and the relaxation process. The longer time-scale process shows a different time constant for the pump on and off and range from about 1.5 s for pumped response to 3 s during relaxation.

In contrast, fibers from previous studies display a single time constant for both decay and recovery between about 0.7 s and 1.0 s [2]. This suggests that the underlying processes in the two sample geometries may be different, which might originate in the surface skin effect, where in thin films, the skin is a much larger fraction of the total volume than in fibers. As such, the thin film measurements reported here most likely are probing a regime where the skin contributes a large fraction of the total response, thus showing two unique processes, while the fiber measurements are mostly probing bulk effects.

The amplitudes of the photomechanical response for the fast and slow processes determined from Equation (2) are plotted in Figure 16. The faster response mechanism amplitude σ_1 shows a large dependence on the pump power, while the slower response associated with the amplitude σ_2 remains approximately unchanged as a function of the intensity.

The fits in Figure 16 to σ_1 and σ_2 as a function of intensity to the power series expansions

$$\sigma_1(I) = -(\kappa_{\sigma_1}^{(0)} + \kappa_{\sigma_1}^{(1)} I + \kappa_{\sigma_1}^{(2)} I^2) \quad (37)$$

and

$$\sigma_2(I) = -(\kappa_{\sigma_2}^{(0)} + \kappa_{\sigma_2}^{(1)} I). \tag{38}$$

Note that $\kappa_{\sigma_1}^{(0)}$ and $\kappa_{\sigma_2}^{(0)}$ are not the material constants but rather related to the pre-stress σ_0 when mounting the samples, which is given by

$$\sigma_0 = \kappa_{\sigma_1}^{(0)} + \kappa_{\sigma_2}^{(0)}. \tag{39}$$

Then, we can define an intensity-dependent photomechanical response as being given by

$$\kappa_{\sigma_1}^{(1)}(I) = \frac{\kappa_{\sigma_1}^{(0)} - \sigma_1(I)}{I} = \kappa_{\sigma_1}^{(1)} + \kappa_{\sigma_1}^{(2)} I. \tag{40}$$

and

$$\kappa_{\sigma_2}^{(1)}(I) = \frac{\kappa_{\sigma_2}^{(0)} - \sigma_2(I)}{I} = \kappa_{\sigma_2}^{(1)}, \tag{41}$$

where $\kappa_{\sigma_i}^{(1)}(I)$ is the long-time photomechanical stress response of process i to light of intensity I.

The parameters $\kappa_{\sigma_1}^{(1)}$ and $\kappa_{\sigma_1}^{(2)}$ are determined for DR1- and DO11-doped PMMA for both vertical and horizontal pump intensities from fits to the data to Equation (37) as shown in the top portion of Figure 16. The same is performed for the long-term stress response for the process associated with time constant t_2 by fitting the data in the bottom portion of Figure 16 to the power series expansion given by Equation (38). In this case, the data are relatively flat, so the fit includes only the linear term.

To summarize, the data given by Figure 16 are used to obtain the photomechanical coefficients $\kappa_{\sigma_i}^{(n)}$ for each process, which are substituted in Equations (37) and (38) to obtain the long-term stress $\sigma_i(I)$. These fits are also used to determine the intensity-dependent photomechanical response $\kappa_{\sigma_i}^{(1)}(I)$ for each process.

The strategy for isolating the mechanisms of the photomechanical response is to compare the photothermal heating model's predictions given by Equation (9) with the magnitude of the fast and slow response determined from the data. The heating model depends on several types of parameters. First, σ_A, σ_B, T_0 and n characterize the thermodynamic properties of the material and are determined from the temperature-dependent stress of a clamped sample as shown in Figure 3. For DO11 and DR1, the data are fit to Equation (4) as shown in Figure 3 to obtain these parameters. Next, the physical properties of each sample, such as their thicknesses w, are directly determined with a micrometer. Properties such as the specific heat capacity of the polymer c and the material density ρ are sourced from the literature. We are thus assuming that the dopant dyes do not contribute substantially to these properties because of their low concentration. This assumption may need to be revisited given the large difference in the data between DO11 and DR1 in Figure 3. Finally, each of the time constants from the fits of the stress response are used to obtain τ at each intensity that is used in Equation (5) to relate the temperature change to the light intensity. Recall that $t_1(I)$ and $t_2(I)$ are defined to be the short and long time constants for pump intensity I. The strategy is to identify which exponential component of the two-exponential fit best agrees with the model's prediction.

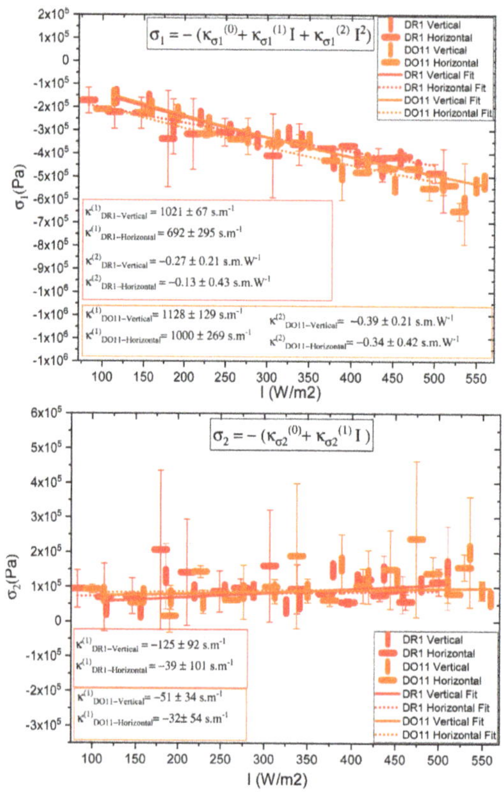

Figure 16. Amplitudes σ_1 (**top**) and σ_2 (**bottom**) for DR1- and DO11-Doped PMMA thin films as a function of pump intensity for both vertically and horizontally polarized pump. The negative stress of the fast process given by σ_1 implies that the length increases under light exposure and the increase gets larger with intensity. σ_2 is positive, implying that for the slow process the sample contracts when pumped with light.

Alternatively, as we saw above, we can eliminate the parameter σ_B in Equation (9) using Equation (4), yielding the semi-empirical model given by Equation (12). In this case, the measured stress $\sigma_i(I)$ at pump intensity I determined from the fits for the process of response time t_i is used in the heating model, de-emphasizing reliance on the phenomenological stress model given by Equation (4). As we will see below, the small differences between these two approaches are insignificant in determining which process is associated with photothermal heating.

Figures 17 and 18 show a summary of the theoretical and experimental intensity-dependent photomechanical constant $\kappa_{\sigma_1}(I)$ for the faster process with response time t_1. The bands are derived from the experimentally determined amplitudes of the fast process and their widths represent the one standard deviation uncertainty band. The vertical and horizonal data markers represent the photothermal heating model's prediction for the intensity-dependent photomechanical response for the two orthogonal pump polarization directions. Their values are determined using the heating theory given by Equation (9), plotted in Figure 17, and Equation (12), plotted in Figure 18. The scatter in the plots originate from the scatter in the measured time constants $t_1(I)$ that are input parameters to the theory. The semi-empirical theory using Equation (12) has an additional contribution to the uncertainty from $\sigma_1(I)$ and σ_A, which leads to larger error bars. Similarly, the error bands of the data are coded with vertical and horizontal hash marks to represent the polarization direction. Figures 19 and 20 show the slower process with the time constant data and theory using $t_2(I)$ and $\sigma_2(I)$ for the analysis.

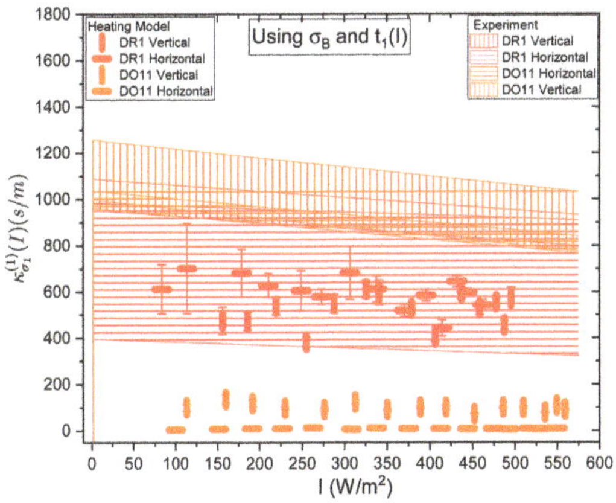

Figure 17. The vertical and horizontal lines (representing the pump polarization relative to the vertical) show the predicted heating contribution to the photomechanical response for DO11 and DR1 thin films using the theory given by Equation (9). The time constants $t_1(I)$ at each intensity used in the model are the measured ones in Figure A1 in Appendix A. The width of the bands show one standard deviation band of the magnitude of the fast component of the experimentally determined intensity dependent photomechanical constant, which is determined from a quadratic fit of the stress data shown in the top graph in Figure 16.

Figures 17 and 18 show that the magnitude of the measured fast linear photomechanical response, as represented by the bands, decreases with pump intensity. The samples are pristine at the start of the experiment and characterized in the same physical spot starting with low intensity and then ending at the highest intensity with uniform intensity increases in between. Measurements of the order parameter, as discussed above, show that the material undergoes long-term changes in axial order of the dopant dyes, which we provisionally attribute to long-term molecular reorientation. This change in the fast response with exposure is consistent with the long-term process of molecular reorientation. Figures 17 and 18 show that the data and the photothermal heating theory for the fast response agree for DR1 but disagree for DO11. As such, we conclude that the fast process is due to heating in DR1 but not in DO11. This is puzzling unless our assumption that the heat capacities are the same for both materials is wrong. The data suggest that the heat capacity of DO11 would need to be an order of magnitude smaller than in neat PMMA to make the heating theory agree with the data, which seems unlikely.

Figures 19 and 20 show the experimental data bands and the theory markers for the slower response given by time constant t_2. For DO11, the photomechanical response of the data and theory seem to marginally agree with some outliers, while for DR1, the data disagree with theory. However, the signs disagree. This suggests that the slow response is not associated with photothermal heating in both materials.

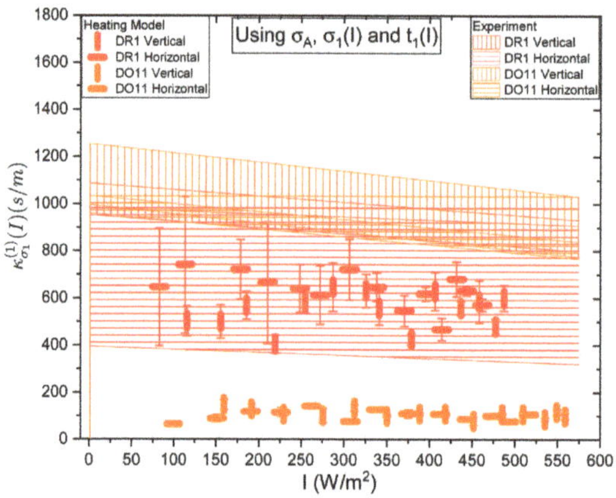

Figure 18. The vertical and horizontal lines (representing the pump polarization relative to the vertical) show the predicted heating contribution to the photomechanical response for DO11 and DR1 thin films using t_1 for the time constant as a function of intensity using the semi-empirical theory given by Equation (12). The one-standard-deviation bands show the experimentally determined intensity-dependent photomechanical constant using a fit of the stress data to a quadratic function in the intensity, as shown in the top graph in Figure 16.

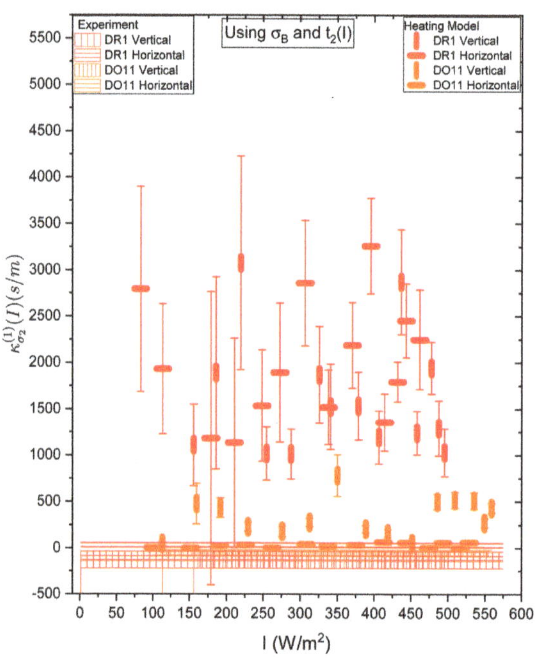

Figure 19. The vertical and horizontal lines (representing the pump polarization relative to the vertical) show the predicted heating contribution to the photomechanical response for DO11 and DR1 thin film using $t_2(I)$ as the time constant in the model given by Equation (9). The bands show the experimentally determined intensity-dependent photomechanical constant determined by the fit of the stress data to a quadratic function in the intensity, as shown in the bottom graph of Figure 16.

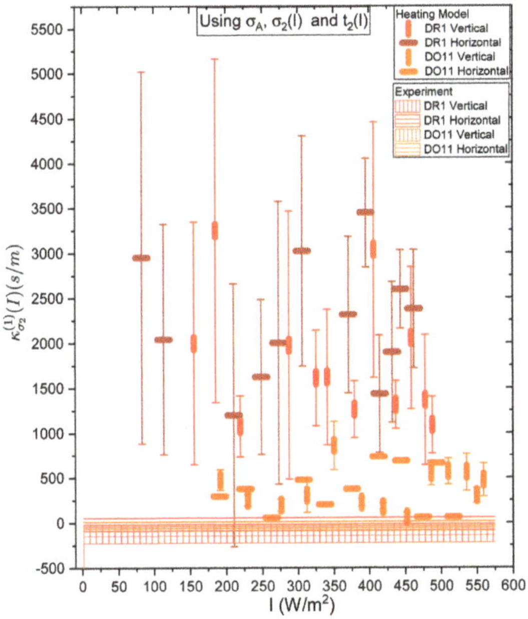

Figure 20. The vertical and horizontal lines (representing the pump polarization relative to the vertical) show the predicted heating contribution to the photomechanical response for DO11 and DR1 thin film using $t_2(I)$ as the time constant in the semi-empirical model given by Equation (12). The one-standard-deviation error bands show the experimentally determined intensity-dependent photomechanical constant determined from the fit of the stress data to a quadratic function in the intensity, as shown in the bottom graph of Figure 16.

Table 4 summarizes the data. The fast response time of both DR1- and DO11-doped PMMA thin films is the same within experimental uncertainty and is consistent with the time constant being determined from the sample geometry, which determines how heat dissipates. For heating, then, the time constants are expected to be similar, as is t_1. The fact that the time constants are independent of polarization is also what one would expect from an isotopic sample. This gives us greater confidence that the fast response originates in photothermal heating. The fact that DO11 has a smaller linear response from heating is also consistent with Figure 3, which shows that heating results in a smaller temperature-dependent force. All of these data together are consistent with photothermal heating being responsible for the fast process.

Table 4. A summary of the photomechanical constants for each process, their time constants and the critical intensity for vertical (V) and horizontal (H) pump polarization.

Material	$\kappa^{(1)}_{\sigma_1}$ (s/m)	$\kappa^{(2)}_{\sigma_1}$ (s · m/W)	$\kappa^{(1)}_{\sigma_2}$ (s/m)	$t_{1,avg}$ (s)	$t_{2,avg}$ (s)	I_{cr_1} (W · m^{-2})
DR1 (V)	1021 ± 67	-0.27 ± 0.21	-125 ± 92	0.500 ± 0.018	1.56 ± 0.17	3785.19 ± 2954
DR1 (H)	692 ± 295	-0.13 ± 0.43	-39 ± 101	0.582 ± 0.018	1.94 ± 0.17	$5323.1 \pm 17{,}752$
DO11 (V)	1128 ± 129	-0.39 ± 0.21	-51 ± 34	0.500 ± 0.021	1.68 ± 0.26	2892.31 ± 1592
DO11 (H)	1000 ± 269	-0.34 ± 0.42	-32 ± 54	0.462 ± 0.024	1.56 ± 0.32	2941.2 ± 3718

The slow response in both materials is of negative sign, consistent with sample contraction as one would expect for molecular hole burning and molecular reorientation away from the pump polarization direction. Molecular reorientation would result in a positive response for a perpendicular pump beam, while orientational hole burning would give a negative response, but of smaller magnitude. The data are consistent with hole burning, but the uncertainties are too large to make a definitive conclusion.

The critical intensity I_{cr_1} due to the fast heating process is defined by

$$I_{\mathrm{cr}_1} = \left| \frac{\kappa_{\sigma_1}^{(1)}}{\kappa_{\sigma_1}^{(2)}} \right|, \tag{42}$$

and quantifies the intensity at which the higher-order term exceeds the linear response. DO11's critical intensity is smaller than DR1's value because its linear response is smaller. For both materials and both polarizations, the first nonlinear correction to the photomechanical response $\kappa_{\sigma_1}^{(2)}$ is the same within experimental uncertainty.

6. Conclusions

This paper presents the results of extensive studies of the time- and polarization-dependent photomechanical response of thin films made of PMMA polymer doped with (1) DR1 and (2) DO11 dyes. In addition, dichroism measurements are used to determine the degree of hole burning and the evolution of molecular orientation due to both the short-time and reversible response to the pump beam and long-term quasi-irreversible reorientation due to prolonged-exposure. Also measured is the temperature-dependent stress of a clamped sample to determine its thermodynamic equation of state, which is used to build a model of the photothermal photomechanical response.

Time-dependent dichroism measurements probe the dopant dye's orientational distribution function in response to a strong pump being turned on and off. In DR1-doped PMMA, these measurements show a fast evolution of dichroism due to hole burning and a slower but stronger signal due to molecular reorientation. These observations are consistent with expectations of a fast initial excitation of a cis isomer followed by the slower process of reorientational diffusion. DO11, which does not photoisomerize, shows no long-term molecular reorientation but does show a smaller degree of hole burning. The underlying mechanism is most likely due to the photo-tautomerization due to a proton hopping from the amine group to the carbonyl group, which is a known process responsible for self-healing [36–38].

Intensity-dependent dichroism measurements are used to determine the axial order parameter $\langle P \rangle$, which is a measure of the axial alignment of the dopant molecules. In DR1, the pristine material after fabrication is slightly anisotropic. The observed dichroism after the material reaches equilibrium under optical pumping increases with increased intensity, as we would expect of molecular reorientation where the reorientational diffusion rate increases with pump intensity. However, the 15 s recovery time between measurements is not enough time to allow for the reorientational alignment to relax. At the higher intensities, the residual axial alignment with the pump off in the time interval that precedes the pump can be as high as 40% of the value induced when the pump is turned on. This residual alignment builds over the course of the experiment. As such, the fraction of the pump light absorbed decreases since fewer molecules have a substantial projection of their axis along the pump polarization. This will lead to a smaller photomechanical response than for a pristine sample. In contrast, DO11 is isotropic from the start and shows very few intensity-dependent effects.

The photomechanical experiments measure the stress induced by light for a clamped sample for two pump polarizations. This can be viewed as the photobaric analogy of photochromism, where light-polarization-dependent stress changes take the place of light-polarization-dependent color change. These measurements show that photothermal heating is the dominant mechanism, with the others falling below experimental uncertainties. The conclusion is that the dye molecules couple weakly to the polymer so that they do not affect its photomechanical properties. In dye-doped polymers, then, the role of the dopant molecules is to absorb light and transfer the energy as heat to the polymer, which reacts through a thermodynamic process of thermal expansion.

In a parallel tract, we have developed a heating model that predicts the pure photothermal heating response. The critical input parameters to this model are the materials' thermodynamic and mechanical properties. The thermally induced stress in the clamped configuration is a parameter that is difficult to determine from first principles. As such, we have measured it directly and fit it to an empirically determined analytical model with three adjustable parameters, which fully characterizes the thermomechanical properties of the material. Surprisingly, the parameters for DO11- and DR1-doped PMMA are vastly different, showing that the small concentration of dopants can vastly change these properties.

Finally, we have used the thermomechanical parameters to predict the photomechanical response. This allows for a more fine-tuned experiment to separate the mechanisms. A fit to the photomechanical data of DR1 to a double exponential separates the fast from the slow response, and a comparison to the heating theory shows that the fast response is due to heating and the slow response is an order of magnitude smaller than the heating response, but within uncertainty of being zero. It is a puzzle why DO11 does not fit the heating theory even though the magnitude of the measured response is the same as in DR1. This may be due to the fact that the photobaric data of DR1 and DO11 are so different. This remains the topic of future research.

Author Contributions: Conceptualization, M.G.K.; Data Curation, Z.G., A.B. and B.Z.; Formal Analysis, Z.G., A.B. and M.G.K.; Funding Acquisition, M.G.K.; Investigation, Z.G. and M.G.K.; Methodology, M.G.K.; Project Administration, M.G.K.; Resources, M.G.K.; Software, B.Z.; Supervision, M.G.K.; Validation, Z.G., A.B., B.Z., M.S.K. and M.G.K.; Visualization, Z.G. and A.B.; Writing—Original Draft, Z.G. and M.G.K.; Writing—Review and Editing, Z.G., A.B., M.S.K. and M.G.K. All authors have read and agreed to the published version of the manuscript.

Funding: This research was funded by the National Science Foundation, Directorate for Engineering (ENG), grant number EFRI-ODISSEI:1332271.

Institutional Review Board Statement: Not applicable.

Data Availability Statement: The original contributions presented in this study are included in the article. Further inquiries can be directed to the corresponding author.

Acknowledgments: We thank Washington State University for additional internal support.

Conflicts of Interest: The authors declare no conflicts of interest.

Abbreviations

The following abbreviations are used in this manuscript:

DR1	Disperse Red 1
DO11	Disperse Orange 11
PMMA	Poly Methyl Methacrylate
FOM	Figure of Merit
PGMEA	Propylene Glycol Methyl Ether Acetate
PBS	Polarizing Beam Splitter

Appendix A

This appendix includes data, in the form of plots, that are used in the main text for the heating model.

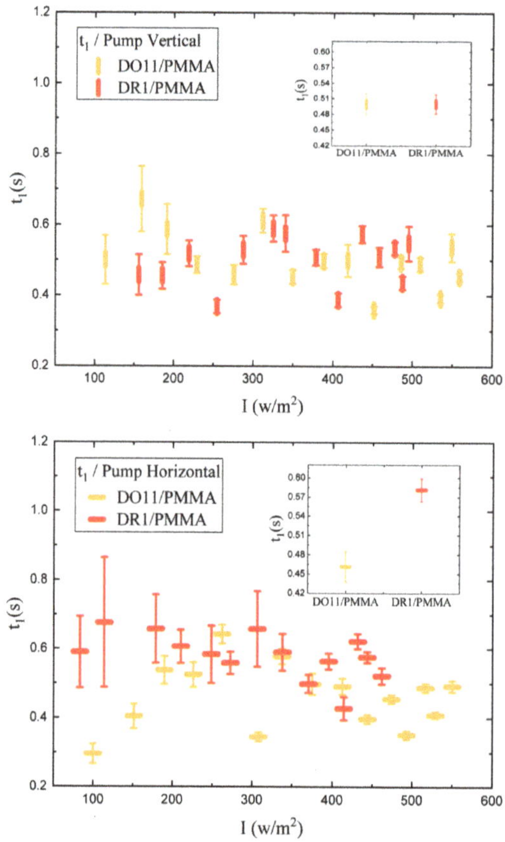

Figure A1. The fast time constant t_1 of the photomechanical response for DR1- and DO11-doped PMMA thin films as a function of intensity. The inset shows the average time constant over the whole intensity range for each material. (**top**) Vertically and (**bottom**) horizontally polarized pump.

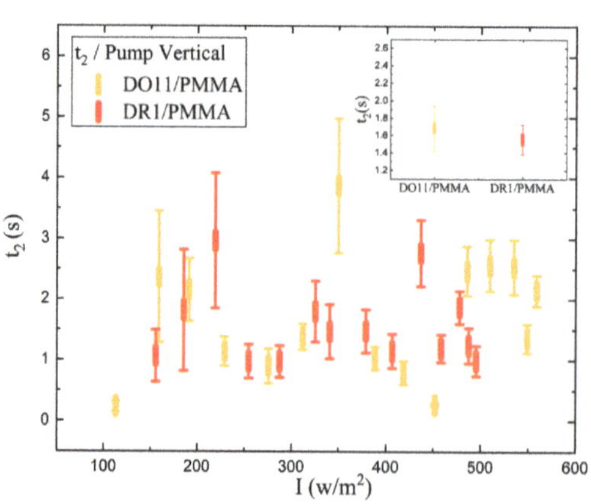

Figure A2. *Cont.*

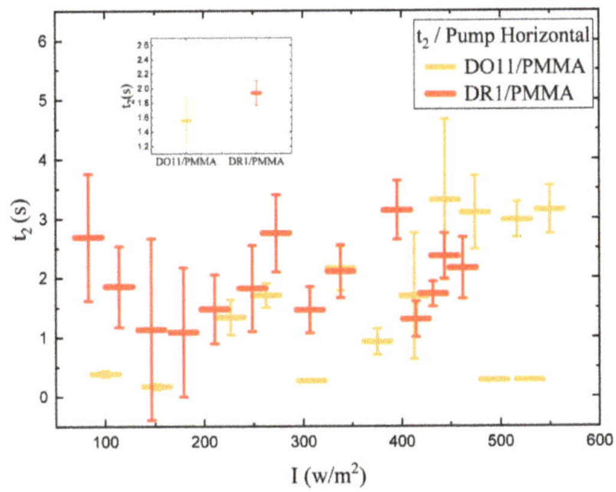

Figure A2. The slow time constant t_2 of the photomechanical response for DR1- and DO11-doped PMMA thin films as a function of intensity. The inset shows the average time constant over the whole intensity range for each material. (**top**) Vertically and (**bottom**) horizontally polarized pump.

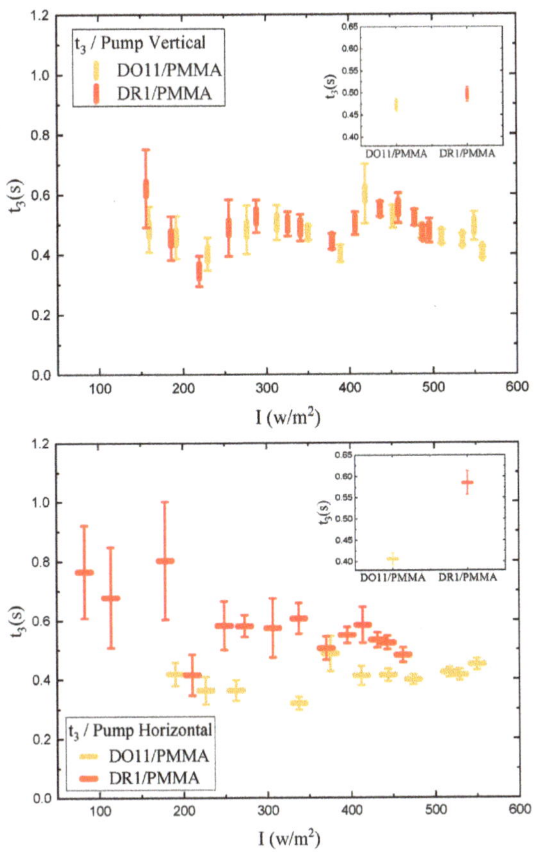

Figure A3. The fast relaxation time constant t_3 for DR1- and DO11-doped PMMA thin films as a function of intensity. The inset shows the average time constant over the whole intensity range for each material. (**top**) Vertically and (**bottom**) horizontally polarized pump.

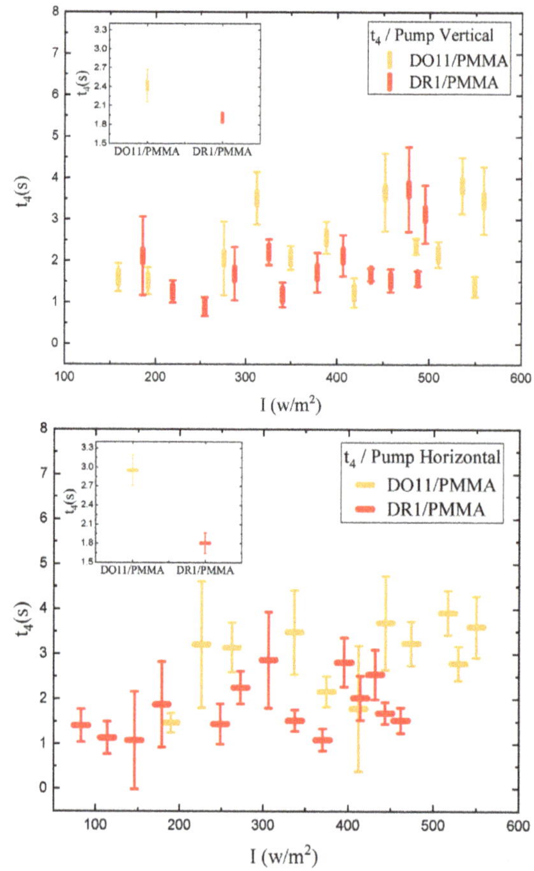

Figure A4. The slow relaxation time constant t_4 for DR1- and DO11-doped PMMA thin films as a function of intensity. The inset shows the average time constant over the whole intensity range for each material. (**top**) Vertically and (**bottom**) horizontally polarized pump.

References

1. Kuzyk, M.G.; Dawson, N.J. Photomechanical materials and applications: A tutorial. *Adv. Opt. Photonics* **2020**, *12*, 847. [CrossRef]
2. Ghorbanishiadeh, Z.; Zhou, B.; Karkhaneh, M.S.; Oehler, R.; Kuzyk, M.G. Photothermal and Reorientational Contributions to the Photomechanical Response of DR1 Azo Dye-Doped PMMA Fibers. *Appl. Sci.* **2021**, *12*, 315. [CrossRef]
3. Mahimwalla, Z.; Yager, K.G.; Mamiya, J.I.; Shishido, A.; Priimagi, A.; Barrett, C.J. Azobenzene photomechanics: Prospects and potential applications. *Polym. Bull.* **2012**, *69*, 967–1006. [CrossRef]
4. Zhou, B.; Bernhardt, E.; Bhuyan, A.; Ghorbanishiadeh, Z.; Rasmussen, N.; Lanska, J.; Kuzyk, M.G. Theoretical and experimental studies of photomechanical materials [Invited]. *J. Opt. Soc. Am. B* **2019**, *36*, 1492. [CrossRef]
5. Hines, L.; Petersen, K.; Lum, G.Z.; Sitti, M. Soft Actuators for Small-Scale Robotics. *Adv. Mater.* **2016**, *29*, 1603483. [CrossRef]
6. Lahikainen, M.; Zeng, H.; Priimagi, A. Reconfigurable photoactuator through synergistic use of photochemical and photothermal effects. *Nat. Commun.* **2018**, *9*, 4148. [CrossRef]
7. Mirfakhrai, T.; Madden, J.D.W.; Baughman, R.H. Polymer artificial muscles. *Mater. Today* **2007**, *10*, 30–38. [CrossRef]
8. Crespi, S.; Simeth, N.A.; König, B. Heteroaryl azo dyes as molecular photoswitches. *Nat. Rev. Chem.* **2019**, *3*, 133–146. [CrossRef]
9. Hill, R.A.; Dreher, S.; Knoesen, A.; Yankelevich, D.R. Reversible optical storage utilizing pulsed, photoinduced, electric-field-assisted reorientation of azobenzenes. *Appl. Phys. Lett.* **1995**, *66*, 2156–2158. [CrossRef]
10. Silva, L.M.; Silva, D.L.; Boas, M.V.; Bretonniere, Y.; Andraud, C.; Vivas, M.G. Probing the high performance of photoinduced birefringence in V-shaped azo/PMMA guest–host films. *RSC Adv.* **2020**, *10*, 40806–40814. [CrossRef] [PubMed]
11. Parthenopoulos, D.A.; Rentzepis, P.M. Three-Dimensional Optical Storage Memory. *Science* **1989**, *245*, 843–845. [CrossRef]

12. Welker, D.J.; Kuzyk, M.G. All-optical switching in a dye-doped polymer fiber Fabry–Perot waveguide. *Appl. Phys. Lett.* **1996**, *69*, 1835–1836. [CrossRef]
13. Moujdi, S.; Bougdid, Y.; Rahmouni, A.; Mahfoud, T.; Nesterenko, D.; Halim, M.; Sekkat, Z. Azo-polymers for holographic recording: Photo-assisted holography and surface relief gratings. In Proceedings of the Practical Holography XXXIII: Displays, Materials, and Applications, San Francisco, CA, USA, 2–7 February 2019; Bjelkhagen, H.I., Bove, V.M., Eds.; SPIE: Bellingham, WA, USA, 2019. [CrossRef]
14. Priimagi, A.; Shevchenko, A. Azopolymer-based micro- and nanopatterning for photonic applications. *J. Polym. Sci. Part B Polym. Phys.* **2013**, *52*, 163–182. [CrossRef]
15. Nesterenko, D.V.; Moujdi, S.; Hayashi, S.; Sekkat, Z. Simulation of photochemically induced motion of matter in gradient light fields. *J. Appl. Phys.* **2020**, *127*, 243106. [CrossRef]
16. Charra, F.; Kajzar, F.; Nunzi, J.M.; Raimond, P.; Idiart, E. Light-induced second-harmonic generation in azo-dye polymers. *Opt. Lett.* **1993**, *18*, 941. [CrossRef]
17. Sekkat, Z.; Dumont, M. Poling of polymer films by photoisomerization of azo dye chromophores. *Mol. Cryst. Liq. Cryst. Sci. Technol. B* **1992**, *2*, 359.
18. Andam, N.; Refki, S.; Ishitobi, H.; Inouye, Y.; Sekkat, Z. Optical Characterization of Ultra-Thin Films of Azo-Dye-Doped Polymers Using Ellipsometry and Surface Plasmon Resonance Spectroscopy. *Photonics* **2021**, *8*, 41. [CrossRef]
19. Bhuyan, A. Mechanism of the Photomechanical Response in Composite Polymeric Materials. Ph.D. Thesis, Washington State University, Pullman, WA, USA, 2020.
20. Elhani, S.; Maouli, I.; Refki, S.; Halim, M.; Hayashi, S.; Sekkat, Z. Quantitative analyses of optically induced birefringence in azo dye containing polymers. *J. Opt.* **2019**, *21*, 115401. [CrossRef]
21. Sekkat, Z.; Yasumatsu, D.; Kawata, S. Pure Photoorientation of Azo Dye in Polyurethanes and Quantification of Orientation of Spectrally Overlapping Isomers. *J. Phys. Chem. B* **2002**, *106*, 12407–12417. [CrossRef]
22. Rodriguez, A.; Vitrant, G.; Chollet, P.A.; Kajzar, F. Photochromic properties of azo dye polymer thin films: Evidence of an additional reversible mechanism of molecular behaviour. *Rev. Mex. Fis.* **2006**, *52*, 135–138.
23. Kawabe, Y.; Okoshi, K. Discrimination of photo-induced isomerization and molecular reorientation processes in azobenzene derivative doped in a polymer. *Opt. Mater. Express* **2018**, *8*, 332. [CrossRef]
24. Kuzyk, M.G.; Moore, R.C.; King, L.A. Second-harmonic-generation measurements of the elastic constant of a molecule in a polymer matrix. *J. Opt. Soc. Am. B* **1990**, *7*, 64–72. [CrossRef]
25. Zhou, Y.; Hauser, A.W.; Bende, N.P.; Kuzyk, M.G.; Hayward, R.C. Waveguiding Microactuators Based on a Photothermally Responsive Nanocomposite Hydrogel. *Adv. Funct. Mater.* **2016**, *26*, 5447–5452. [CrossRef]
26. Abbas, B.; Khalis, M.A. Dichroism of poly (methyl methacrylate) thin films doped with disperse orange 11 molecules. *Acta Phys. Pol. A* **2010**, *115*, 857–863. [CrossRef]
27. Embaye, N.B.; Ramini, S.K.; Kuzyk, M.G. Mechanisms of reversible photodegradation in disperse orange 11 dye doped in PMMA polymer. *J. Chem. Phys.* **2008**, *129*, 054504. [CrossRef]
28. Dumont, M.; Sekkat, Z.; Loucif-Saibi, R.; Nakatani, K.; Delaire, J.A. Photoisomerization, photo-induced orientation and orientational relaxation of azo dyes in polymeric films. *Nonlinear Opt.* **1993**, *5*, 395–406.
29. Sekkat, Z.; Dumont, M. Polarization effects in photoisomerization of azo dyes in polymeric films. *Appl. Phys. B Photophys. Laser Chem.* **1991**, *53*, 121–123. [CrossRef]
30. Delaire, J.A.; Nakatani, K. Linear and Nonlinear Optical Properties of Photochromic Molecules and Materials. *Chem. Rev.* **2000**, *100*, 1817–1846. [CrossRef]
31. Bernhardt, E.A.; Garrison, C.M.; Rasmussen, N.F.; Lanska, J.T.; Kuzyk, M.G. An apparatus for measuring a material's photomechanical response. *Am. J. Phys.* **2018**, *86*, 943–952. [CrossRef]
32. Dawson, N.J.; Kuzyk, M.G.; Neal, J.; Luchette, P.; Palffy-Muhoray, P. Experimental studies of the mechanisms of photomechanical effects in a nematic liquid crystal elastomer. *J. Opt. Soc. Am. B* **2011**, *28*, 1916. [CrossRef]
33. Ghorbanishiadeh, Z.; Zhou, B.; Sheibani Karkhaneh, M.; Oehler, R.; Kuzyk, M.G. Effect of Disperse Red 1 Azobenzene Dye Doping and Annealing on the Thermomechanical and Photomechanical Properties of PMMA Fibers. *Appl. Sci.* **2022**, *12*, 7991. [CrossRef]
34. Singer, K.D.; Kuzyk, M.G.; Sohn, J.E. Second-order nonlinear-optical processes in orientationally ordered materials: Relationship between molecular and macroscopic properties. *J. Opt. Soc. Am. B* **1987**, *4*, 968. [CrossRef]
35. Chu, W.; Tsui, S.M. Photoreductive Model of Disperse Orange 11 in Aqueous Acetone and Triethylamine. *J. Environ. Eng.* **2001**, *127*, 741–747. [CrossRef]
36. Howell, B.; Kuzyk, M.G. Amplified Spontaneous Emission and Recoverable Photodegradation in Disperse-Orange-11-Doped-Polymer. *J. Opt. Soc. Am. B* **2002**, *19*, 1790. [CrossRef]

37. Ramini, S.; Kuzyk, M. A self healing model based on polymer-mediated chromophore correlations. *J. Chem. Phys.* **2012**, *137*, 054705. [CrossRef]
38. Hung, S.T.; Bhuyan, A.; Schademan, K.; Steverlynck, J.; McCluskey, M.D.; Koeckelberghs, G.; Clays, K.; Kuzyk, M.G. Spectroscopic studies of the mechanism of reversible photodegradation of 1-substituted aminoanthraquinone-doped polymers. *J. Chem. Phys.* **2016**, *144*, 114902. [CrossRef]

MDPI AG
Grosspeteranlage 5
4052 Basel
Switzerland
Tel.: +41 61 683 77 34

Polymers Editorial Office
E-mail: polymers@mdpi.com
www.mdpi.com/journal/polymers

www.ingramcontent.com/pod-product-compliance
Lightning Source LLC
Chambersburg PA
CBHW040838250926
51850CB00007BA/1352
9783725849277